Complex fluid flows are encountered widely in nature, in living beings, and in engineering practice. These flows often involve both geometric and dynamic complexity and present problems that are difficult to analyze because of their wide range of length and time scales, as well as their geometric configuration.

This book describes some newly developed computational techniques and modeling srategies for analyzing and predicting complex transport phenomena. It summarizes advances in the context of a pressure-based algorithm. Among methods discussed are discretization schemes for treating convection and pressure, parallel computing, multigrid methods, and composite, multiblock techniques. With respect to physical modeling, the book addresses issues of turbulence closure and multiscale, multiphase transport from an engineering viewpoint. Both fundamental and practical issues are considered, along with the relative merits of competing approaches.

Numerous examples are given throughout the text. The final chapter is devoted to practical applications that illustrate the advantage of various numerical and physical tools.

Mechanical, aerospace, chemical, and materials engineers can use the techniques presented in this book to tackle important, practical problems more effectively.

T0215262

Computational Techniques for
Complex Transport Phenomena

To the memory of
Chia-Shun Yih
(1918–1997)

Computational Techniques for Complex Transport Phenomena

W. Shyy
University of Florida

S. S. Thakur
University of Florida

H. Ouyang
Atchison Casting Corp.

J. Liu
Western Atlas Logging Services

E. Blosch
Northrop-Grumman

CAMBRIDGE
UNIVERSITY PRESS

CAMBRIDGE UNIVERSITY PRESS
Cambridge, New York, Melbourne, Madrid, Cape Town, Singapore, São Paulo

Cambridge University Press
The Edinburgh Building, Cambridge CB2 2RU, UK

Published in the United States of America by Cambridge University Press, New York

www.cambridge.org
Information on this title: www.cambridge.org/9780521592680

First published 1997
This digitally printed first paperback version 2005

A catalogue record for this publication is available from the British Library

Library of Congress Cataloguing in Publication data
Computational techniques for complex transport phenomena / W. Shyy...[et al.].
p.   cm.
Includes bibliographical references (p.   ).
ISBN 0-521-59268-2 (hb)
1. Fluid dynamics – Data processing.   2. Multiphase flow – Mathematical models.
3. Turbulence – Mathematical models
4. Transport theory – Mathematical
models.   I. Shyy. W. (Wei)
TA357.5.D37C66   1997
620.1'04 – dc21   97-4471
                                                                                CIP

ISBN-13 978-0-521-59268-0 hardback
ISBN-10 0-521-59268-2 hardback

ISBN-13 978-0-521-02360-3 paperback
ISBN-10 0-521-02360-2 paperback

Contents

Preface

This book deals with complex fluid flow and heat/mass transport phenomena encountered in natural and human-made environments. These problems, especially those encountered in engineering practice, are often difficult to analyze due to geometric and dynamic complexities, among other factors. Dynamic complexity results from nonlinear physical mechanisms, and geometric complexity is common to virtually all practical devices. To analyze such physical systems, one needs to resolve the wide range of length and time scales as well as the details of the geometric configuration. Consequently, the computational capability required to facilitate such analyses far exceeds that which is available at the present time. Before such a capability can, if ever, become available, one must resort to a more pragmatic approach.

To develop a predictive capability, it is essential to view the complex physical system in its entirety and to strive for appropriate computational and modeling techniques. Quite often, this means that fundamental rigor and engineering approximation need to be reconciled. In the present book, we address a set of computational techniques suitable for analyzing and predicting complex transport phenomena. What we envision is that a design engineer or an analyst can use the techniques presented to conduct investigations and obtain useful information for a given engineering problem in a matter of days rather than weeks or months. In order to achieve this goal, we have purposely avoided taking a deductionist's viewpoint, which tends to concentrate on a specific aspect of the originally complex system in a rigorous manner but with much simplification in terms of the overall complexities. While such an approach is appealing from the viewpoint of fundamental research, it seldom addresses the issue of interaction among multiple competing mechanisms. Furthermore, while it is necessary to present the various techniques individually, we ensure that these techniques are compatible with one another and can be integrated to form a tool for predicting complex transport phenomena. Individual techniques are only as effective as the overall numerical capability within which they operate and as the physical model into which they are integrated.

To accomplish our goal, instead of presenting a broad range of the materials, we present only selected topics with sufficient detail. Issues related to both numerical computation and physical modeling are presented. In numerical techniques, we first summarize recent progress made in the context of a pressure-based algorithm and discuss the discretization schemes for convection and pressure terms, parallel computing and multigrid methods, and composite, multiblock techniques. We present fundamental as well as practical issues and the relative merits of the competing approaches. All the topics presented are still under intensive investigation. In particular, in the area of parallel computing, the combination of commercial competition and limited market size has caused the available computer architectures to experience a very high rate of turnover, sometimes even before they are thoroughly evaluated. To accommodate such a dynamic situation, we have chosen to stress the basic underlying issues, such as parallel efficiency and scalability, using the existing architectures to highlight the points made.

In physical modeling, we discuss the issues of turbulence modeling and multiphase transport, with limited and sharply focused scopes. For turbulence modeling, the K-ε two-equation closure is discussed, with particular emphasis on the possible extensions of the original model, as well as on the implementation issues in generalized curvilinear coordinates. Needless to say, the K-ε model is not the only one available for engineering computations; however, we don't want to turn the present text into another book on turbulence modeling. Even though the K-ε turbulence model has documented difficulties for complex fluid flows, it is practically important and hence worth further refinement. Furthermore, models such as second-moment closure are still not as thoroughly developed and tested. To do justice to these more complex models, additional research and development will be needed. The implementational aspects discussed in conjunction with the K-ε model are also applicable to other two-equation-based turbulence models. Similarly, for multiphase dynamics, we concentrate on the volume-averaged technique and highlight its computational implications.

The present book is based on the collaborative research we have conducted over the past several years. It complements the two previous books authored and coauthored by one of us (Shyy 1994, Shyy et al. 1996). Together, these three books offer a comprehensive view of approaching fluid flow problems with complexities arising from geometrical configurations, boundary conditions, and various intrinsic physical mechanisms. We thank our families for moral support during the course of preparing this book and our co-workers in the computational thermo-fluids group, in particular, Dr. Venkata Krishnamurthy and Dr. Jeffrey Wright, who offered insight and collaborated with us to develop some of the ideas put forth in this book. We are grateful to Florence Padgett of Cambridge University Press for her enthusiasm, support and encouragement throughout this endeavor. We also thank Ellen Tirpak of TechBooks, Inc. for her substantial contributions during typesetting. The Department of Aerospace Engineering, Mechanics & Engineering Science of the University of Florida has given us a very supportive environment to conduct our research. It is our good fortune to be associated with this fine institution.

April, 1997
Gainesville

1 Introduction

1.1 Dynamic and Geometric Complexity

Complex fluid flow and heat/mass transfer problems encountered in natural and human-made environments are characterized by both *geometric and dynamic complexity*. For example, the geometric configuration of a jet engine, a heart, or a crystal growth device is irregular; to analyze the heat and fluid flow in these devices, geometric complexity is a major issue. From the analytical point of view, dynamic complexity is a well-established characteristic of fluid dynamics and heat/mass transfer. The combined influence of dynamic and geometric complexities on the transport processes of heat, mass, and momentum is the focus of the present work. A computational framework, including both numerical and modeling approaches, will be presented to tackle these complexities. In this chapter, we will first present basic background to help identify the issues involved and to highlight the state of our current knowledge.

1.1.1 Dynamic Complexity

Dynamic complexity results from the disparities of the length, time, and velocity scales caused by the presence of competing mechanisms, such as convection, conduction, body forces, chemical reaction, and surface tension; these mechanisms are often coupled and nonlinear. A case in point is the classical boundary layer theory originated by Prandtl (Schlichting 1979, Van Dyke 1975), whose foundation is built on the realization that the ratio of viscous and convective length and time scales can differ by orders of magnitude for high Reynolds number flows. With sufficiently high Reynolds number or other appropriate nondimensional parameters, such as Rayleigh or Marangoni number (Shyy 1994), turbulence appears, and there exists a wide spectrum of scales, ranging from the global one dictated by the flow configuration, to the smallest one where turbulent energy is dissipated into heat (Landhal and Mollo-Christensen 1992, Monin and Yaglom 1971 and 1975, Tennekes and Lumley 1972). In addition to these standard textbook materials, relevant examples abound in nature

and in engineering practice. A timely example can be drawn from materials solidification and processing. It is well established that the interaction of transport mechanisms
in the melt with the solidification process can have a strong effect on the resulting
structure and properties of the material, the main reason being that these mechanisms
create multiple length and time scales in the physical system (Kessler et al. 1988,
Kurz and Fisher 1984, Langer 1980, Shyy 1994, Shyy et al. 1996b). For example,
based on dimensional analysis, capillary, convective, and diffusive (including heat,
mass, and momentum) length scales are important in a system containing solidifying
binary species.

Most of the work published in this area deals with the issues confined within
a particular range of scales. At the microscopic level, work has been conducted by
many researchers to analyze and simulate the formation, growth, and evolution of the
solid–liquid interface where the mechanisms at the morphological length scale are
considered (DeGregoria and Schwartz 1986, Huang and Glicksman 1981, Kobayashi
1993, Lynch 1982, McFadden and Coriell 1987, Tiller 1991, Udaykumar and Shyy
1995, Wheeler et al. 1993). Physical mechanisms such as capillarity and diffusion
are important to consider. At the macroscopic level, where convective and diffusive
effects on momentum, heat, and mass transport are important, limited analyses have
been made to study the solidification problem, without accounting for the smaller
length scales due to capillarity (Beckermann and Viskanta 1988, Bennon and Incropera 1987a, Canright and Davis 1991, Christensen et al. 1989, Crowley 1983,
Dantzig 1989, Garandet et al. 1990, Heinrich et al. 1989, Hurle et al. 1983, Schwabe
1988, Shyy et al. 1992c, Szekely 1979). In order to treat the physical processes
involving different scales, one needs to devise separate scaling procedures for each
physical regime, as well as to find ways to couple these scales in a coherent manner
so as to optimize the utilization of computing resources.

Based on the above discussion, it appears that two classes of multiscale problems
exist. In the first class, small scales occupy distinct regions in space so that the large
and small scales need only be matched or patched together; the boundary layer theory
is a good example where the viscous, small-scale effect is confined to in the vicinity of
the solid boundary. Methods such as singular perturbation and matched asymptotic
expansion have been developed during the last several decades to deal with such
problems (Meyer and Parter 1980, Van Dyke 1975). In the second class, the disparate
scales are not confined in thin regions, and these scales interact with one another.
Turbulence is a classical example where the macroscopic and microscopic scales are
present simultaneously and exchange energy via the cascading process (Tennekes and
Lumley 1972). Physical systems undergoing phase change (Langer 1980, Shyy 1994)
also form such an example. Analytical techniques such as the renormalization group
(RNG) method, originally developed for treating the critical phenomena involving
phase change, have also been recently applied to treat turbulent fluid flows (McComb
1990, Orszag et al. 1993), yielding some success.

1.1.2 Geometric Complexity

In addition to the dynamic complexity, for many problems involving complex
geometries (including multibody configurations and bodies with complex shapes),

substantial difficulties still remain in generating a reasonable single grid to cover the entire flow domain. There are several available strategies for handling complex geometries, including composite structured grids (Chesshire and Henshaw 1990, Rai 1985, 1986, Shyy 1994, Thakur et al. 1996, Wright and Shyy 1993), unstructured grids (Löhner and Parikh 1988, Morgan et al. 1991, Venkatakrishnan 1996), hybrid structured/unstructured grid methods (Kao and Liou 1995, Wang and Yang 1994), and the cut-cell approach with regular, Cartesian grid (Coirier and Powell 1996, Shyy et al. 1996a, Udaykumar et al. 1996). Each of these strategies has its own advantages and disadvantages. Among them, the unstructured grid technique has been thoroughly reviewed by Venkatakrishnan (1996), and the cut-cell approach has been extensively discussed in Shyy et al. (1996a). Composite structured grids can be broadly classified as either patched grids (also known as abutting grids; see Rai 1985, 1986) or overlaid grids (such as Chimera grids; see Steger 1991). Overlaid (or overlapping) grids have the advantage that the neighboring subgrids (blocks) need not match perfectly at their internal boundaries, and hence, individual blocks of desirable quality (in terms of skewness) and orientation can be employed (Steger and Benek 1987, Steger 1991, Tu and Fuchs 1992). Issues such as mesh interface generation (Wey 1994) and multigrid computations (Johnson and Belk 1995) in the context of overset grids have been discussed. Interesting applications based on such an approach have been reported, for example, by Shih et al. (1993) and Chyu et al. (1995). On the other hand, recent work of Braaten and Connell (1996), applying an unstructured grid method, has shown substantial promise for solving propulsion-related fluid flow problems. While the power and flexibility of the algorithm in treating complex geometries with accurate resolution of the flow features makes the unstructured grid approach very attractive, drawbacks have also been identified. For example, the use of purely tetrahedral meshes makes the computed solutions somewhat noisy in regions of high flow gradients and requires an excessive number of points to be inserted by the refiner in the viscous layer. Both of these aspects are undesirable for certain transport problems; for example, since the conduction heat transfer rate depends on an accurate calculation of the temperature gradient, a noisy solution can seriously compromise the accuracy of the computation.

With overlaid grids, information transfer between neighbors is more complicated and perhaps more ambiguous than with patched grids, and it is not very straightforward to maintain conservation across interfaces. Patched grids, though somewhat more restrictive in terms of grid quality, are attractive from the point of view of information transfer across internal boundaries. Honoring flux conservation at interfaces appears to be more straightforward.

There are several important issues associated with an interface scheme for discontinuous grids. One is the accuracy of the interface treatment. This issue can be addressed from the point of view of the type (order) of interpolation employed for estimating fluxes and or variables at an interface from the information in the neighboring block (Rai 1986) or from the point of view of the stretching error in the estimation of gradients resulting from the sudden change in grid size across the interface (Kallinderis 1992). Another issue is that of maintaining flux conservation across the interfaces. Fluxes from the neighboring blocks should cancel each other in order to maintain conservation.

In addition to the structured composite grid method, a hybrid structured/unstructured grid method has also been proposed, which uses the Chimera overlapped structured grids as the initial grid layout, replacing the grid overlapping region by an unstructured grid. This method, called the DRAGON grid (Kao and Liou 1995), makes limited use of the unstructured grid in the region sandwiched between the structured grids, resulting in a more modest increase in memory and computational effort than that required by a completely unstructured grid approach.

The multiblock method will be the primary focus of the present text, because (1) it can reduce the topological complexity of a single grid system by employing several grid blocks, permitting each individual grid block to be generated independently so that both geometry and resolution in the desired region can be treated more satisfactorily; (2) grid lines need not be continuous across grid interfaces, and local grid refinement and adaptive redistribution can be conducted more easily to accommodate different physical length scales present in different regions; and (3) the multiblock method also provides a natural route for parallel computations. In order to make good use of the multiblock method, one needs to handle the grid interface within the flow solver in a careful manner.

1.2 Computational Complexity

In the context of a computational approach, there is an additional issue related to complexity, namely, the interaction between computation and physics. We will discuss this aspect using an adaptive grid computation as an illustration. Using a simple one-dimensional, convective-diffusive equation, it will be shown that if an adaptive grid technique is employed to help better resolve the solution variation, complex patterns of grid movement can form without reaching a final equilibrium state. These patterns result strictly from the interaction between numerical algorithms and dimensionless parameters, such as Reynolds number, deduced from the governing equation.

At a fundamental level, many physical systems display irregular dynamic or "chaotic" behavior. Mathematicians have used the name "strange attractor" to denote bounded, chaotic, nonperiodic solutions of deterministic, nonlinear differential equations in contrast to more predictable motions such as those near equilibrium points and limit cycles. A remarkable fact is that despite the deterministic nature of the governing equations, the string of numbers so produced is in any practical sense unpredictable; and yet they can also be completely repeatable. It is now well known that solutions of these systems are distinguished from regular ones by at least three features: (1) they are exponentially sensitive to small changes in the initial conditions, (2) their power spectra are irregular and comprise broadband components; and (3) the surfaces on which they lie in the space of the dependent variables are of fractal dimensions.

Many scholarly and popular writings have been published on this subject (e.g., Gleick 1987, Grebogi et al. 1987, Guckenheimer and Holmes 1983, Lichtenberg and Lieberman 1982, Miles 1984, Moon 1987, Thompson and Stewart 1986). By studying problems arising from diverse applications, qualitatively we see that with modest values of a dimensionless parameter, the solution first converges from some initial conditions to a static equilibrium value. Then, as the value of this parameter increases, the solution displays periodic motion. Further increasing this parameter beyond a critical value, the solution makes a transition into a chaotic regime. These characteristics prompt many people to suggest the relevance between the chaos theory and the onset and subsequent development of turbulence in fluid dynamics and related transport phenomena. Aided by such a viewpoint, new insight has been obtained, notably by Lumley, Holmes, and their co-workers (e.g., Aubry et al. 1988, Holmes et al. 1996). Nevertheless, a solid mathematical link has yet to be forged between, say, the Navier-Stokes equations and chaos. Furthermore, it is not clear in practice how much additional insight can be gained from analyzing chaos to truly help solve the classic problem of turbulance.

In the following, we employ the adaptive grid method to solve a simple problem to help illustrate the dynamic (and chaotic) characteristics displayed in a course of computation. The method utilized is based on that proposed by Dwyer (1984), where grid points are distributed along a given arc length in space depending on the relative importance of the following three factors: (1) total arc length (smoothing term), (2) dependent variable function variation (first derivative), and (3) dependent variable slope variation (second derivative).

The mathematical expression for the technique is

$$\xi = \xi_{max} \frac{\int_0^x W\,dx}{\int_0^{x_{max}} W\,dx} \tag{1.1}$$

where ξ is the general coordinate, W is the weighting function used to adapt the grid, x is the arc length, and x_{max} is the maximum arc length. Equation (1.1) can also be written as

$$\int_0^x W\,dx = \frac{\xi}{\xi_{max}} \int_0^{x_{max}} W\,dx. \tag{1.2}$$

It is noted that if ξ is incremented with a constant value, Eq. (1.2) also implies

$$W_i \Delta x_i = \text{constant} \tag{1.3}$$

where x_i is the interval along the given arc and W_i is the corresponding weighting function in the interval.

W may have different forms according to different physical problems. For example, W may assume the form (Dwyer 1984) of

$$W = 1 + b_1 \left| \frac{\partial \phi}{\partial x} \right| + b_2 \left| \frac{\partial^2 \phi}{\partial x^2} \right| \tag{1.4}$$

where ϕ is the solution of a dependent variable. In Eq. (1.4), b_1 and b_2 are the "normalizing factors," and their determination depends on the relative importance of each term measured by R_1 and R_2, where

$$R_1 = \frac{b_1 \int_0^{x_{max}} \left| \frac{\partial \phi}{\partial x} \right| dx}{\int_0^{x_{max}} W \, dx}, \tag{1.5}$$

$$R_2 = \frac{b_2 \int_0^{x_{max}} \left| \frac{\partial \phi^2}{\partial x^2} \right| dx}{\int_0^{x_{max}} W \, dx}. \tag{1.6}$$

By assigning values to R_1 and R_2 for each term, b_1 and b_2 are determined accordingly. The values of R_1 and R_2 are decided by the desired relative importance of each term.
W takes the following form:

$$W = 1 + b_1 \left| f \left(\frac{\partial \phi}{\partial x} \right) \right| + b_2 \left| g \left(\frac{\partial^2 \phi}{\partial x^2} \right) \right| \tag{1.7}$$

where f and g are numerical functions of first- and second-order derivatives, respectively, defined as:

$$f \left(\frac{\partial \phi}{\partial x} \right) = \frac{\Delta \phi}{\Delta x}, \qquad g \left(\frac{\partial^2 \phi}{\partial x^2} \right) = \left(\frac{\Delta \phi^+}{\Delta x^+} - \frac{\Delta \phi^-}{\Delta x^-} \right) \bigg/ (\Delta x / 2) \tag{1.8}$$

where

$$
\begin{aligned}
\Delta \phi &= \phi_{i+1} - \phi_{i-1}, & \Delta x &= x_{i+1} - x_{i-1}, \\
\Delta \phi^+ &= \phi_{i+1} - \phi_i, & \Delta \phi^- &= \phi_i - \phi_{i-1}, \\
\Delta x^+ &= x_{i+1} - x_i, & \Delta x^- &= x_i - x_{i-1}, \\
\Delta \phi &= \phi_{i+1} - \phi_{i-1}, & \Delta x &= x_{i+1} - x_{i-1}.
\end{aligned}
$$

What we are interested in finding out here is the characteristics of the dynamical responses caused by, for example, the variations of the parameters R_1 and R_2. Are there distinctive patterns that one can identify for oscillatory and apparently chaotic behavior of grid distribution in the course of iterative adaption? In view of the influences of the free parameters R_1 and R_2, what role does the Reynolds number have in the present context?
The model problem is

$$
\begin{aligned}
u\phi_x &= v\phi_{xx} + S(x), & u, v &= \text{constant} > 0 \\
\phi(0) &= 0, & (\phi_x)_{x=L} &= 0,
\end{aligned} \tag{1.9}
$$

and the source $S(x)$ has the piecewise-linear form:

$$S(x) = \begin{cases} ax + b & 0 \leq x \leq x_1 \\ -\dfrac{(ax + b)}{x_2} x + \dfrac{(ax_1 + b)}{x_2}(x_1 + x_2) & x_1 \leq x \leq x_1 + x_2. \end{cases} \tag{1.10}$$

The values of a, b, x_1, and x_2 used here are $a = -2.0$, $b = 3.0$, $x_1 = 2h$, $x_2 = h$, and $h = \frac{15}{16}$. The value of L is 15.

The numerical scheme chosen for both the convection and diffusion terms in the present example is second-order central differencing. Since the grids become nonuniform in the adaptive procedure, the nonuniformity factor β should be included in the formulation. Here β is defined as:

$$\beta_i = \frac{x_i - x_{i-1}}{x_{i+1} - x_i} = \frac{\Delta x^-}{\Delta x^+}.$$

In the following, we shall use β and Δx to designate β_i and Δx_i, respectively. The Taylor series expansion then gives

$$\phi_{i+1} = \phi_i + \Delta x \cdot \left.\frac{\partial \phi}{\partial x}\right|_i + \frac{1}{2}\Delta x^2 \cdot \left.\frac{\partial^2 \phi}{\partial x^2}\right|_i + \cdots, \tag{1.11}$$

$$\phi_{i-1} = \phi_i - \beta \Delta x \cdot \left.\frac{\partial \phi}{\partial x}\right|_i + \frac{1}{2}(\beta \Delta x)^2 \cdot \left.\frac{\partial^2 \phi}{\partial x^2}\right|_i + \cdots \tag{1.12}$$

and hence the first derivative and second derivative at grid (i) can be written as:

$$\left.\frac{\partial \phi}{\partial x}\right|_i = \frac{1}{\Delta x}\left[-\frac{1}{\beta(1+\beta)}\phi_{i-1} + \frac{1-\beta^2}{\beta(1+\beta)}\phi_i + \frac{\beta}{1+\beta}\phi_{i+1}\right] + \mathcal{O}(\Delta x^2) \tag{1.13}$$

$$\left.\frac{\partial \phi^2}{\partial x^2}\right|_i = \frac{1}{\Delta x^2}\left[\frac{2}{\beta(1+\beta)}\phi_{i-1} - \frac{2}{\beta}\phi_i + \frac{2}{1+\beta}\phi_{i+1}\right] + \mathcal{O}(\Delta x). \tag{1.14}$$

The numerical approach starts with the uniform grid solution. This uniform grid solution is used to estimate the weighting function in Eq. (1.4), and the new grid position is determined from Eq. (1.2). These new grid positions are in turn used to recalculate the numerical solution of the model problem. The adaptation continues until the convergence criterion is met. The criterion employed is that the maximum distance between two consecutive grid adaptions should be smaller than 10^{-3}. If this criterion is not met, the adaptation procedure is repeated. The history of grid positions during these iterations will then be processed to study the dynamic structure.

Even for a simple one-dimensional problem such as the present model problem and with a given numerical discretization scheme, four free parameters can influence the performance of the adaptive grid computation. They are R_1 and R_2, the number of grid points, and the overall Reynolds number. The overall Reynolds number is defined as:

$$Re = \frac{uL}{v}. \tag{1.15}$$

Here no attempt has been made to cover all the possible combinations of these four parameters; this is neither possible nor necessary. Following the work of Shyy (1991a,b), selected cases will be presented to depict the salient features of the dynamic structure of the present adaptive grid computation.

First, we demonstrate the expected usefulness of the adaptive grid method. Figure 1.1 shows the comparison of exact and numerical solutions on both the uniform

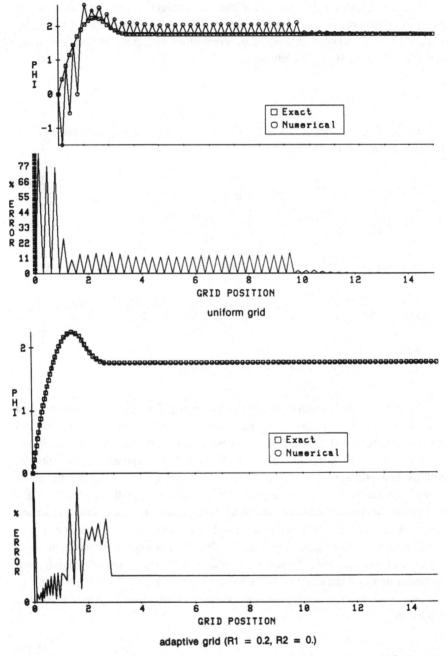

Figure 1.1 Comparison of exact and numerical solutions for $Re = 1.5 \times 10^9$ with 100 nodes.

grid and adaptive grid (with $R_1 = 0.2$, $R_2 = 0$) of 100 nodes for a very high Reynolds number case, $Re = 1.5 \times 10^9$. Wiggles of large magnitude appear in the uniform grid resolution, while the adaptive grid solution reduces the error by more than 100 times. For this case, it takes only a couple of adaptive iterations to reach the converged equilibrium grid distribution. For the case of smaller Re, $Re = 1.5 \times 10^4$, both the uniform grid and adaptive grid solutions (with $R_1 = 0.2$, $R_2 = 0$) are very accurate. However, not all the adaptive grid solutions with any values of R_1 and R_2 can produce good results. In particular, it has been found that for the class of problem under study, the second-derivative term in the weight function can cause extreme sensitivity of the performance of the adaptive grid method. For example, for the identical problem of $Re = 1.5 \times 10^4$ and 100 nodes as shown in Fig. 1.2, the adaptive grid method with $R_1 = 0$ and $R_2 = 0.9$ exhibits a persistently oscillatory pattern in the course of iteration without being able to reach an equilibrium grid distribution, as shown in Figs. 1.3 and 1.4. This observation is more remarkable if one notices that the uniform grid solution is already very accurate. For the parameters of $R_1 = 0$ and $R_2 = 0.9$, even a very accurate initial numerical solution cannot yield an equilibrium grid distribution. This means that the oscillatory pattern of the adaptive grid procedure has nothing to do with the accuracy of the original numerical solution; it is the internal mechanism of the adaptive grid method itself that produces this oscillatory behavior. Figure 1.3 shows three snapshots of the comparisons between the exact and adaptive grid solutions after 10, 100, 200, and 1900 adaptive iterations. Figure 1.4 shows the grid distribution in the first 200 adaptive iterations, as well as a more detailed view of the adaptive regridding of the first 10 grid points.

In the following, we shall use the set of parameters $R_1 = 0$ and $R_2 = 0.9$ to explore the dynamic structure of the present adaptive grid method. A standard tool used in the area of dynamical chaos, which proves to be very useful here, is the so-called one-dimensional map. The map is produced by plotting the grid positions x of the same index i of the two consecutive adaptive iterations, n and $n + 1$. Hence, the map to be shown is plotted as $x_i(n + 1)$ versus $x_i(n)$ for all the adaptive iterations. Figure 1.5 shows such a map for the case of $Re = 1.5 \times 10^4$, $R_1 = 0.2$, $R_2 = 0$, and with 100 nodes, for $i = 20$. It shows that the grid position of $i = 20$ quickly stabilizes at the equilibrium position of 0.9489. Shown in Fig. 1.6 are the structures of grid distribution in the course of adaptive iteration for a wide span of Reynolds numbers, ranging from $Re = 1.5 \times 10^{-2}$ to $Re = 1.5 \times 10^9$ with $R_1 = 0$ and $R_2 = 0.9$ and 100 grid points. First, what immediately appears obvious is the similarity of the structure of these one-dimensional maps of different Re. Even for the Reynolds number as low as $Re = 1.5 \times 10^{-2}$ and with the use of 100 grid points, the large contribution of the second derivative in the weight function can drive the present adaptive grid method to depict a chaotic and yet a well-structured pattern.

As the Reynolds number becomes higher, the basic pattern still remains the same, albeit the randomness becomes larger, and some substructures start to emerge. It should be emphasized that the patterns shown in Fig. 1.6 constitute many repetitious trajectories, that is, at some stages of adaptive iterations, the grid positions seemingly approach equilibrium distributions (which should be the intersection point of the map

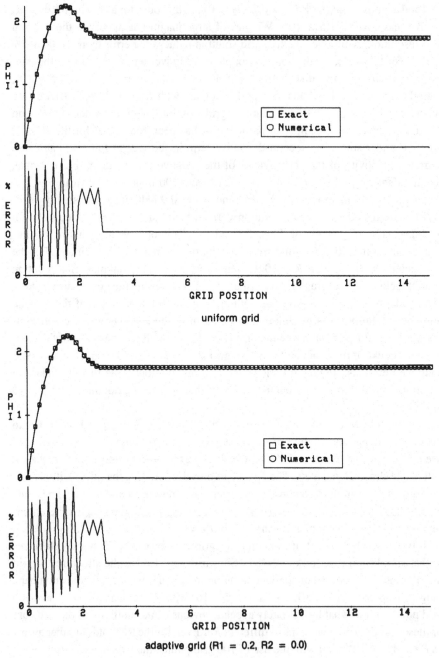

uniform grid

adaptive grid (R1 = 0.2, R2 = 0.0)

Figure 1.2   Comparison of exact and numerical solutions for $Re = 1.5 \times 10^4$ with 100 nodes.

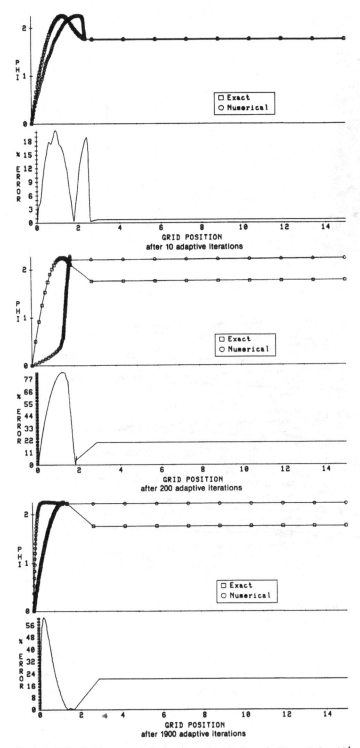

Figure 1.3 Performance of adaptive grid solutions for $Re = 1.5 \times 10^4$ with 100 nodes, $R_1 = 0.0$, $R_2 = 0.9$.

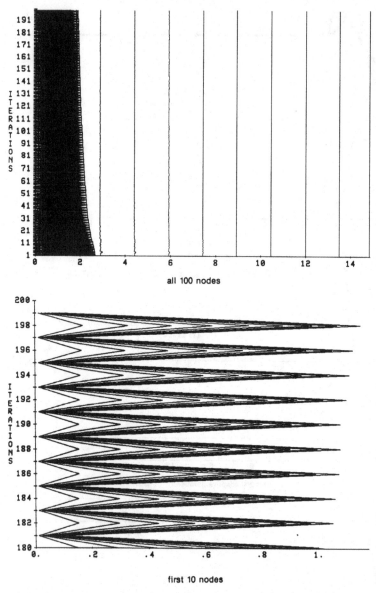

Figure 1.4   Grid positions with respect to adaptive iterations for $Re = 1.5 \times$ 100 nodes, $R_1 = 0.0$, $R_2 = 0.9$.

and the straight line issued at 45° from the origin), and they are never able to achieve a total equilibrium state without abruptly being pushed away by small disturbances. To demonstrate this feature more clearly, a series of more detailed maps is presented in Figs. 1.7 and 1.8 for $Re = 1.5 \times 10^{-2}$ and $Re = 1.5 \times 10^4$, respectively. In each of the plots, $x_i(n + 1)$ versus $x_i(n)$ for 100 consecutive iterations are shown at several different adaptive stages. It is clear that self-repeating patterns emerge and that

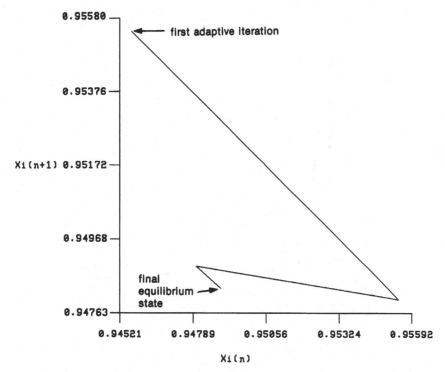

Figure 1.5   One-dimensional map of $x_i(n+1)$ versus $x_i(n)$, $i = 20$, for the case $Re = 1.5 \times 10^4$ with 100 nodes, $R_1 = 0.2$, $R_2 = 0.0$.

solutions never settle down to the equilibrium state. This characteristic is consistent with the well-established findings of the chaotic behavior of dynamic systems.

Similar observations can be made from many other cases with different values of parameters. Figure 1.9 shows the cases of $R_1 = 0.2$ and $R_2 = 0.7$, and with varying Re essentially all of the discussions made above are applicable here, too.

The above results indicate that both the equilibrium state and the chaotic state of the grid distribution can be observed, depending on the choice of R_1 and R_2. For the fixed values of R_1 and R_2, the dynamic structure of the present adaptive grid method remains unchanged for a wide range of Reynolds number. However, the randomness of the structure increases with Reynolds number. The dynamic structure illustrated here is not caused by the inaccuracy of the numerical solution of the model equation. In fact, it has been demonstrated that even a highly accurate numerical solution on a uniform grid can result in a chaotic pattern of the adaptive grid distribution. The choice of R_1 and R_2 is the key to the phenomenon observed here, that is, it is the internal mechanism of the adaptive grid method itself that causes the chaos of grid distribution. The materials presented in this section mainly serve to point out the potential difficulties one may encounter when using computational tools to analyze complex transport problems.

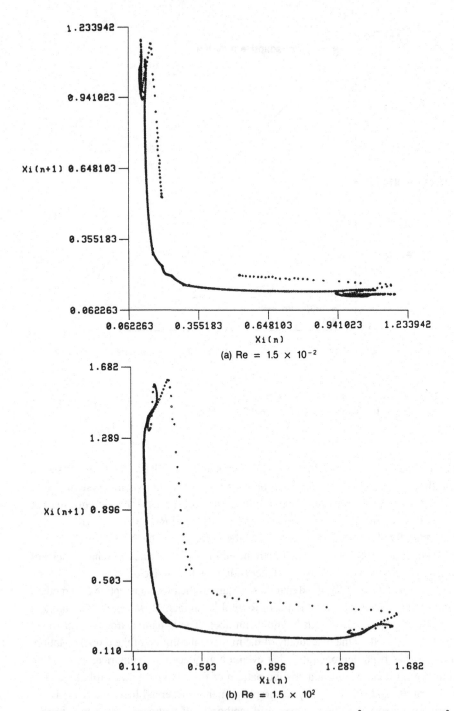

Figure 1.6   One-dimensional maps for the cases (a) $Re = 1.5 \times 10^{-2}$, (b) $Re = 1.5 \times 10^{2}$, (c) $Re = 1.5 \times 10^{3}$, (d) $Re = 1.5 \times 10^{4}$, (e) $Re = 1.5 \times 10^{5}$, and (f) $Re = 1.5 \times 10^{9}$ with 100 nodes, $R_1 = 0.0$, $R_2 = 0.9$. Shown are the maps for the 20th node.

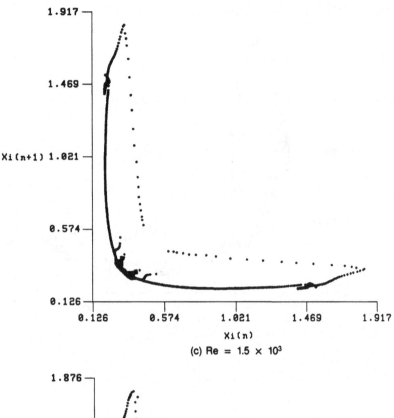

(c) Re = 1.5 × 10³

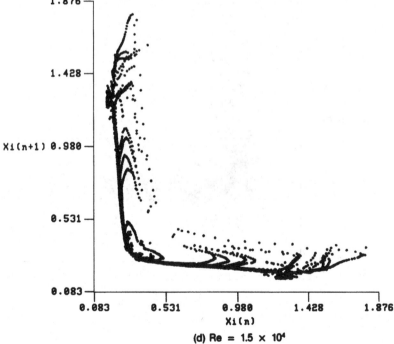

(d) Re = 1.5 × 10⁴

Figure 1.6   (cont.)

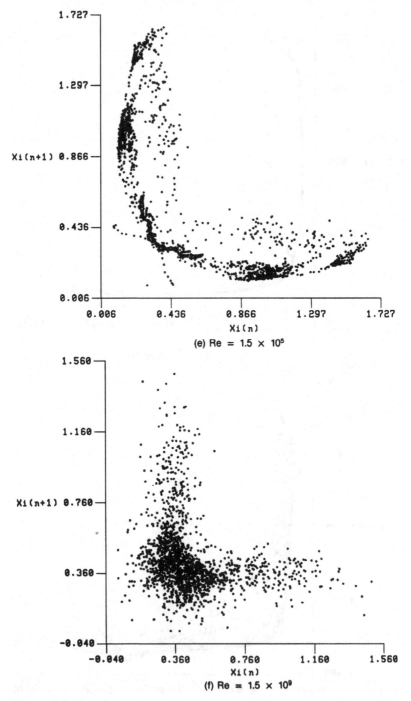

(e) Re = 1.5 × 10⁵

(f) Re = 1.5 × 10⁹

Figure 1.6   (cont.)

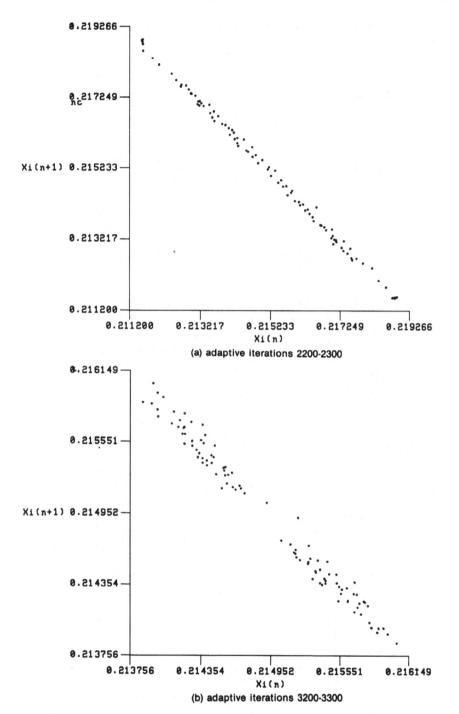

Figure 1.7 Detailed maps for the case $Re = 1.5 \times 10^{-2}$, with 100 nodes, $R_1 = 0.0$, $R_2 = 0.9$. (a) Adaptive iterations 1100–2300; (b) adaptive iterations 3200–3300; (c) adaptive iterations 4200–4300; (d) adaptive iterations 4800–4900.

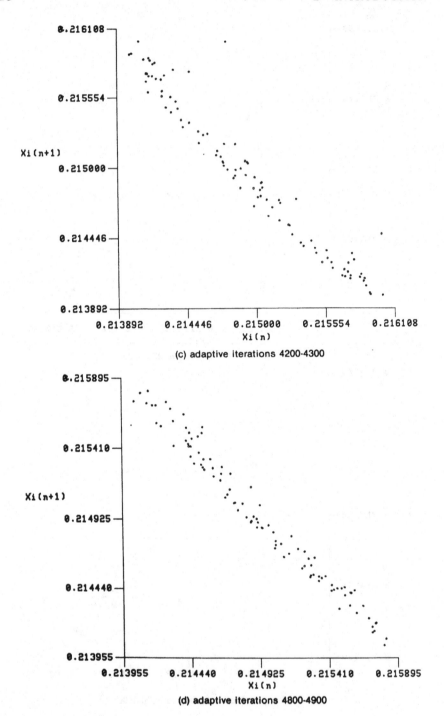

(c) adaptive iterations 4200-4300

(d) adaptive iterations 4800-4900

Figure 1.7　(cont.)

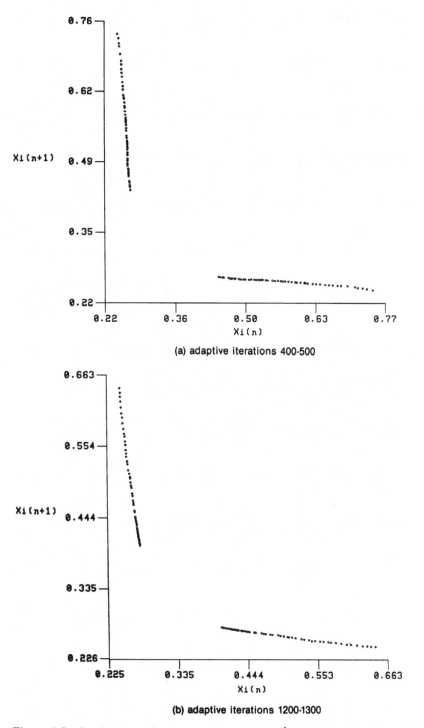

(a) adaptive iterations 400-500

(b) adaptive iterations 1200-1300

Figure 1.8 Detailed maps for the case $Re = 1.5 \times 10^4$, with 100 nodes, $R_1 = 0.0$, $R_2 = 0.9$. (a) Adaptive iterations 400–500; (b) adaptive iterations 1200–1300; (c) adaptive iterations 3250–3350; (d) adaptive iterations 4500–4600.

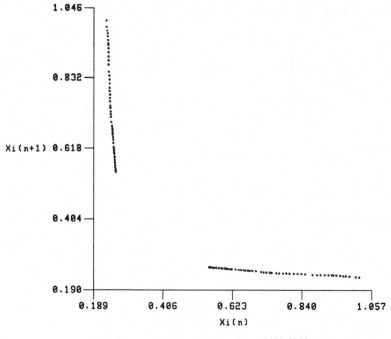

(c) adaptive iterations 3250-3350

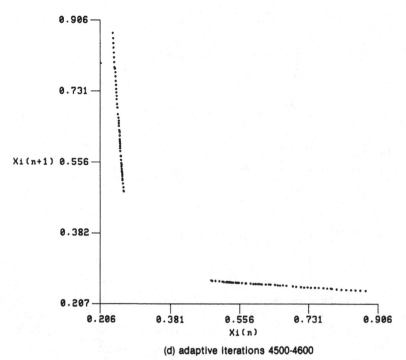

(d) adaptive iterations 4500-4600

Figure 1.8   (cont.)

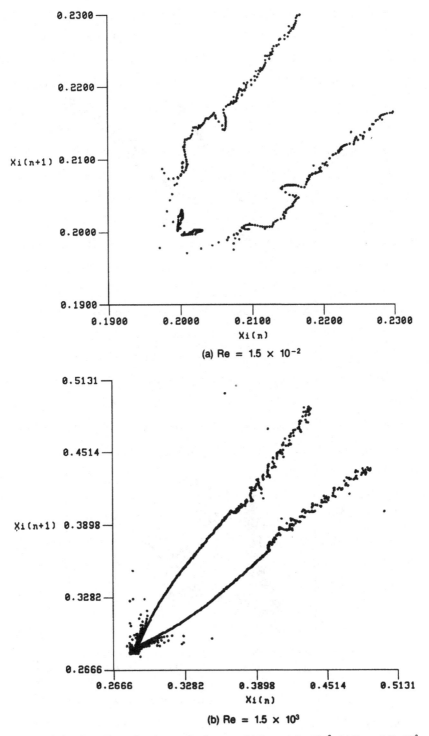

(a) Re = 1.5 × 10⁻²

(b) Re = 1.5 × 10³

Figure 1.9   One-dimensional maps for the cases (a) $Re = 1.5 \times 10^{-2}$, (b) $Re = 1.5 \times 10^{3}$, (c) $Re = 1.5 \times 10^{4}$, and (d) $Re = 1.5 \times 10^{5}$ with 100 nodes, $R_1 = 0.2$, $R_2 = 0.7$. Shown are the maps for the 20th node.

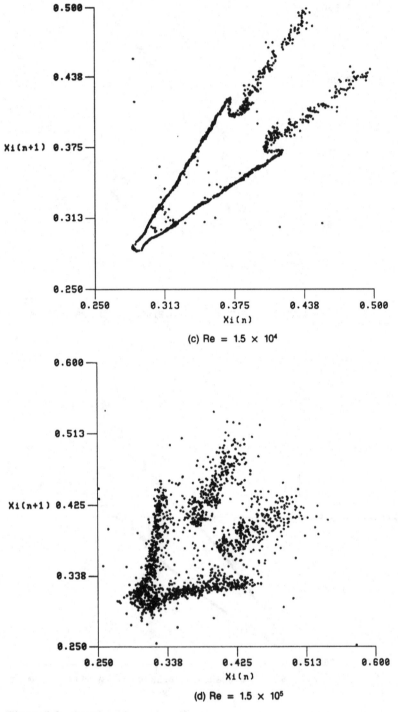

(c) Re = 1.5 × 10⁴

(d) Re = 1.5 × 10⁵

Figure 1.9   (cont.)

1.3 Scope of the Present Book

In the present effort, two types of complexity, namely dynamic complexity and geometric complexity, will be addressed using computational tools. With the rapid progress made in both computer hardware and software, we are now at a point that many practical problems can be handled virtually routinely. However, for a large number of truly difficult problems, fundamental difficulties in both physical modeling and numerical computations remain. In the present text, a computational framework will be presented to address both dynamic and geometric complexities in a comprehensive manner. The materials presented are guided by the viewpoint that a pragmatic, multiple-resolution modeling and computational strategy capable of handling the existence of distinguishable competing mechanisms, acting at disparate scales and speeds, is very desirable. Specifically, the challenge in the present context is to develop a *computational capability* for solving general transport equations to simulate the physical phenomena at both macroscopic and microscopic scales; a modeling strategy capable of handling the mutual influence between different scales in the form of, say, modified source terms, effective transport properties such as diffusivities, or boundary conditions; and a *resolution patching* capability based on multiblock techniques to allow different scales, or geometrical details, to be patched together if different numerical resolutions are needed in different spatial domains.

In addition to gridding issues, one needs to discuss formulation of the field equations solver, treatment of the conservation laws in a discrete form, convection and diffusion treatments, and solution procedure and convergence. These aspects have been presented in some detail in a recent book (Shyy 1994) and will be further expanded. Specifically, we will present recent development of high-accuracy convection and pressure treatments. Recent development in the context of the pressure-based algorithm will be highlighted with applications to both incompressible and compressible flows. These aspects are presented in Chapter 2. With a large set of differential equations system discretized on a large number of meshes, substantial computing effort is frequently needed to obtain convergent solutions. In order to expedite the runtime of a given computation, both convergence acceleration employing the multigrid technique and parallel computing can be highly useful. These two aspects will be discussed in Chapter 3. The multiblock technique will then be addressed in detail in Chapter 4, including the issue of conservative treatment across discontinuous grid interface and data structure.

On the physical side, topics to be presented are turbulence modeling and macroscopic transport equations in the context of engineering computations. In Chapter 5, the turbulence closure issues will be addressed with emphasis on the nonequilibrium, rotational, and compressibility effects. In Chapter 6, a multiscale model based on the concept of volume averaging will be developed, using phase change problems as the main illustration. Finally, in Chapter 7, we present several practical applications to help illustrate the benefit of employing the numerical and physical tools discussed.

2 Numerical Scheme for Treating Convection and Pressure

In this chapter we present some recent developments in the treatment of convective and pressure fluxes. There is an abundance of information on convection schemes for Euler and Navier-Stokes equations in the literature. No attempt is made here to summarize the various schemes proposed to date. Instead we focus on a controlled variation scheme (CVS) developed in the context of pressure-based algorithms that treat the convective and pressure fluxes separately. The attention of this chapter is on developing the formalism for TVD-based higher-order schemes for pressure-based methods and on demonstrating the accuracy of such schemes. We first begin with a summary of the basic methodology of pressure-based methods (Patankar 1980, Shyy 1994), which constitute one of several classes of successful techniques for handling complex transport problems (Ferziger and Peric 1996, Fletcher 1988, Hirsch 1990). The discussion presented in this chapter is equally applicable to other numerical algorithms as well.

2.1 Summary of Pressure-Based Algorithms

All the computations to be presented in this book have been conducted with a pressure-based algorithm that has been very widely used for several engineering applications. A more detailed discussion of pressure-based methods for complex geometries can be found in Shyy (1994). The algorithm we have developed can handle complex three-dimensional flow domains. However, in this chapter, we will present the governing equations in two dimensions for the sake of clarity of development of the controlled variation scheme presented in the next section. The extension to three-dimensional cases is straightforward (Shyy 1994).

2.1.1 Governing Equations and Numerical Algorithm

The two-dimensional steady-state, incompressible, constant-property Navier-Stokes equations are used:

$$\frac{\partial}{\partial x}(\varrho u) + \frac{\partial}{\partial y}(\varrho v) = 0, \tag{2.1}$$

$$\frac{\partial}{\partial x}(\varrho u u) + \frac{\partial}{\partial y}(\varrho v u) = -\frac{\partial p}{\partial x} + \frac{\partial}{\partial x}\left(\mu\frac{\partial u}{\partial x}\right) + \frac{\partial}{\partial y}\left(\mu\frac{\partial u}{\partial y}\right), \tag{2.2}$$

$$\frac{\partial}{\partial x}(\varrho u v) + \frac{\partial}{\partial y}(\varrho v v) = -\frac{\partial p}{\partial y} + \frac{\partial}{\partial x}\left(\mu\frac{\partial v}{\partial x}\right) + \frac{\partial}{\partial y}\left(\mu\frac{\partial v}{\partial y}\right). \tag{2.3}$$

With the introduction of the coordinate transformation $\xi = \xi(x, y)$, $\eta = \eta(x, y)$, the equations above are then cast into the curvilinear coordinates (Shyy 1994),

$$\frac{\partial}{\partial \xi}(\varrho U) + \frac{\partial}{\partial \eta}(\varrho V) = 0, \tag{2.4}$$

$$\frac{\partial}{\partial \xi}(\varrho U u) + \frac{\partial}{\partial \eta}(\varrho V u) = -y_\eta\frac{\partial p}{\partial \xi} + y_\xi\frac{\partial p}{\partial \eta} + \frac{\partial}{\partial \xi}\left[\frac{\mu}{J}(q_1 u_\xi - q_2 u_\eta)\right]$$
$$+ \frac{\partial}{\partial \eta}\left[\frac{\mu}{J}(-q_2 u_\xi + q_3 u_\eta)\right], \tag{2.5}$$

$$\frac{\partial}{\partial \xi}(\varrho U v) + \frac{\partial}{\partial \eta}(\varrho V v) = +x_\eta\frac{\partial p}{\partial \xi} - x_\xi\frac{\partial p}{\partial \eta} + \frac{\partial}{\partial \xi}\left[\frac{\mu}{J}(q_1 v_\xi - q_2 v_\eta)\right]$$
$$+ \frac{\partial}{\partial \eta}\left[\frac{\mu}{J}(-q_2 v_\xi + q_3 v_\eta)\right], \tag{2.6}$$

where

$$U = u y_\eta - v x_\eta, \qquad V = v x_\xi - u y_\xi, \tag{2.7}$$

$$q_1 = x_\eta^2 + y_\eta^2, \qquad q_3 = x_\xi^2 + y_\xi^2, \tag{2.8}$$

$$q_2 = x_\xi x_\eta + y_\xi y_\eta, \qquad J = x_\xi y_\eta - x_\eta y_\xi. \tag{2.9}$$

There are several choices of the grid and variable arrangements (Ferziger and Peric 1996, Shyy 1994, Shyy and Vu 1991), depending on whether the flow variables are arranged in a staggered or collocated manner. There are several issues involved in grid arrangement, including (i) boundary treatment; (ii) implicitly or explicitly added numerical smoothing and impact on numerical accuracy, especially under the condition of large pressure and density variations; and (iii) programming and data structures. These issues have not been settled conclusively; however, suffice it to say that with due care and consistency, success can be obtained with any reasonable choice. In the following, a staggered grid system is adopted, as shown in Fig. 2.1, for discretization of the above equations. Scalar variables, such as pressure (p), are located at the center of the control volume. Both u and U are located at the midpoints of the east and west faces of the control volume. Both v and V are located at the midpoints of the north and south faces of the control volume. In terms of the notation shown in Fig. 2.1, for a node P enclosed in its cell and surrounded by its neighbors N, S, E, and W, the finite-difference approximation to the momentum equations can be obtained by taking the integral of the momentum equations over the control volume. By arbitrarily taking $\Delta\xi = \Delta\eta = 1$, the resulting momentum equations yield:

$$\left[\varrho U u + y_\eta p - \frac{\mu}{J}(q_1 u_\xi - q_2 u_\eta)\right]\Big|_w^e$$
$$+ \left[\varrho V u - y_\xi p - \frac{\mu}{J}(-q_2 u_\xi + q_3 u_\eta)\right]\Big|_s^n = 0 \tag{2.10}$$

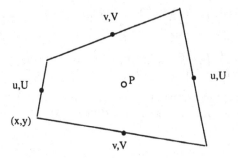

(i) Configuration of a staggered grid system

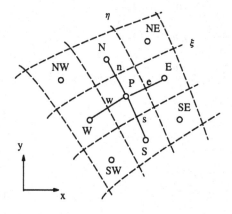

(ii) Curvilinear grid system

Figure 2.1   Staggered grid and notation for curvilinear grid system.

$$\left[\varrho U v - x_\eta p - \frac{\mu}{J}(q_1 v_\xi - q_2 v_\eta)\right]\Big|_w^e$$

$$+ \left[\varrho V v + x_\xi p - \frac{\mu}{J}(-q_2 v_\xi + q_3 v_\eta)\right]\Big|_s^n = 0. \tag{2.11}$$

Furthermore, the above equations can be put into a generalized form as:

$$\left[\varrho U \Phi - \frac{\mu}{J}(q_1 \Phi_\xi - q_2 \Phi_\eta)\right]\Big|_w^e + \left[\varrho V \Phi - \frac{\mu}{J}(-q_2 \Phi_\xi + q_3 \Phi_\eta)\right]\Big|_s^n = S \cdot J \tag{2.12}$$

where Φ is the general dependent variable and S is the source term. With appropriate finite-difference schemes representing the convective and diffusive terms at the control volume boundaries, the discretized equation relating the variable at a central point P to its neighboring values is obtained (Shyy et al. 1985):

$$A_P \Phi_P = A_E \Phi_E + A_W \Phi_W + A_S \Phi_S + A_N \Phi_N + S \tag{2.13}$$

where the coefficients result from different numerical schemes chosen in the course of discretization. The pressure terms and the cross-derivative portion of the viscous terms, due to the nonorthogonal grid effects, are taken into the source term, S. Specifically, for example, if the central-difference scheme is used for the convection and diffusion terms, the coefficients are

$$A_E = \left.\frac{\mu}{J}q_1\right|_e - \left.\frac{1}{2}\varrho U\right|_e \tag{2.14}$$

$$A_W = \left.\frac{\mu}{J}q_1\right|_w + \left.\frac{1}{2}\varrho U\right|_w \tag{2.15}$$

$$A_N = \left.\frac{\mu}{J}q_3\right|_n - \left.\frac{1}{2}\varrho V\right|_n \tag{2.16}$$

$$A_S = \left.\frac{\mu}{J}q_3\right|_s + \left.\frac{1}{2}\varrho V\right|_s \tag{2.17}$$

$$A_P = A_E + A_W + A_N + A_S. \tag{2.18}$$

The continuity equation can be written in a similar discretized form over each control volume as follows:

$$(y_\eta \varrho u - x_\eta \varrho v)_e - (y_\eta \varrho u - x_\eta \varrho v)_w + (-y_\xi \varrho u + x_\xi \varrho v)_n$$
$$- (-y_\xi \varrho u + x_\xi \varrho v)_s = 0. \tag{2.19}$$

Let u^* and v^* be the intermediate velocity components with a given distribution of pressure, p^*. Since in general the mass continuity equation is not satisfied by this intermediate velocity field, the pressure p^* must be corrected. The corrected pressure is obtained from (Patankar 1980)

$$p = p^* + p' \tag{2.20}$$

where p' is called the pressure correction. The corresponding velocity corrections u' and v' can be introduced in a similar manner:

$$u = u^* + u', \qquad v = v^* + v'. \tag{2.21}$$

To derive the pressure correction equation, u^* and v^* are obtained from the momentum equations as follows:

$$u_P^* = \sum_{i=E,W,N,S} \frac{A_i^u}{A_P^u} u_i^* + D^u + (B^u p_\xi^* + C^u p_\eta^*) \tag{2.22}$$

$$v_P^* = \sum_{i=E,W,N,S} \frac{A_i^v}{A_P^v} v_i^* + D^v + (B^v p_\xi^* + C^v p_\eta^*) \tag{2.23}$$

where D^u and D^v are, respectively, the cross-derivative viscous terms in the u- and v-momentum equations, and

$$B^u = -\frac{y_\eta}{A_P^u}, \qquad C^u = \frac{y_\xi}{A_P^u},$$
$$ \tag{2.24}$$
$$B^v = \frac{x_\eta}{A_P^v}, \qquad C^v = -\frac{x_\xi}{A_P^v}.$$

The velocity components are thus corrected by the following formulas:

$$u = u^* + (B^u p_\xi' + C^u p_\eta') \tag{2.25}$$

$$v = v^* + (B^v p_\xi' + C^v p_\eta'). \tag{2.26}$$

Subsequently, the corresponding correction forms for U and V are obtained by substituting Eqs. (2.25)–(2.26) into Eq. (2.7):

$$U = U^* + (B^u y_\eta - B^v x_\eta) p'_\xi + (C^u y_\eta - C^v x_\eta) p'_\eta \tag{2.27}$$

$$V = V^* + (B^v x_\xi - B^u y_\xi) p'_\xi + (C^v x_\xi - C^v y_\xi) p'_\eta \tag{2.28}$$

where U^* and V^* are calculated based on u^* and v^*. It is noted that the continuity equation, Eq. (2.19), can also be written in the following finite-difference form:

$$(\varrho U)_e - (\varrho U)_w + (\varrho V)_n - (\varrho V)_s = 0. \tag{2.29}$$

To retain a five-point approximation in the pressure-correction equation, the p'_η term in Eq. (2.27) and the p'_ξ term in Eq. (2.28) are dropped, leading to the following simplified correction equations for U and V:

$$U = U^* + (B^u y_\eta - B^v x_\eta) p'_\xi \tag{2.30}$$

$$V = V^* + (C^v x_\xi - C^v y_\xi) p'_\eta. \tag{2.31}$$

These equations are then substituted into Eq. (2.29) to obtain the following pressure-correction equations:

$$a_P p'_P = a_E p'_E + a_W p'_W + a_N p'_N + a_S p'_S + S_P \tag{2.32}$$

$$a_P = a_E + a_W + a_N + a_S \tag{2.33}$$

$$S_P = (\varrho U^*)_w - (\varrho U^*)_e + (\varrho V^*)_s - (\varrho V^*)_n \tag{2.34}$$

with the coefficients given by:

$$a_E = \varrho \left(\frac{y_\eta^2}{A_P^u} + \frac{x_\eta^2}{A_P^v} \right)_e \tag{2.35}$$

$$a_W = \varrho \left(\frac{y_\eta^2}{A_P^u} + \frac{x_\eta^2}{A_P^v} \right)_w \tag{2.36}$$

$$a_N = \varrho \left[\frac{x_\xi^2}{A_P^v} + \frac{y_\xi^2}{A_P^u} \right]_n \tag{2.37}$$

$$a_S = \varrho \left[\frac{x_\xi^2}{A_P^v} + \frac{y_\xi^2}{A_P^v} \right]_s. \tag{2.38}$$

2.1.1.1 PRESSURE BOUNDARY CONDITIONS

Usually, two types of boundary conditions can be adopted for the pressure-correction equation. If the pressure is known at the boundary, the pressure correction is zero there. If the velocity is known at the boundary, then, according to Eq. (2.30) or Eq. (2.31),

$$U = U^* \quad \text{or} \quad V = V^* \tag{2.39}$$

or

$$p'_\xi = 0 \quad \text{or} \quad p'_\eta = 0, \tag{2.40}$$

that is, the Neumann type of boundary condition is applied to the pressure correction equation. At outlets, where usually the velocity is not known, the Neumann type of boundary condition is applied iteratively by updating the exit velocity based on the global mass conservation (Shyy 1994).

2.1.2 Solution Procedure

In single-block grid computations, the solution procedure is as follows: (a) The momentum equations are solved to obtain the Cartesian velocity components with the guessed pressure field. When solving the momentum equations, the contravariant velocity components U and V are calculated after updating each of the Cartesian velocity components. (b) With the updated U and V, the pressure-correction equation is solved to obtain p'. (c) The Cartesian velocity and pressure field are updated with the solution of the pressure-correction equation, employing the Dyakonov iteration procedure (Braaten and Shyy 1986, Shyy 1994). Steps (a)–(c) are repeated until the momentum and continuity equations are simultaneously satisfied to the required degree of accuracy.

2.2 Treatment of Convection and Pressure Splitting

An enormous variety of discretization schemes have been proposed for the inviscid fluxes in the Navier-Stokes equations (see, for example, Hirsch 1990). For incompressible flows, several alternatives have been commonly used, including first- and second-order upwinding, central differencing and QUICK (Quadratic Upwind Interpolation for Convective Kinematics), among others (see, e.g., Shyy 1994). With an intention of resolving sharp gradients in compressible flows, several TVD (total variation diminishing) and ENO (essentially nonoscillatory) schemes have been developed. Our objective in this book is to combine the various components in the framework of a pressure-based method (SIMPLE) to form an algorithm for accurately and efficiently handling complex transport phenomena, of which the first component is the discretization scheme. Thus, in this section we focus on developing a scheme called the controlled variation scheme (CVS) with an intent for implementing it in a pressure-based algorithm, which is discussed in a later section. We begin the discussion with a brief review of available high-resolution schemes, followed by the motivation and the need for a scheme like the CVS for pressure-based methods.

For the hyperbolic system of Euler equations, a number of high-resolution schemes have been proposed. Most of these schemes are designed to satisfy the TVD property for scalar conservation laws and systems of equations with constant coefficients, whereby spurious oscillations are suppressed. Several different approaches can be found in the literature, such as the modified flux approach (scalar diffusion) of Harten (1983), flux vector splittings (Steger and Warming 1981, van Leer 1982), flux difference splittings (Osher 1984, Roe 1981), and so on. All these schemes developed for the Euler equations can be directly extended to the Navier-Stokes equations.

The main motivation behind the various approaches is to achieve high accuracy and efficiency in numerical computations, especially for complex flows that may involve strong convective effects, sharp gradients, recirculation, chemical reactions, and turbulence models. Different schemes have different accuracy and efficiency characteristics. For example, flux vector splitting schemes are quite efficient and relatively simple but produce excessive smearing. Moreover, the Steger–Warming splitting (Steger and Warming 1981) produces glitches at points where eigenvalues change sign, such as sonic points. The van Leer splitting (van Leer 1982) is designed to remedy this, but it suffers from excessive numerical diffusion in viscous regions. Subsequent efforts have been made to reduce this diffusion (Hänel and Schwane 1989). On the other hand, flux difference splittings, such as Roe and Osher splittings, have substantially lesser numerical diffusion. However, they too are known to yield inaccurate results in some simple flows. For example, Roe splitting produces nonphysical "carbuncle" shocks in supersonic flows over blunt bodies (Liou and Steffen 1993).

Taking all the above factors into account, there is continued interest and ongoing effort in the development of new schemes that are robust in terms of accuracy as well as efficiency. Towards this end, one promising approach is the treatment of convective and pressure fluxes as two separate entities. Employing this idea, Thakur and Shyy (Shyy and Thakur 1994a,b, Thakur and Shyy 1992, 1993) have developed a controlled variation scheme (CVS) in which the convective flux is estimated using Harten's second-order TVD scheme (modified flux approach) where the local characteristic speeds of the different equations are coordinated by assigning them the values of the local convective speeds; the pressure terms are treated as source terms and are central differenced or treated in a special manner by employing Strang's time-splitting technique. The main objective in these studies has been to develop these higher-resolution schemes for sequential solvers using pressure-based algorithms. Liou and Steffen (1993) have also proposed a scheme, called the Advection Upstream Splitting Method (AUSM), which treats the convective terms and the pressure terms separately. In the AUSM scheme, the interface convective velocity is obtained by an appropriate splitting, and the convected variable is upwinded based on the sign of the interface convective velocity. The pressure terms are also handled using an appropriate splitting formula.

In the CVS presented in this section, guided by the eigenvalues of the total flux as well as the individual convective and pressure fluxes, the treatment is as follows. The convective flux is fully upwinded, whereas the pressure flux is split yielding contributions from upstream and downstream neighbors. Two different formulations, which lead to different pressure fluxes, are discussed. The eigenvalues of the respective pressure fluxes are used to interpret the physical significance of the two formulations. It is shown that the most desirable formulation is perhaps that which is consistent with the physical mechanism that the convective fluxes get transported at the mean convection speed and the pressure signals propagate both upstream and downstream in subsonic flows. Two one-dimensional test cases – the standard shock tube problem and a longitudinal combustion instability problem previously investigated by Shyy et al. (1993) – are used to demonstrate that the CVS and the AUSM scheme yield accuracy comparable to the Roe scheme. The results for the combustion instability problem, in particular, will illustrate that the approach of treating convective and pressure fluxes separately can indeed coordinate signal propagation, even in the presence of source terms such as heat release.

2.2.1 Estimation of the Fluxes for the CVS and AUSM Schemes

We use the one-dimensional system of conservation laws for an ideal gas as an illustration:

$$\frac{\partial W}{\partial t} + \frac{\partial F}{\partial x} = 0 \tag{2.41a}$$

where

$$W = \begin{bmatrix} \varrho \\ m \\ E \end{bmatrix} \qquad F = \begin{bmatrix} \varrho u \\ mu + p \\ (E + p)u \end{bmatrix} = \begin{bmatrix} \varrho u \\ mu + p \\ Hu \end{bmatrix}. \tag{2.41b}$$

Here, m is the momentum, E is the total energy, $E = \varrho(e + u^2/2)$, and $H = E + p$ is the total enthalpy. A numerical scheme for Eq. (2.41a) can be written as, for example,

$$W_i^{n+1} + \lambda\theta\left(F_{i+\frac{1}{2}}^{n+1} - F_{i-\frac{1}{2}}^{n+1}\right) = W_i^n - \lambda(1 - \theta)\left(F_{i+\frac{1}{2}}^n - F_{i-\frac{1}{2}}^n\right) \tag{2.42}$$

where $\lambda = \Delta t / \Delta x$, $F_{i\pm\frac{1}{2}}$ are the numerical fluxes at the control volume interfaces, the superscripts n and $n + 1$ represent time levels, and θ is a measure of implicitness of the scheme. We obtain explicit, fully implicit, and Crank–Nicolson schemes for $\theta = 0$, 1, and 1/2, respectively.

A recent approach is to treat convection and acoustic wave propagation as physically distinct (but coupled) mechanisms. The breakup of the total flux into convective and pressure fluxes can be done in at least two different ways, as presented next.

Formulation 1: Based on Total Enthalpy. One way of breaking up the total flux into convective and pressure fluxes is to treat the total energy flux (Hu) as part of the convective flux. Thus, the pressure flux consists of just the p term in the momentum flux:

$$F = F^c + F^p = \begin{bmatrix} \varrho u \\ mu \\ Hu \end{bmatrix} + \begin{bmatrix} 0 \\ p \\ 0 \end{bmatrix} \tag{2.43a}$$

$$= M\Phi + F^p = M\begin{bmatrix} \varrho a \\ \varrho u a \\ \varrho H a \end{bmatrix} + \begin{bmatrix} 0 \\ p \\ 0 \end{bmatrix}. \tag{2.43b}$$

Such a breakup of the flux has been used, for example, in the AUSM scheme of Liou and Steffen (1993).

Formulation 2: Based on Total Energy. Another way of breaking up the total flux into convective and pressure fluxes is to treat the energy flux (Eu) as part of the convective flux. Thus, the pressure flux now consists of p and pu terms:

$$F = F^c + F^p = \begin{bmatrix} \varrho u \\ mu \\ Eu \end{bmatrix} + \begin{bmatrix} 0 \\ p \\ pu \end{bmatrix}. \tag{2.44}$$

We first present the treatment of the convective flux for either of the above two formulations.

2.2.2 Convective Fluxes

2.2.2.1 CONTROLLED VARIATION SCHEME (CVS)

In this chapter, we will present the development and assessment of the CVS for segregated (or sequential) solution techniques in which the governing equations are numerically treated as a collection of scalar conservation laws instead of a simultaneous system. Even though these laws are not independent, they can be treated individually in numerical procedures. Such has been the approach taken in previous studies (Shyy and Thakur 1994a,b, Thakur and Shyy 1992, 1993). With this in view, let w represent the dependent variable of each of the scalar conservation laws comprising the system (2.41). A typical scalar conservation law using an explicit scheme can be written as

$$w_i^{n+1} = w_i^n - \lambda\left(f_{i+\frac{1}{2}}^n - f_{i-\frac{1}{2}}^n\right). \tag{2.45}$$

Using a TVD formulation originally proposed by Harten (1983), the numerical flux f can be written as

$$f_{i+\frac{1}{2}} = \frac{1}{2}\left\{f_i + f_{i+1} - Q\left(b_{i+\frac{1}{2}}\right)\Delta_{i+\frac{1}{2}}w\right\} \tag{2.46}$$

where Q is the convective dissipation function given by

$$Q_{i+\frac{1}{2}} \equiv Q\left(b_{i+\frac{1}{2}}\right) = \begin{cases} \frac{1}{2}\left(\frac{b^2}{\delta} + \delta\right), & \text{if } |b| < \delta \\ |b|, & \text{if } |b| \geq \delta \end{cases} \tag{2.47a}$$

and

$$\Delta_{i+\frac{1}{2}}w = w_{i+1} - w_i. \tag{2.47b}$$

The parameter δ in Eq. (2.47a) is used to eliminate the violation of the entropy condition for characteristic speeds close to zero (Harten 1983), and $b_{i+\frac{1}{2}}$ is the local characteristic speed on the right interface of the control volume. The local characteristic speed $b_{i+\frac{1}{2}}$ is conventionally defined as follows (Harten 1983):

$$b_{i+\frac{1}{2}} = \begin{cases} \frac{f_{i+1}-f_i}{\Delta_{i+\frac{1}{2}}w}, & \text{if } \Delta_{i+\frac{1}{2}}w \neq 0 \\ \frac{\partial f}{\partial w}, & \text{if } \Delta_{i+\frac{1}{2}}w = 0 \end{cases} \tag{2.48a}$$

and

$$\Delta_{i+\frac{1}{2}}w = w_{i+1} - w_i. \tag{2.48b}$$

The CVS utilizes the above form of TVD type schemes while defining the characteristic speeds in a different way. In the CVS, the local characteristic speed $b_{i+\frac{1}{2}}$ for the system (2.41) is defined as the local convective speed:

$$b_{i+\frac{1}{2}} = \frac{1}{2}\left(u_i + u_{i+1}\right). \tag{2.49}$$

It has been shown previously (Shyy and Thakur 1994a,b, Thakur and Shyy 1992, 1993) that while solving the system of Euler or Navier-Stokes equations in a sequential manner, the use of Eq. (2.48a) to yield a different characteristic speed for each of the individual equations leads to nonphysical numerical oscillations. On the other hand, the use of Eq. (2.49) as the common interface characteristic speed for all the equations leads to a coordination of signal propagation eliminating any oscillations.

In the present work, we employ the explicit scheme ($\theta = 0$) for one-dimensional unsteady-flow problems and the fully implicit scheme (i.e., $\theta = 1$) for multidimensional steady-flow cases as the basis for development of the CVS. For the latter, the implicit and highly nonlinear equations would require iterations at every timestep if a time-stepping approach to steady state is employed. If an infinite timestep is chosen to solve for steady state, as in the present study, the number of iterations required to achieve convergence will be very large. Consequently, some linearized versions of implicit TVD schemes have been devised (Yee 1986, 1987, Yee et al. 1985). We base the CVS on the linearized nonconservative implicit (LNI) scheme described by Yee (1986, 1987), following which $f_{i+\frac{1}{2}} - f_{i-\frac{1}{2}}$ can be written as

$$
\begin{aligned}
\left[f_{i+\frac{1}{2}} - f_{i-\frac{1}{2}}\right]^{n+1} &= \frac{1}{2}\left[f_{i+1} + f_i - Q\left(b_{i+\frac{1}{2}}\right)\Delta_{i+\frac{1}{2}}w - f_i - f_{i-1}\right. \\
&\quad \left. + Q\left(b_{i-\frac{1}{2}}\right)\Delta_{i-\frac{1}{2}}w\right]^{n+1} \\
&= \frac{1}{2}\left[b_{i+\frac{1}{2}} - Q_{i+\frac{1}{2}}\right]^{n+1}(w_{i+1} - w_i)^{n+1} \\
&\quad - \frac{1}{2}\left[-b_{i-\frac{1}{2}} - Q_{i-\frac{1}{2}}\right]^{n+1}(w_i - w_{i-1})^{n+1}.
\end{aligned} \tag{2.50}
$$

The superscripts n and $n+1$ signify the previous and current iteration levels at steady state, respectively. The above nonlinear equation can be linearized by dropping the superscripts of the coefficients of $\Delta_{i\pm\frac{1}{2}}w^{n+1}$ from $n+1$ to n. This form can be shown to be TVD (Yee 1987). This form of the implicit scheme cannot be expressed in the conservation form and thus it is nonconservative except at steady state, where it has been shown that it does reduce to a conservative form (Yee 1987).

2.2.2.2 THE AUSM SCHEME

We now briefly present the treatment of convective flux in the AUSM scheme proposed by Liou and Steffen (1993), which also treats the convective terms and pressure terms separately. For the AUSM scheme, the numerical convective flux at an interface is written as

$$
F^c_{i+\frac{1}{2}} = M_{i+\frac{1}{2}}\left(\Phi_{i+\frac{1}{2}}\right)_{L/R} \tag{2.51}
$$

where $M_{i+\frac{1}{2}}$ is the interface convective velocity and Φ is the convected variable.

The interface convective Mach number is expressed as the sum of the split values of the positive and negative contributions from the left and right states of the interface:

$$
M_{i+\frac{1}{2}} = \left(M_{i+\frac{1}{2}}\right)^+_L + \left(M_{i+\frac{1}{2}}\right)^-_R. \tag{2.52}
$$

For first-order accuracy, the left (L) and the right (R) states on the $\left(i + \frac{1}{2}\right)$ interface are obtained by a first-order extrapolation of the nodal values of the left and the right neighbors of the interface:

$$\left(M_{i+\frac{1}{2}}\right)_L^+ = M_i^+, \qquad \left(M_{i+\frac{1}{2}}\right)_R^- = M_{i+1}^-. \tag{2.53}$$

The spitting chosen here (based on the van Leer splitting for the Euler equations) is the following:

$$M^\pm = \begin{cases} \pm\frac{1}{4}(M \pm 1)^2, & \text{if } |M| \leq 1, \\ \frac{1}{2}(M \pm |M|), & \text{otherwise.} \end{cases} \tag{2.54}$$

The convected variable is upwinded, depending on the sign of the interface velocity, as follows:

$$\left(\Phi_{i+\frac{1}{2}}\right)_{L/R} = \begin{cases} \left(\Phi_{i+\frac{1}{2}}\right)_L, & \text{if } M_{i+\frac{1}{2}} \geq 0, \\ \left(\Phi_{i+\frac{1}{2}}\right)_R, & \text{otherwise.} \end{cases} \tag{2.55}$$

For first-order accuracy, a first-order extrapolation using one upwind nodal value is employed:

$$\left(\Phi_{i+\frac{1}{2}}\right)_L = \Phi_i, \qquad \left(\Phi_{i+\frac{1}{2}}\right)_R = \Phi_{i+1}. \tag{2.56}$$

2.2.2.3 FORMULATION OF THE FLUXES OF THE CVS AND AUSM SCHEMES

Wada and Liou (1994) have proposed a version of the AUSM scheme based on flux difference splitting, labeled as AUSMD. It is interesting to note that the estimation of the convective fluxes in the momentum equations by the AUSMD scheme is identical to that of the linearized CVS scheme for steady-state computations, except that the interface convective velocities are estimated differently in the two schemes. The net convective flux along the x direction, for example, using the CVS scheme, Eq. (2.50), with $\delta = 0$, can be expressed as follows:

$$\begin{aligned}
\left[(\varrho u^2)_{i+\frac{1}{2}} - (\varrho u^2)_{i-\frac{1}{2}}\right]^{\text{CVS}} &= \frac{1}{2}\varrho_{i+\frac{1}{2}}\left(u_{i+\frac{1}{2}} - \left|u_{i+\frac{1}{2}}\right|\right) \cdot (u_{i+1} - u_i) \\
&\quad - \frac{1}{2}\varrho_{i-\frac{1}{2}}\left(-u_{i-\frac{1}{2}} - \left|u_{i-\frac{1}{2}}\right|\right) \cdot (u_i - u_{i-1}) \\
&= \frac{1}{2}\varrho_{i+\frac{1}{2}}\left(u_{i+\frac{1}{2}} - \left|u_{i+\frac{1}{2}}\right|\right)(u_{i+1} - u_i) \\
&\quad - \frac{1}{2}\varrho_{i-\frac{1}{2}}\left(-u_{i-\frac{1}{2}} - \left|u_{i-\frac{1}{2}}\right|\right)(u_i - u_{i-1}). \tag{2.57}
\end{aligned}$$

In the AUSMD scheme, the convective flux is expressed as follows:

$$\begin{aligned}
\left[(\varrho u^2)_{i+\frac{1}{2}} - (\varrho u^2)_{i-\frac{1}{2}}\right]^{\text{AUSMD}} \\
= \frac{1}{2}\left[(\varrho u)_{i+\frac{1}{2}}(u_i + u_{i+1}) - \left|(\varrho u)_{i+\frac{1}{2}}\right|(u_{i+1} - u_i)\right] \\
- \frac{1}{2}\left[(\varrho u)_{i-\frac{1}{2}}(u_{i-1} + u_i) - \left|(\varrho u)_{i-\frac{1}{2}}\right|(u_i - u_{i-1})\right]. \tag{2.58a}
\end{aligned}$$

In terms of actual implementation, several different options are possible based on the expression for the mass flux (or m). For example, the original AUSM scheme is obtained if the mass flux is defined by:

$$(\varrho u)_{i+\frac{1}{2}} = \frac{1}{2}\left[u_{i+\frac{1}{2}}(\varrho_{i+1} + \varrho_i) - \left|u_{i+\frac{1}{2}}\right|(\varrho_{i+1} - \varrho_i)\right].$$

Similarly, a variant of the AUSM scheme can be obtained by expressing the mass flux in terms of the Mach number. Here, in order to demonstrate a basic similarity between the overall flux expressions for the CSV and the AUSMD schemes, the interface mass flux is expressed as the product of interface density (which is upwinded from the nodal values) and interface velocity. Thus, the right side of Eq. (2.58a) becomes

$$\frac{1}{2}\varrho_{i+\frac{1}{2}}\left(u_{i+\frac{1}{2}} - \left|u_{i+\frac{1}{2}}\right|\right)(u_{i+1} - u_i) - \frac{1}{2}\varrho_{i-\frac{1}{2}}\left(-u_{i-\frac{1}{2}} - \left|u_{i-\frac{1}{2}}\right|\right)(u_i - u_{i-1})$$
$$+ u_i\left(\varrho_{i+\frac{1}{2}}u_{i+\frac{1}{2}} - \varrho_{i-\frac{1}{2}}u_{i-\frac{1}{2}}\right). \tag{2.58b}$$

From Eqs. (2.57) and (2.58), it can be seen that

$$\left[(\varrho u^2)_{i+\frac{1}{2}} - (\varrho u^2)_{i-\frac{1}{2}}\right]^{\text{AUSMD}} = \left[(\varrho u^2)_{i+\frac{1}{2}} - (\varrho u^2)_{i-\frac{1}{2}}\right]^{\text{CVS}}$$
$$+ u_i\left(\varrho_{i+\frac{1}{2}}u_{i+\frac{1}{2}} - \varrho_{i-\frac{1}{2}}u_{i-\frac{1}{2}}\right). \tag{2.59}$$

The last term in the above expression is nothing but the nodal value of the dependent variable multiplied by the net mass flux in the x direction. A similar expression results from the convective fluxes along the y direction. Thus the difference between the numerical convective fluxes between the CVS and the AUSMD schemes is the net mass flux term integrated over a control volume, which must be zero at steady state (from the continuity equation). Thus, for steady state applications, the two flux estimations are identical. The only difference is the method of estimation of the interfacial velocities – the CVS scheme just averages the nodal point values (Eq. 2.49) whereas the AUSMD scheme uses splitting based on the local Mach number (Eq. 2.52).

2.2.3 Treatment of the Pressure Flux

As mentioned earlier, the pressure flux is treated separately in both CVS and AUSM schemes. Different approaches can be taken as described next.

2.2.3.1 SPLITTING OF THE p TERM

The p term in the momentum equation for both the formulations given by Eqs. (2.43) and (2.44) can be treated by splitting as follows:

$$F^p_{i+\frac{1}{2}} = p_{i+\frac{1}{2}} = \left(p_{i+\frac{1}{2}}\right)^+_L + \left(p_{i+\frac{1}{2}}\right)^-_R. \tag{2.60}$$

For first-order accuracy, the left (L) and the right (R) states on the $\left(i + \frac{1}{2}\right)$ interface are obtained by a first-order extrapolation of the nodal p values of, respectively, the left and the right neighbors of the interface (Fig. 2.2).

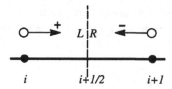

Figure 2.2   Schematic of the contributions from split pressures at an interface.

The splitting of pressure can be achieved in a manner similar to the van Leer splitting for the fluxes of the Euler equations (van Leer 1982). The van Leer splitting is based on the requirements that the split fluxes and their first derivatives be continuous and that the split fluxes be polynomials of the lowest possible degree. This leads to a splitting of the fluxes in terms of factors $(M \pm 1)^2$. Liou et al. (1990) have suggested a similar splitting for pressure:

$$
\begin{aligned}
p &= \alpha[(M+1)^2 - (M-1)^2] + \beta[(M+1)^2 + (M-1)^2] \\
&= 4\alpha M + 2\beta(M^2 + 1).
\end{aligned}
\tag{2.61}
$$

By choosing $2\beta = p$ (which is true for $M = 0$), we obtain

$$
\alpha = -\frac{1}{4}pM
\tag{2.62}
$$

and hence p can be written as

$$
p = \frac{p}{4}(M+1)^2(-M+2) + \frac{p}{4}(M+1)^2(M+2).
\tag{2.63}
$$

Thus, as suggested by Liou and Steffen (1993) and Liou et al. (1990), the following splitting is employed:

$$
p^{\pm} =
\begin{cases}
\frac{p}{4}(M \pm 1)^2(2 \mp M), & \text{if } |M| \le 1, \\
\frac{p}{2}(M \pm |M|)/M, & \text{otherwise.}
\end{cases}
\tag{2.64}
$$

Thus, for supersonic flow, the above formulation leads to full upwinding of pressure, namely,

$$
p_{i+\frac{1}{2}} = p_i, \quad \text{if } |M| > 1 \quad \text{and} \quad M > 0.
\tag{2.65}
$$

The splitting takes place only for subsonic flow where contributions from both upwind and downwind neighbors are taken into account:

$$
\begin{aligned}
p_{i+\frac{1}{2}} &= (p_i)^+ + (p_{i+1})^-, \quad \text{if } |M| \le 1 \\
&= \frac{1}{4}(M_i + 1)^2(2 - M_i)p_i + \frac{1}{4}(M_{i+1} - 1)^2(2 + M_{i+1})p_{i+1}.
\end{aligned}
\tag{2.66}
$$

This is consistent with the fact that pressure signal propagates only in the upwind direction for supersonic flows and in both upwind and downwind directions for subsonic flows.

An alternative form of dissipation for the pressure flux can also be devised. For example, the following form for the pressure flux at the interface has been utilized by some researchers (Jameson 1993, Tatsumi et al. 1995):

$$p_{i+\frac{1}{2}} = \frac{1}{2}(p_i + p_{i+1}) - \frac{1}{2}Q^p_{i+\frac{1}{2}}\Delta_{i+\frac{1}{2}}p \qquad (2.67a)$$

with

$$Q^p_{i+\frac{1}{2}} = \frac{1}{2}M_{i+\frac{1}{2}}\left(3 - M^2_{i+\frac{1}{2}}\right) \qquad (2.67b)$$

where $Q^p_{i+\frac{1}{2}}$ is the numerical diffusion introduced by the pressure splitting to the central difference flux and $M_{i+\frac{1}{2}}$ is the interface Mach number. It can be observed that as $M_{i+\frac{1}{2}}$ increases from 0 to 1, $Q^p_{i+\frac{1}{2}}$ varies from 0 to 1 smoothly, changing the nature of the pressure splitting from central differencing $\left(\text{for } M_{i+\frac{1}{2}} = 0\right)$ to full upwinding (for $M_{i+\frac{1}{2}} = 1$) in a continuous manner.

2.2.3.2   SPLITTING OF THE pu TERM

For Formulation 2, given by Eq. (2.44), the pressure flux has the pu term in the energy equation, which can also be split in a manner similar to the p term in the momentum equation. As suggested by Liou et al. (1990), the pu term can be expressed as consisting of $(M \pm 1)^2$ factors occurring in the splitting for p and a quadratic function in u:

$$pu = \frac{\varrho a}{4}(M + 1)^2(Au^2 + 2Bua + Ca^2) - \frac{\varrho a}{4}(M - 1)^2(Au^2 - 2Bua + Ca^2)$$

$$= (A + B)\varrho u^3 + (B + C)\varrho ua^2. \qquad (2.68)$$

The ϱu^3 term can be eliminated by enforcing the condition

$$A + B = 0 \qquad (2.69a)$$

to obtain

$$B + C = \frac{p}{\varrho a^2}. \qquad (2.69b)$$

Thus, a family of infinite choices for splitting pu are possible based on the parameter B. From the consideration of the total energy flux in the Euler equations, van Leer has proposed the following choice of B:

$$B = \frac{h/a^2}{1 + 2h/a^2} \qquad (2.70)$$

where $h \equiv H/\varrho$. The simplest choice, as proposed by Hänel et al. (1987) is

$$B = 0. \qquad (2.71)$$

Thus, as suggested by Liou et al. (1990), in the present study, the following splitting is employed:

$$
pu^{\pm} = \begin{cases} \pm\frac{1}{4}\varrho a(M \pm 1)^2(Au^2 \pm 2Bua + Ca^2), & \text{if } |M| \leq 1, \\ \frac{pu}{2}(M \pm |M|)/M, & \text{otherwise.} \end{cases} \tag{2.72}
$$

2.2.4   Analysis of Eigenvalues

2.2.4.1   FORMULATION 1

We next present a brief analysis of the eigenvalues associated with the convective and pressure fluxes. It must be stated at the outset that one has to examine the eigenvalues of the total combined flux (convective and pressure) in order to interpret the true nature of signal propagation in the gas dynamic system. However, the following analysis gives an idea of the nature of the convective and pressure fluxes in the two formulations given by Eqs. (2.43) and (2.44). In particular, from the viewpoint of operator splitting, the following analysis will be relevant to the individual components of convective and pressure fluxes, respectively. Also, this breakup of the total flux is expected to be more critical for the CVS, since in this scheme the coupling between the convective and pressure fluxes is not explicitly coordinated as a function of the local Mach number, as in the AUSM scheme. For the convective and pressure fluxes given by Eq. (2.43), the Jacobians of the convective and pressure fluxes are given by

$$
A^c \equiv \frac{\partial F^c}{\partial W} = \begin{bmatrix} 0 & 1 & 0 \\ -u^2 & 2u & 0 \\ -\gamma\frac{E}{\varrho}u + (\gamma - 1)u^3 & \gamma\frac{E}{\varrho} - \frac{3}{2}(\gamma - 1)u^2 & \gamma u \end{bmatrix} \tag{2.73a}
$$

$$
A^p \equiv \frac{\partial F^p}{\partial W} = \begin{bmatrix} 0 & 0 & 0 \\ (\gamma - 1)\frac{u^2}{2} & -(\gamma - 1)u & (\gamma - 1) \\ 0 & 0 & 0 \end{bmatrix} \tag{2.73b}
$$

where γ is the ratio of specific heats of the gas. The eigenvalues of the above Jacobians can be found by solving the equations

$$
A^c - \Lambda^c I = 0, \qquad A^p - \Lambda^p I = 0 \tag{2.74}
$$

where I is the identity matrix, resulting in the following eigenvalues for the convective and pressure fluxes:

$$
\Lambda^c = \begin{bmatrix} 0 \\ 2u \\ \gamma u \end{bmatrix}, \qquad \Lambda^p = \begin{bmatrix} 0 \\ -(\gamma - 1)u \\ 0 \end{bmatrix}. \tag{2.75}
$$

These seem to indicate that the convective flux has an upwind character and that the pressure flux has a downwind character only. In accordance with the physical characteristics, it is desirable that the convective fluxes are completely upwinded and that pressure fluxes are split based on the local Mach number. It is this thought that prompts us to investigate Formulation 2 (Eq. 2.44), which is perhaps more consistent with the numerical treatment of the individual (convective and pressure) fluxes.

If the three equations – continuity, momentum, and energy – are looked upon as a collection of three scalar conservation laws, as in the case of sequential solvers (Thakur and Shyy 1992, 1993), then the characteristic speeds that one obtains for the convective terms in the three equations are $(u, u, \gamma u)$. Likewise, for the pressure terms in the three equations, the characteristic speeds are $(0, -(\gamma - 1)u, 0)$. A similar analysis has also been performed by Tatsumi et al. (1995).

2.2.4.2 FORMULATION 2

For the convective and pressure fluxes given by Eq. (2.44), the Jacobians of the convective and pressure fluxes are given by

$$
A^c \equiv \frac{\partial F^c}{\partial W} =
\begin{bmatrix}
0 & 1 & 0 \\
-u^2 & 2u & 0 \\
-\frac{E}{\varrho}u & \frac{E}{\varrho} & u
\end{bmatrix}
\tag{2.76a}
$$

$$
A^P \equiv \frac{\partial F^P}{\partial W}
$$

$$
=
\begin{bmatrix}
0 & 0 & 0 \\
(\gamma - 1)\frac{u^2}{2} & -(\gamma - 1)u & (\gamma - 1) \\
(\gamma - 1)\left(-\frac{E}{\varrho}u + u^3\right) & (\gamma - 1)\left(\frac{E}{\varrho} - \frac{3}{2}u^2\right) & (\gamma - 1)u
\end{bmatrix}.
\tag{2.76b}
$$

The eigenvalues of the above Jacobians are:

$$
\Lambda^c =
\begin{bmatrix}
0 \\
2u \\
u
\end{bmatrix},
\qquad
\Lambda^P =
\begin{bmatrix}
0 \\
+\sqrt{\frac{\gamma-1}{\gamma}}a \\
-\sqrt{\frac{\gamma-1}{\gamma}}a
\end{bmatrix},
\tag{2.77}
$$

which indicate that the convective flux has an upwind character and that the pressure flux has both a downwind and an upwind character. The eigenvalues of the pressure flux in this formulation also suggest that the speed of propagation of pressure signals is dependent on the acoustic speed. Thus, it appears that Formulation 2 is more consistent with the dynamics of the system if one treats convection and acoustic wave propagation as two separate entities. Once again, it must be stated that Formulations 1 and 2 are expected to make a greater difference for the CVS than the AUSM scheme due to the reasons stated earlier.

If the three equations in the Euler system (continuity, momentum, and energy equations) are looked upon as a collection of three scalar conservation laws, then the characteristic speeds that one obtains for the convective terms in the three equations are (u, u, u). Likewise, for the pressure terms in the three equations, the characteristic speeds are $(0, -(\gamma - 1)u, (\gamma - 1)u)$.

2.2.5 Numerical Dissipation of the Various Schemes

The net interfacial flux for the schemes discussed in the previous sections, along with the Roe scheme (Roe 1981) for comparison, can be expressed as follows:

$$\textbf{CVS:} \quad f_{i+\frac{1}{2}} = \frac{1}{2}\left[f_i^c + f_{i+1}^c - Q_{i+\frac{1}{2}}\Delta_{i+\frac{1}{2}}w\right] + p_{i+\frac{1}{2}} \tag{2.78}$$

$$\textbf{AUSM:} \quad F_{i+\frac{1}{2}} = \frac{1}{2}\left[M_{i+\frac{1}{2}}(\Phi_i + \Phi_{i+1}) - \left|M_{i+\frac{1}{2}}\right|\Delta_{i+\frac{1}{2}}\Phi\right] + P_{i+\frac{1}{2}} \tag{2.79}$$

$$\textbf{Roe:} \quad F_{i+\frac{1}{2}} = \frac{1}{2}\left[F_i + F_{i+1} - \left|\bar{\bar{A}}_{i+\frac{1}{2}}\right|\Delta_{i+\frac{1}{2}}W\right]. \tag{2.80}$$

Note that the CVS essentially treats the system of equations as a set of scalar conservation laws. In all of the above schemes, the last term in the square brackets represents the numerical dissipation added to the central difference scheme. A significant contrast among the above schemes is that the Roe scheme involves the computation of the linearized matrix $\bar{\bar{A}}(W_{i+\frac{1}{2}})$, unlike both CVS and AUSM schemes. In the AUSM scheme, the coupling between the numerical convective and pressure fluxes via the splitting formulas for both velocity and pressure at the control volume interfaces is expected to coordinate signal propagation, thus yielding no spurious oscillations. In the CVS, too, there is such a coupling but perhaps to a lesser degree, since interfacial velocity is directly estimated by two-point averaging. However, the parameter δ in the CVS can be used to regulate the amount of numerical dissipation in order to suppress spurious oscillations should they occur.

2.2.6 Extension to Second-Order Spatial Accuracy

The net flux for each of the schemes discussed in the previous sections is spatially first-order accurate. One can extend the net flux formally to second-order accuracy by employing a variable extrapolation or a flux extrapolation approach. The latter is chosen here.

Let the total first-order flux at an interface be given by $f_{i\pm\frac{1}{2}}$ where the various quantities at the interfaces are defined by first-order extrapolations. We define $f_i = f(u_i)$ as point-valued fluxes at cell centers.

A general expression for a higher-order flux at the interface can be written as (Hirsch 1990):

$$f_{i+\frac{1}{2}}^{(2)} = f_{i+\frac{1}{2}} + \frac{1}{2}\left[\frac{1-\varkappa}{2}(f_i - f_{i-\frac{1}{2}}) + \frac{1+\varkappa}{2}(f_{i+1} - f_{i+\frac{1}{2}})\right]$$
$$+ \frac{1}{2}\left[\frac{1+\varkappa}{2}(f_i - f_{i+\frac{1}{2}}) + \frac{1-\varkappa}{2}(f_{i+1} - f_{i+\frac{3}{2}})\right]. \tag{2.81}$$

For $\varkappa = 1$, we obtain the second-order central difference scheme, whereas $\varkappa = -1$ yields the fully upwind second-order scheme. Choosing $\varkappa = -1$, the second-order interface flux becomes

$$f_{i+\frac{1}{2}}^{(2)} = f_{i+\frac{1}{2}} + \frac{1}{2}\left[(f_i - f_{i-\frac{1}{2}}) + (f_{i+1} - f_{i+\frac{3}{2}})\right], \tag{2.82}$$

which can be expressed with the use of limiters (for suppression of oscillations) as follows:

$$f_{i+\frac{1}{2}}^{(2)} = f_{i+\frac{1}{2}} + \frac{1}{2}\psi\left(r_{i-\frac{1}{2}}^{+}\right)\cdot\left(f_i - f_{i-\frac{1}{2}}\right) - \frac{1}{2}\psi\left(r_{i+\frac{3}{2}}^{-}\right)\cdot\left(f_{i+1} - f_{i+\frac{1}{2}}\right) \quad (2.83)$$

where

$$r_{i+\frac{1}{2}}^{+} = \frac{f_{i+2} - f_{i+\frac{3}{2}}}{f_{i+1} - f_{i+\frac{1}{2}}}, \qquad r_{i+\frac{1}{2}}^{-} = \frac{f_{i-1} - f_{i-\frac{1}{2}}}{f_i - f_{i+\frac{1}{2}}}. \quad (2.84)$$

The function $\Psi(r)$ is the flux limiter mentioned above. The minmod flux limiter has been employed in the present study, namely:

Minmod limiter : $\Psi(r) = \max[0, \min(1, r)].$ \quad (2.85)

Other limiters such as van Leer's monotonic limiter or Roe's superbee limiter can also be used (Hirsch 1990).

For the implicit version of the CVS, the second-order interfacial flux can be written as

$$f_{i+\frac{1}{2}}^{(2)} = f_{i+\frac{1}{2}} + \frac{1}{4}\psi\left(r_{i-\frac{1}{2}}^{+}\right)\cdot\left\{f_i - f_{i-1} + Q_{i-\frac{1}{2}}\Delta_{i-\frac{1}{2}}w\right\}$$
$$- \frac{1}{4}\psi\left(r_{i+\frac{3}{2}}^{-}\right)\cdot\left\{f_{i+2} - f_{i+1} - Q_{i+\frac{3}{2}}\Delta_{i+\frac{3}{2}}w\right\}, \quad (2.86)$$

which can be further simplified, using Eq. (2.48a):

$$f_{i+\frac{1}{2}}^{(2)} = f_{i+\frac{1}{2}} + \frac{1}{4}\psi\left(r_{i-\frac{1}{2}}^{+}\right)\cdot\left\{b_{i-\frac{1}{2}} + Q_{i-\frac{1}{2}}\right\}\cdot(w_i - w_{i-1})$$
$$+ \frac{1}{4}\psi\left(r_{i+\frac{3}{2}}^{-}\right)\cdot\left\{-b_{i-\frac{3}{2}} + Q_{i+\frac{3}{2}}\right\}\cdot(w_{i+2} - w_{i+1}). \quad (2.87)$$

Similarly, $f_{i-\frac{1}{2}}^{(2)}$ can be written as

$$f_{i-\frac{1}{2}}^{(2)} = f_{i-\frac{1}{2}} + \frac{1}{4}\psi\left(r_{i-\frac{3}{2}}^{+}\right)\cdot\left\{b_{i-\frac{3}{2}} + Q_{i-\frac{3}{2}}\right\}\cdot(w_{i-1} - w_{i-2})$$
$$+ \frac{1}{4}\psi\left(r_{i+\frac{1}{2}}^{-}\right)\cdot\left\{-b_{i+\frac{1}{2}} + Q_{i+\frac{1}{2}}\right\}\cdot(w_{i+1} - w_i). \quad (2.88)$$

Finally, from Eqs. (2.50), (2.87), and (2.88), we get the following net flux for the linearized implicit version of the CVS:

$$f_{i+\frac{1}{2}}^{(2)} - f_{i-\frac{1}{2}}^{(2)} = \frac{1}{2}\left\{\left[b_{i+\frac{1}{2}} - Q_{i+\frac{1}{2}}\right]\left[1 + \frac{1}{2}\psi\left(r_{i+\frac{1}{2}}^{-}\right)\right]\right\}^{n}(w_{i+1} - w_i)^{n+1}$$
$$- \frac{1}{2}\left\{\left[-b_{i-\frac{1}{2}} - Q_{i-\frac{1}{2}}\right]\left[1 + \frac{1}{2}\psi\left(r_{i-\frac{1}{2}}^{+}\right)\right]\right\}^{n}(w_i - w_{i-1})^{n+1}$$
$$+ \frac{1}{4}\left\{\psi\left(r_{i+\frac{3}{2}}^{-}\right)\left[-b_{i+\frac{3}{2}} + Q_{i+\frac{3}{2}}\right](w_{i+2} - w_{i+1})\right\}^{n}$$
$$- \frac{1}{4}\left\{\psi\left(r_{i-\frac{3}{2}}^{+}\right)\left[b_{i-\frac{3}{2}} + Q_{i-\frac{3}{2}}\right](w_{i-1} - w_{i-2})\right\}^{n}. \quad (2.89)$$

2.2.7   Results of One-Dimensional Test Cases

Two one-dimensional cases are presented in this section to demonstrate the accuracy of the controlled variation scheme. The first one is the standard shock tube problem. The second case is simulation of a longitudinal combustion instability which involves thermoacoustic coupling due to the interplay between pressure oscillations and periodic heat release in the combustor.

2.2.7.1   SHOCK TUBE PROBLEM

The shock tube problem presented here has been previously investigated in Thakur and Shyy (1992, 1993). The initial conditions on the left and the right of the diaphragm are reported there. In the present work, we study the CVS and AUSM schemes in terms of their capacity to resolve discontinuities such as shock waves and contact surfaces. The total length of the tube is 14 units with the initial location of the diaphragm in the middle of the tube. There are 141 grid points used, and the value of $\lambda \equiv \Delta t / \Delta x$ is 0.1. Results are presented in the form of total energy profiles after 200 timesteps. For all the results presented in the following, the fluxes of all the schemes employed have been extrapolated to second order using the minmod limiter.

Figures 2.3 and 2.4 show the total energy profiles obtained with Formulations 1 and 2 of the CVS, respectively (as classified in Section 2.2.1), using two values of the parameter δ, which regulates the amount of numerical dissipation. It can be observed that Formulation 1, which treats only the p term as part of the pressure flux, yields a slight overshoot near the shock location, as seen in Fig. 2.3(a). For both values of δ, Formulation 2 yields solution profiles that are qualitatively better than those obtained with Formulation 1, consistent with the interpretation of the eigenvalues of the Jacobian matrices for the two formulations, as discussed in Section 2.2.4. Comparing these profiles with those obtained using the second-order Roe scheme, shown in Fig. 2.6, it can be observed that the CVS, especially with Formulation 2, yields accuracy comparable to the Roe scheme.

The solution profiles obtained with the second-order AUSM scheme are shown in Fig. 2.5. Again, both formulations for the convective and pressure fluxes are investigated. It is seen that the AUSM scheme yields results with Formulation 1 that are comparable with those obtained with Formulation 2. A possible explanation is that the AUSM scheme uses splittings for both the convective interface velocity and the pressure flux along with upwinding for the convected variables. Thus, for the AUSM scheme, the u-velocity of the eigenvalues already exhibits directional bias according to the local Mach number, which is not the case for the CVS. The AUSM scheme, like the CVS, also yields results of accuracy comparable to those obtained with the Roe scheme (Fig. 2.6).

2.2.7.2   LONGITUDINAL COMBUSTION INSTABILITY PROBLEM

This test case has been devised by Shyy et al. (1993) to investigate the interaction of convection and a source term in the form of heat release. It involves pressure oscillations in a one-dimensional model of a combustor that are sustained by the oscillations of heat release. The heat release in the combustor is specified using a

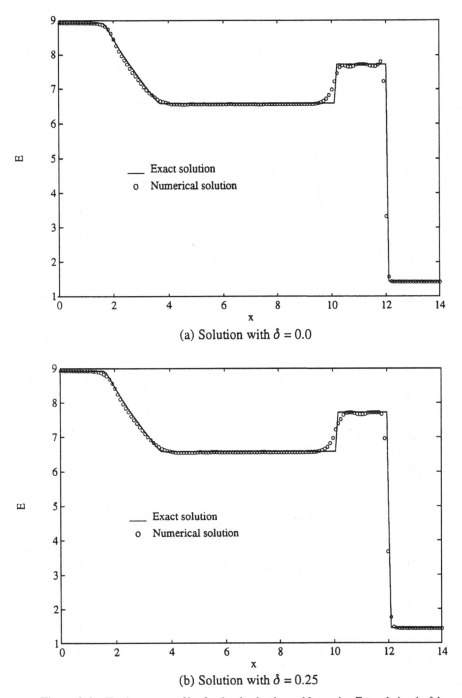

(a) Solution with $\delta = 0.0$

(b) Solution with $\delta = 0.25$

Figure 2.3 Total energy profiles for the shock tube problem using Formulation 1 of the CVS (p term only in the pressure flux) with two values of δ; minmod limiter is used.

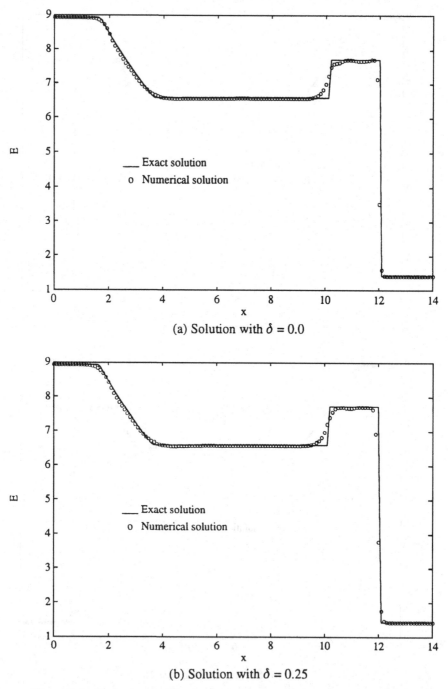

(a) Solution with $\delta = 0.0$

(b) Solution with $\delta = 0.25$

Figure 2.4   Total energy profiles for the shock tube problem using Formulation 2 of the CVS (p and pu terms in the pressure flux) with two values of δ; minmod limiter is used.

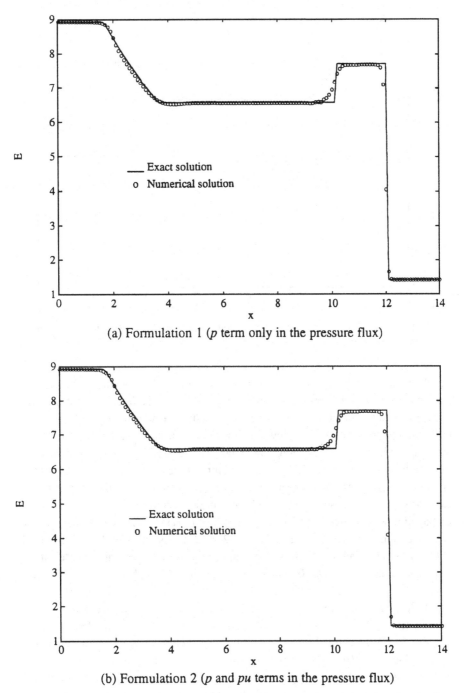

(a) Formulation 1 (p term only in the pressure flux)

(b) Formulation 2 (p and pu terms in the pressure flux)

Figure 2.5 Total energy profiles for the shock tube problem using the second-order AUSM scheme with the two formulations for the pressure flux; minmod limiter is used.

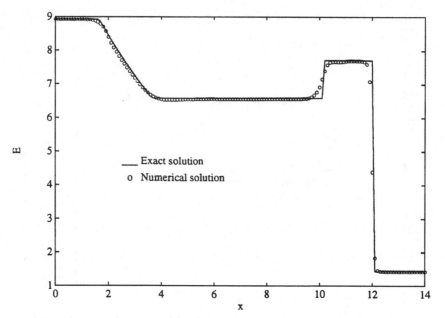

Figure 2.6   Total energy profiles for the shock tube problem using the second-order Roe scheme; minmod limiter is used.

simple model based on some experimental observations. The details of the heat release model have been presented in Shyy et al. (1993). The interaction of the heat
release source term and the convective and acoustic mechanisms in the system can
lead to nonphysical high-frequency oscillations in some solution profiles. These numerical oscillations, besides being fundamentally undesirable from the point of view
of numerical accuracy, may lead to an instability of the computation or even trigger
nonlinear instabilities in the system. In this regard, this problem is quite a stringent test
case for any numerical scheme that seeks to coordinate convection and acoustic wave
propagation with source-term effects such as heat release. The value of $\lambda \equiv \Delta t/\Delta x$
is 0.03, and the simulation is carried on for 2×10^6 timesteps. Results are presented
in the form of the following: (i) pressure and temperature mode shapes plotted at the
last ten instants, which are 10^4 timesteps apart (Fig. 2.7); and (ii) pressure and heat
release time series at the location $x = 0.75$ for the last 4×10^4 timesteps (Fig. 2.8).
The following schemes are used: (a) CVS with $\delta = 0$ and pressure splitting with
Formulation 2, (b) second-order AUSM scheme, and (c) second-order Roe scheme.

For the CVS, central differencing of the pressure flux along with a lower amount
of damping ($\delta = 0$) leads to spurious oscillations in some of the pressure mode
shapes (see the results in Shyy et al. 1993). These oscillations can be suppressed
with a sufficiently high dose of dissipation by increasing the value of δ to 0.8 or
higher (Shyy et al. 1993). This extra dissipation, of course, leads to a smearing of
solution profiles, which is manifested in the form of reduced magnitudes of all the
mode shapes. This can be resolved by treating pressure as a source term along with
heat release and by imparting a special treatment such as Strang's operator splitting
method (Thakur and Shyy 1992, 1993). Such a special source-term treatment yields
an improved accuracy with no spurious oscillations (Shyy et al. 1993). However,

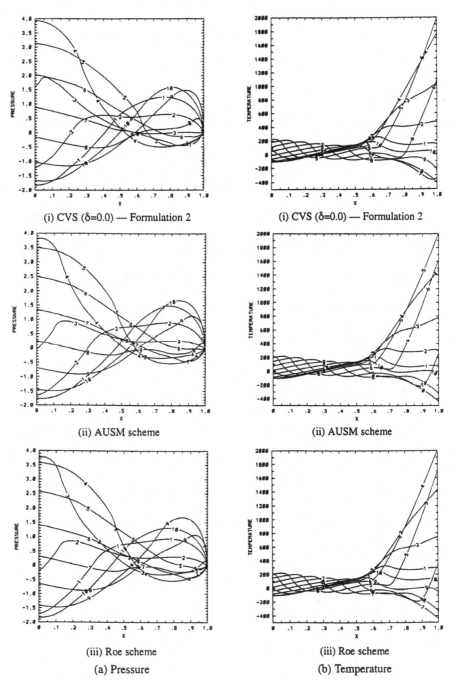

(i) CVS (δ=0.0) — Formulation 2   (i) CVS (δ=0.0) — Formulation 2

(ii) AUSM scheme   (ii) AUSM scheme

(iii) Roe scheme   (iii) Roe scheme

(a) Pressure   (b) Temperature

Figure 2.7 Ten pressure and temperature mode shapes for the combustion instability problem using the second-order CVS, AUSM, and Roe schemes (with minmod limiter).

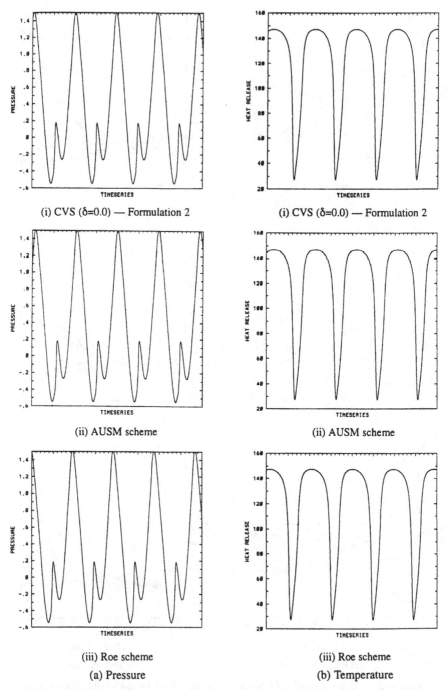

(i) CVS (δ=0.0) — Formulation 2   (i) CVS (δ=0.0) — Formulation 2

(ii) AUSM scheme   (ii) AUSM scheme

(iii) Roe scheme   (iii) Roe scheme

(a) Pressure   (b) Temperature

Figure 2.8   Pressure and heat release time series at $x = 0.75$ for the combustion instability problem using second-order CVS, AUSM, and Roe schemes (with minmod limiter).

this is at an increased computational expense, which can be avoided by splitting the pressure flux in an appropriate manner as discussed earlier in this chapter.

As seen from Figs. 2.7 and 2.8, the results obtained with the CVS along with splitting of the pressure flux, employing Formulation 2 of the convective and pressure fluxes, are quite satisfactory and comparable in accuracy to those obtained with the second-order Roe scheme. The CVS is able to coordinate the interaction of pressure oscillations and the heat release in an appropriate manner, and no spurious oscillations are observed. The CVS yields pressure mode shapes with slightly larger magnitudes compared to those resulting from the second-order AUSM and Roe schemes.

The second-order AUSM scheme also yields results comparable in accuracy to those obtained with the second-order Roe scheme. However, slight oscillations are observed in mode 3 (Fig. 2.7) of the pressure mode shapes in the region given by $x = 0.6$ to $x = 0.8$. Overall, the CVS, AUSM, and Roe schemes yield results very comparable in accuracy. However, one can observe from the mode shapes of the variables (especially pressure) that the results obtained by these three schemes are not exactly in phase. This is to be expected because there is a difference in the dispersive and dissipative characteristics among the schemes, which, however small, will result in a slight difference in the phase characteristics of the solution, especially after 2×10^6 timesteps.

The approach of treating the convective and pressure fluxes in the Euler and Navier-Stokes equations as two distinct, though coupled, entities appears to be very promising, as demonstrated by the results in this section and by other workers (Jameson 1993, Liou and Steffen 1993). The upwinding of the convective flux and the splitting of the pressure fluxes (based on local Mach number) achieve the proper propagation of signals in the system, yielding high resolution in the solution profiles with no spurious oscillations. Such an approach can also be effective in the presence of source terms, as demonstrated by the results of the longitudinal combustion instability problem. Overall, the CVS yields accuracy comparable to the Roe scheme and the AUSM scheme.

2.3 Implementation of the CVS in the Pressure-Based Algorithm

In the previous section, a high-resolution scheme, namely the controlled variation scheme (CVS), is developed for the Euler equations treated as a set of scalar conservation laws in one dimension for clarity of development of the scheme. The governing equations are treated as separate conservation laws since the overall objective is to incorporate such high-resolution convection and pressure treatments in pressure-based algorithms that are essentially sequential in nature. In this section, the implementation of the CVS in a pressure-based algorithm for curvilinear coordinates and its performance for two-dimensional test cases are detailed and evaluated. It is demonstrated that via the CVS and AUSM type schemes, pressure-based algorithms can yield accurate simulations of complex compressible flows, including high resolution of shock waves.

2.3.1 Momentum Equations

We now formulate the CVS for the momentum equations given by Eq. (2.12) by formally extending the scheme from one dimension to the present two-dimensional case. It should be recalled that the convective fluxes of the CVS and AUSMD schemes are identical for steady-state applications except for the computation of the interface convective velocities. Using the form of the CVS presented in Eq. (2.89), independently along the x and y directions, Eq. (2.12) can be expressed for steady flow as

$$C_{i+\frac{1}{2}}^n(\phi_E - \phi_P)^{n+1} - C_{i-\frac{1}{2}}^n(\phi_P - \phi_W)^{n+1} + C_{j+\frac{1}{2}}^n(\phi_N - \phi_P)^{n+1}$$

$$- C_{j-\frac{1}{2}}^n(\phi_P - \phi_S)^{n+1} = D_{i+\frac{1}{2}}^n(\phi_E - \phi_P)^{n+1} - D_{i-\frac{1}{2}}^n(\phi_P - \phi_W)^{n+1}$$

$$+ D_{j+\frac{1}{2}}^n(\phi_N - \phi_P)^{n+1} - D_{j-\frac{1}{2}}^n(\phi_P - \phi_S)^{n+1} + S \tag{2.90}$$

where the various coefficients are given by

$$C_{i\pm\frac{1}{2}} = \frac{1}{2}\varrho_{i\pm\frac{1}{2}}\left[\pm b_{i\pm\frac{1}{2}} - Q_{i\pm\frac{1}{2}}\right]\left[1 + \frac{1}{2}\psi\left(r_{i\pm\frac{1}{2}}^{\mp}\right)\right] \tag{2.91a}$$

$$D_{i\pm\frac{1}{2}} = \left(\frac{\mu}{J}q_1\right)_{i\pm\frac{1}{2}} \equiv \left(\frac{\mu}{J}q_1\right)_{e/w}, \text{ etc.} \tag{2.91b}$$

and $b_{i+\frac{1}{2}}$ etc. are the interface velocities. S is the source term consisting of pressure gradient and viscous cross-derivative terms as well as the remaining higher-order contributions from the convective fluxes. Note that the subscript i is used to denote the ξ direction, and j to denote the η direction.

Using the conventional notation for the SIMPLE algorithm expressed in Eq. (2.13), Eq. (2.90) can be written as

$$A_W = \left[\frac{\mu}{J}q_1\right]_{i-\frac{1}{2}} - \frac{1}{2}\varrho_{i-\frac{1}{2}}\left[-b_{i-\frac{1}{2}} - Q_{i-\frac{1}{2}}\right]\left[1 + \frac{1}{2}\psi\left(r_{i-\frac{1}{2}}^{+}\right)\right] \tag{2.92a}$$

$$A_E = \left[\frac{\mu}{J}q_1\right]_{i+\frac{1}{2}} - \frac{1}{2}\varrho_{i+\frac{1}{2}}\left[b_{i+\frac{1}{2}} - Q_{i+\frac{1}{2}}\right]\left[1 + \frac{1}{2}\psi\left(r_{i+\frac{1}{2}}^{-}\right)\right] \tag{2.92b}$$

$$A_S = \left[\frac{\mu}{J}q_3\right]_{j-\frac{1}{2}} - \frac{1}{2}\varrho_{j-\frac{1}{2}}\left[-b_{j-\frac{1}{2}} - Q_{j-\frac{1}{2}}\right]\left[1 + \frac{1}{2}\psi\left(r_{j-\frac{1}{2}}^{+}\right)\right] \tag{2.92c}$$

$$A_N = \left[\frac{\mu}{J}q_3\right]_{j+\frac{1}{2}} - \frac{1}{2}\varrho_{j+\frac{1}{2}}\left[b_{j+\frac{1}{2}} - Q_{j+\frac{1}{2}}\right]\left[1 + \frac{1}{2}\psi\left(r_{j+\frac{1}{2}}^{-}\right)\right] \tag{2.92d}$$

where

$$A_P = A_W + A_E + A_S + A_N \tag{2.92e}$$

$$S = -\frac{1}{4}\varrho_{i+\frac{3}{2}} \cdot \psi\left(r_{i+\frac{3}{2}}^{-}\right) \cdot \left\{-b_{i+\frac{3}{2}} + Q_{i+\frac{3}{2}}\right\} \cdot (\phi_{i+2} - \phi_{i+1})$$

$$+ \frac{1}{4}\varrho_{i-\frac{3}{2}} \cdot \psi\left(r_{i-\frac{3}{2}}^{+}\right) \cdot \left\{b_{i-\frac{3}{2}} + Q_{i-\frac{3}{2}}\right\} \cdot (\phi_{i-1} - \phi_{i-2})$$

$$- \frac{1}{4}\varrho_{j+\frac{3}{2}} \cdot \psi\left(r_{j+\frac{3}{2}}^{-}\right) \cdot \left\{-b_{j+\frac{3}{2}} + Q_{j+\frac{3}{2}}\right\} \cdot (\phi_{j+2} - \phi_{j+1})$$

$$+ \frac{1}{4} \cdot \varrho_{j-\frac{3}{2}} \cdot \psi\left(r_{j-\frac{3}{2}}^{+}\right) \cdot \left\{b_{j-\frac{3}{2}} + Q_{j-\frac{3}{2}}\right\} \cdot (\phi_{j-1} - \phi_{j-2})$$

$$- \left[\left(\frac{\mu}{J}q_2\phi_\eta\right)_{i+\frac{1}{2}} - \left(\frac{\mu}{J}q_2\phi_\eta\right)_{i-\frac{1}{2}}\right]$$

$$- \left[\left(\frac{\mu}{J}q_2\phi_\xi\right)_{j+\frac{1}{2}} - \left(\frac{\mu}{J}q_2\phi_\xi\right)_{j-\frac{1}{2}}\right] + P^* \tag{2.92f}$$

where $b_{i-\frac{1}{2}} \equiv U_{i-\frac{1}{2}}$ etc., and the ratios $r_{i\pm\frac{1}{2}}^{\pm}$ are the same as in Eq. (2.84). Also, in the above, $\boldsymbol{P}^*$ represents terms involving pressure. It can be observed that the above form is spatially a five-point scheme along both directions which can be conveniently solved using the ADI method along with a tridiagonal matrix solver. Also, it should be noted that the coefficient matrix has a dominant diagonal.

For the boundary control volumes, first-order numerical fluxes are employed to obtain the coefficients A_E, A_W, etc.

2.3.2 Pressure Correction Equation

In the pressure correction equation, for compressible flows, density corrections are related to pressure corrections through the equation of state. This changes the nature of the pressure correction equation from a pure diffusion equation (for incompressible flows) to a convection–diffusion equation (for compressible flows). For details the reader is referred to Shyy (1994).

Conventionally, the estimation of the mass flux terms is done by utilizing the normal velocity components located at the faces of the pressure correction control volumes (due to the staggered grid) and by upwinding the density based on the direction of the interfacial normal velocity (Shyy 1989), that is,

$$
\begin{aligned}
(\varrho U)_{i+\frac{1}{2}}^{(1)} &= \frac{1}{2}\left\{\left[1 + sign(U_{i+\frac{1}{2}})\right]\varrho_i + \left[1 - sign(U_{i+\frac{1}{2}})\right]\varrho_{i+1}\right\}U_{i+\frac{1}{2}} \\
&= \frac{1}{2}\left\{(\varrho_i + \varrho_{i+1}) \cdot U_{i+\frac{1}{2}} - \left|U_{i+\frac{1}{2}}\right| \cdot (\varrho_{i+1} - \varrho_i)\right\}
\end{aligned} \tag{2.93}
$$

where $\left(i + \frac{1}{2}\right)$ refers to the east face of the pressure correction control volume. The above conventional flux estimation is only first-order accurate. It should be noted that the above estimation of the mass flux is identical to that by the AUSMD scheme (Wada and Liou 1994) except, again, the estimation of the interface velocity is different.

The results presented here employ a second-order estimation of the mass flux

$$
\begin{aligned}
(\varrho U)_{i+\frac{1}{2}}^{(2)} &= (\varrho U)_{i+\frac{1}{2}}^{(1)} + \frac{1}{4}\psi\left(r_{i-\frac{1}{2}}^{+}\right) \cdot \left\{b_{i-\frac{1}{2}} + Q_{i-\frac{1}{2}}\right\} \cdot (\varrho_i - \varrho_{i-1}) \\
&\quad + \frac{1}{4}\psi\left(r_{i+\frac{1}{2}}^{-}\right) \cdot \left\{-b_{i+\frac{3}{2}} + Q_{i+\frac{3}{2}}\right\} \cdot (\varrho_{i+2} - \varrho_{i+1}) \tag{2.94}
\end{aligned}
$$

where $b_{i-\frac{1}{2}} \equiv U_{i-\frac{1}{2}}$ etc., and the ratios $r_{i\pm\frac{1}{2}}^{\pm}$ are the same as in Eq. (2.84).

It should be noted that the remaining terms in the pressure correction equation, that is, the ones involving velocity and density corrections, contribute only to the stability, not accuracy, of the overall algorithm. These terms vanish when overall convergence is achieved since we obtain a continuity-satisfying velocity field at convergence (for which pressure correction is zero everywhere, up to machine accuracy).

2.3.3 Additional Issues Due to the Staggered Grid Layout

An important issue in the CVS and AUSM schemes is the estimation of interfacial convective velocities and pressures. Due to the staggered grid arrangement conventionally used in pressure-based algorithms, additional issues have to be addressed as follows.

2.3.3.1 CVS

As mentioned earlier, in the CVS, a straightforward two-point averaging is used. Thus, we have

$$b_{i+\frac{1}{2}} \equiv b_{i+\frac{1}{2},j} = \frac{1}{2}(U_{i,j} + U_{i+1,j}) \tag{2.95a}$$

$$b_{j+\frac{1}{2}} \equiv b_{i,j+\frac{1}{2}} = \frac{1}{2}(V_{i,j} + V_{i,j+1}). \tag{2.95b}$$

It should be noted that the above interpolations result from the staggered nature of the grids employed for the velocity components. Due to the staggered location of the scalar variables such as pressure and density, these variables are readily available at some of the control volume interfaces (east and west faces for u-control volumes, and north and south faces for v-control volumes). Wherever density is required at a point away from the scalar nodes, it is upwinded based on the direction of the local normal velocity component. The pressure values at the u- and v-velocity nodes are obtained by the pressure splitting given by Eq. (2.64). The local Mach numbers required for this splitting along the ξ and η directions are obtained from the local normal velocity and the local speed of sound:

$$M_{\xi} = \frac{0.5(U_{i,j} + U_{i+1,j})/\sqrt{q_1}}{\sqrt{\gamma p_{i,j}/\varrho_{i,j}}}, \qquad M_{\eta} = \frac{0.5(V_{i,j} + V_{i,j+1})/\sqrt{q_3}}{\sqrt{\gamma p_{i,j}/\varrho_{i,j}}}. \tag{2.96}$$

On a staggered grid, for a u-control volume along the ξ direction, for example, the values of pressure at the interfaces $\left(i + \frac{1}{2} \text{ and } i - \frac{1}{2}\right)$ are known from the existing pressure values at the interfaces (due to the staggered grid, shown in Fig. 2.9). Pressure itself is computed from the pressure correction equation in which the mass fluxes are computed by upwinding density based on the interface convective velocity. Thus, there is no need for splitting pressure in order to get the interface pressures for either u-control volumes along the ξ direction or v-control volumes along the η direction. In the remaining direction for each of the control volumes, however, one can employ the splitting formula to obtain interface pressures.

2.3.3.2 AUSM

The AUSM-type schemes use a splitting formula based on the local Mach number to compute the interface velocities as well as interface pressures. For a staggered grid layout, such as that employed in the present algorithm, the velocity components and scalar variables (pressure and density) are not located at the same node. This brings up another issue (for the AUSM scheme), namely, the estimation of Mach number at the locations corresponding of u, v, and p, which are half a cell length apart from each other. For example, along the ξ direction, for the u-control volumes, p is located on the control volume interfaces, and for the p-control volumes, u is located on the interfaces. Thus, we need to estimate local Mach numbers along the ξ and η directions at the nodes as well as the interfaces of u-, v-, and p-control volumes. This requires, for example, the values of p and ϱ (to compute the speed of sound for the Mach number computation for u-splitting) at the u location and values of u

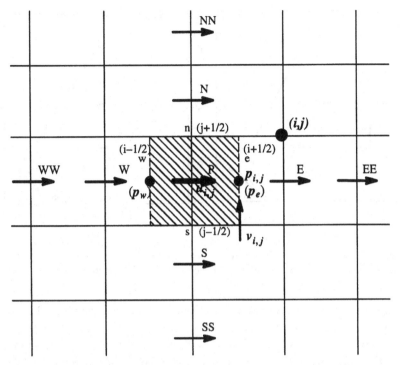

Figure 2.9 Illustration of the staggered locations of u, v, and p and the nomenclature for a typical u-control volume.

(the local convection speed of the Mach number computation for p-splitting) at the p location. This can be handled via an iterative process. Considering the ξ direction, for example, the following steps are performed:

Step 1. The Mach number at the location of p is estimated initially by Eq. (2.54).

Step 2. This Mach number is used to split p, and these split values of p are used to estimate p at the locations of u. The density values at the locations of u are obtained by upwinding based on the local contravariant velocity U (normal velocity along the ξ direction).

Step 3. Based on these p and ϱ, the Mach number at the location of u is obtained.

Step 4. This Mach number is used to split U, and these split values of U are used to estimate an averaged value of U at the p locations.

Step 5. Steps 1 through 4 are repeated a few times (typically five) until convergence is achieved.

2.3.4 Results of Two-Dimensional Computations

2.3.4.1 INCOMPRESSIBLE DRIVEN CAVITY FLOW

As a first example, consider the incompressible flow in a square cavity that is driven by a sliding lid. This is the most well-known benchmark case for incompressible flow solvers. Figure 2.10 shows the u velocity component profiles along the vertical

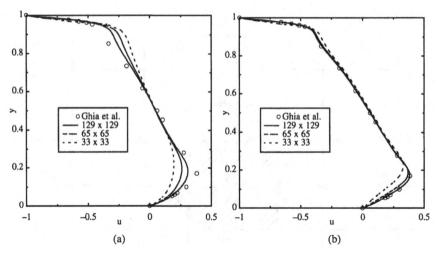

Figure 2.10   *U* profiles along vertical centerline for lid-driven cavity (*Re* = 1,000). (a) First-order CVS for all equations. (b) Second-order CVS for all equations.

centerline of the cavity for the case *Re* = 1,000. This Reynolds number is based on the width of the cavity and the sliding velocity of the lid. Both the first-order CVS and the second-order CVS were employed with three grid resolutions (shown in the figure). The benchmarks used for comparison are the commonly used results of Ghia et al. (1982). It is clear from the figure that the first-order CVS is rather diffusive, as expected, and approaches grid independence slowly. Even with 129 × 129 nodes, the solution still differs considerably from the grid-independent benchmark results, which were also obtained on a 129 × 129 grid but with a second-order scheme. The second-order CVS performs much better than the first-order CVS by comparison. The solution on the coarsest grid (33 × 33) with the second-order CVS is seen to be comparable to the finest grid solution (129 × 129) with the first-order CVS. With the finest grid, the second-order CVS solution compares very favorably with the benchmark results.

To demonstrate the performance of the CVS in the presence of artificial internal boundaries for composite grids, a two-block horizontally divided cavity flow is also computed. The technique to handle multiple blocks with discontinuous grid interfaces is discussed in Chapter 4. The case presented here serves to illustrate the accuracy of the CVS for multiblock domains. The computations were performed on two grids, one twice the resolution of the other. Both grid systems consist of a top block with twice the horizontal grid size as the bottom block. Details of the composite grid computations will be discussed in Chapter 4. Figure 2.11 shows the *u* velocity component profiles along the vertical centerline of the cavity for both computations. The fine grid case uses only 3/4 of the number of nodes as the benchmark, yet the comparison is very good.

2.3.4.2   SUPERSONIC WEDGE FLOW

In order to investigate the performance of the CVS and AUSM schemes for two-dimensional compressible flows involving shocks, a supersonic flow over a wedge is

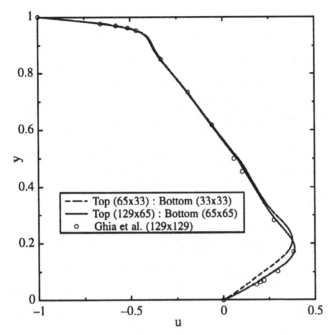

Figure 2.11 u velocity profiles along the vertical centerline for a horizontally divided cavity with a discontinuous grid interface; top block is 65×33 and the bottom block is 33×33.

chosen as a test case. Both first- and second-order CVS and AUSM schemes (minmod limiter is used for the second-order fluxes) for the momentum equations, as well as the mass flux estimation in the pressure correction equation, are investigated. Results obtained using the Roe scheme (implemented in a density-based algorithm) are also presented for comparison with the CVS and AUSM schemes.

This problem consists of an oblique shock generated by a supersonic flow over a wedge and its subsequent reflections by a solid flat plate underneath the wedge and by the wedge surface itself; the problem has been investigated by Wang and Widhopf (1987), among others. A schematic of the flow is depicted in Fig. 2.12. The inlet

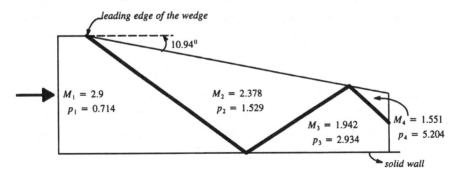

Figure 2.12   Schematic of the supersonic flow over a wedge.

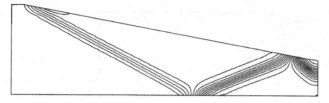

(a) First-order CVS for all the equations on the 101 × 21 grid.

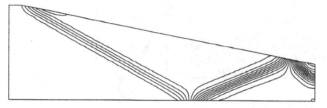

(b) First-order AUSM for all the equations on the 101 × 21 grid.

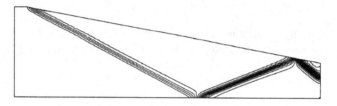

(c) First-order CVS for all the equations on the 201 × 41 grid.

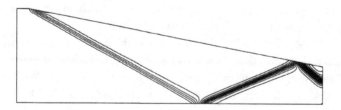

(d) First-order AUSM for all the equations on the 201 × 41 grid.

Figure 2.13   Pressure contours for a supersonic flow (inlet Mach number = 2.9) over a wedge (angle 10.94°) on the 101 × 21 and 201 × 41 grids using first-order CVS and AUSM schemes.

Mach number is 2.9, and the wedge angle is 10.94°. Two grid systems are used for the computations, namely those consisting of 101 × 21 and 201 × 41 uniformly distributed nodes. The location of the leading edge of the wedge is at the discontinuity in the slope of the top boundary of the grid layout.

The upstream boundary condition specifies the incoming flow at the given Mach number, whereas a zero-order extrapolation is used for the downstream boundary

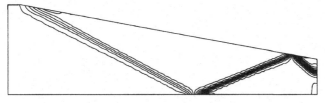

(a) Second-order CVS for all the equations on the 101×21 grid.

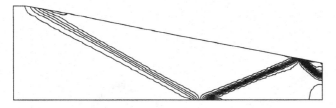

(b) Second-order AUSM for all the equations on the 101×21 grid.

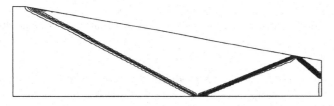

(c) Second-order CVS for all the equations on the 201×41 grid.

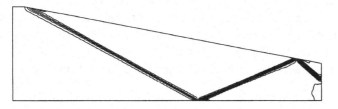

(d) Second-order AUSM for all the equations on the 201×41 grid.

Figure 2.14 Pressure contours for a supersonic flow (inlet Mach number = 2.9) over a wedge (angle 10.94°) on the 101 × 21 and 201 × 41 grids using second-order CVS and AUSM schemes.

condition (at the exit). The entire bottom boundary and the wedge part of the top boundary are reflecting surfaces, and thus the normal velocity components there are specified as zero.

The results are presented in Figs. 2.13–2.15 in the form of thirty pressure contours with equal increments between the minimum and maximum pressure values. For all the cases using the CVS, AUSM, and Roe schemes, on both the grids, the

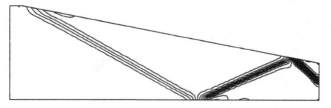

(a) Second-order Roe scheme on the 101×21 grid.

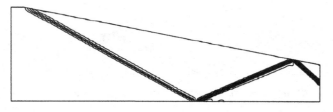

(b) Second-order Roe scheme on the 201×41 grid.

Figure 2.15   Pressure contours for a supersonic flow (inlet Mach number = 2.9) over a wedge (angle 10.94°) on the 101 × 21 and 201 × 41 grids using a second-order Roe scheme.

correct pressure jump and shock angles are predicted. However, using the first-order schemes on the 101 × 21 grid, the shock is excessively smeared, as seen in Fig. 2.13. Even with the refined grid (201 × 41 nodes), the first-order flux estimation does not yield a grid-independent solution. The accuracy improves when the momentum fluxes and the mass fluxes in the pressure correction equation are estimated using the second-order CVS along with the minmod limiter, as seen in Fig. 2.14. On the refined grid (201 × 41 nodes), for example, a crisp shock structure can be observed (Fig. 2.14). It can be observed from the results that both the CVS and AUSM schemes implemented in the pressure-based solver yield accuracy comparable to the Roe scheme (Fig. 2.15).

2.4   Concluding Remarks

The separate treatment of convective and pressure fluxes is a key feature of all pressure-based algorithms for multidimensional fluid flows. Some recently developed schemes based on separate treatment of convective and pressure fluxes – such as the controlled variation scheme (CVS) and the AUSM-type schemes – are thus very naturally amenable for application in these algorithms. The approach of treating the convective and pressure fluxes in the Euler and Navier-Stokes equations as two distinct though coupled entities appears to be very promising, as demonstrated by the results in this chapter. The upwinding of the convective flux and the splitting of the pressure fluxes (based on local Mach number) achieve the proper propagation of signals in the system, yielding high resolution in the solution profiles with no spurious oscillations. Such an approach can also be effective in the presence of source terms, as demonstrated by the results of the longitudinal combustion instability problem. Overall, both the CVS and the AUSM schemes yield accuracy comparable to the Roe scheme.

Both the CVS and the AUSM scheme yield accurate results for two-dimensional compressible flows using a pressure-based algorithm. It has been demonstrated that, with these schemes, the pressure-based algorithms can indeed be very robust and accurate for compressible flows involving shocks, in addition to their well-established robustness for incompressible flows.

3 Computational Acceleration with Parallel Computing and Multigrid Method

3.1 Introduction

To tackle the complexities of many practically relevant transport problems, we often encounter mathematical systems involving more than a few partial differential equations, and discretized on a large number of grid points. In order to find ways to solve such large number of matrix systems efficiently, two approaches can be devised, namely (i) to improve the convergence rate of an iterative method employed to solve large linear matrix systems and/or to handle the nonlinearity of the physical phenomena, and (ii) to expedite the rate of the floating-point computation. In this chapter we will present both the multigrid method to accelerate convergence rate and the parallel computing technique to expedite the data processing speed. We will discuss parallel computing first, then the multigrid technique in the context of a parallel computing environment.

Parallel computing introduces new issues distinct from the physical modeling issues and the numerical accuracy, stability, and consistency issues that are fundamental to all scientific computing problems. The goal of this chapter is to illustrate some of the important current issues that arise in the context of pressure-based methods for incompressible fluid dynamics, in particular for data-partitioned problems.

Data-partitioned problems arise naturally in many scientific computing problems because the laws of physics can be cast in terms of differential equations for scalar and vector field quantities, and the same equations apply concurrently at all points of the domain. Of course there are many problems that contain other forms of parallelism, such as tree-searching algorithms in computer science, which are best attacked by a "pool of tasks" parallelization. The literature on parallel computing tends to be closely tied to the problem domain, and therefore some of the research issues that are important for the pool of tasks–type of problems, for example, are not directly of interest to those working on problems that are parallelized by data partitioning. Nevertheless, all problems present two fundamental challenges in terms of parallelization: the selection of algorithms that are suitably matched to the target parallel architecture(s) (or the other way around), and the programming challenge of selecting appropriate data structures and expressing the parallelism in an efficient implementation. In the next overview section we review these fundamental issues. The remaining discussion then focuses on the performance of the nonlinear pressure-based multigrid method parallelized by data partitioning. The issues discussed have a broader context, though.

Kumar et al. (1994) cover issues in parallel algorithm development, using examples from a broad cross section of application areas. Golub and Ortega (1993) and Bertsekas and Tsitsiklis (1989) give good overviews of parallel computing concepts and discuss parallel algorithms for numerical linear algebra. Thorough coverage of the wide variety of parallel architectures available is given in Hwang (1993), Hwang and Briggs (1984), and Stone (1993). Parallel programming is discussed in Ben-Ari (1982), Carriero and Gelernter (1990), Hatcher and Quinn (1991), and Lester (1993). Even though the specific computer hardware discussed below will surely be dated with time, we believe that the technical discussion and specific ideas will remain largely valid.

3.2 Overview of Parallel Computing

3.2.1 Motivation

In view of the progress made in the algorithms and numerical techniques areas, we have seen a greatly increased interest in the issue of the robustness of algorithms and techniques (MacCormack 1991, Shyy 1994), and on the application of computational fluid dynamics (CFD) to complex transport problems that idealize neither the physics nor the geometry and that take full account of the coupling between fluid dynamics and other areas of physics (Roe 1991, Shyy 1994). Such applications require formidable resources, particularly in the areas of computing speed, memory, storage, and input/output bandwidth (Simon et al. 1992). Thus, parallel computing is receiving much attention right now from the CFD community.

3.2.2 Classification of Parallel Machines

Parallel computers are generally categorized according to their *parallel architecture*, which includes the type and layout of multiple processors, the parallel flow control technique, the mechanism by which communication is accomplished, and the machine granularity. Both the program developer and the user need to be aware of the parallel architecture because considerations such as performance, ease of programming, portability, and cost are affected. Every stage of program development, from choosing the type of parallelization to be employed, to the implementation in a particular programming language, depends on the architecture in a manner that reflects the relative importance placed on these factors.

Flynn (1972) gave the popular categorization of parallel machines according to the style of program control, into single-instruction-stream/multiple-data-stream (SIMD) models and multiple-instruction-stream/multiple-data-stream (MIMD) models. As these terms are still widely used, we will explain their meaning.

In the SIMD model, a master processor feeds individual instructions or blocks of instructions to a bunch of slave processors that execute the instructions concurrently. Strictly speaking, there is only one copy of the program, and it resides on the master processor. Thus, the benefits of the SIMD model are that *synchronization* is automatically enforced because there is only one program counter, and each slave processor does not need to store the program. The tradeoff is that each of the slave processors must do exactly the same thing.

SIMD machines can handle some data-parallel problems particularly well. Data parallelism is parallelism that is derived from the problem data. Each processor processes its own data concurrently. The local data can be independent of every other processor's data, or can be a subset of the global problem data, in which case the local computations will usually have some dependence on remotely held data, necessitating interprocessor communication. Important problems that can execute efficiently on SIMD computers include the analysis of seismic data sets, pattern recognition, and finite-difference/finite-element-type numerical simulations. Examples of SIMD machines include Thinking Machines' CM-2 and MasPar Corporation's MP-1 and MP-2. Thinking Machines' CM-5 can also execute programs in a SIMD fashion.

SIMD machines cannot handle all data-parallel problems well, however. Whether a data-parallel algorithm is appropriate for SIMD execution depends on the needs of the algorithm. If the degree of parallelism present in some instructions is less than the number of available processors, some processors must sit idle, since in the SIMD execution style every processor does the same instruction. Actually all processors participate in the computation, but the "idle" processors neglect to store the result. This problem is known as the *load-balancing* problem for SIMD computers. Load imbalance translates directly to less efficient performance. A common source of idle processors is the implementation of boundary conditions.

In contrast, MIMD computers are composed of independent processors that each execute a copy of the program concurrently. The Intel Paragon is an example of a MIMD-style parallel machine. Like SIMD computers, MIMD computers can also be used to solve problems where the parallelism is derived from the data. However, since the multiple processes each have their own program counter, special treatment of certain data elements can be accommodated while other processors continue doing useful work. Thus, in the MIMD model, synchronization is enforced only when it is needed, not after every instruction. However, the automatic synchronization of SIMD execution lowers interprocessor communication latencies. Thus there are tradeoffs with regard to performance. Furthermore, if the MIMD machine is built from physically distributed processor-memory pairs, that is, a *multicomputer*, then each processor must run a more or less complete operating system and store a copy of the program.

The major difference between SIMD and MIMD execution is flexibility. SIMD machines are efficient only for highly data-parallel problems that involve all or most of the data in all operations. In contrast, the MIMD execution style can accommodate data parallelism as well as parallelism that derives from having multiple, independent tasks, each working on the same data. Such *task-parallel* algorithms can be constructed by having different processors solve different differential equations in a coupled system, or handle different zones within the overall geometric configuration. Computers that support MIMD execution can make use of off-the-shelf components, can therefore be constructed easier and cheaper than SIMD machines, and can more quickly respond to cost/performance improvements.

For data-parallel problems, the use of MIMD machines commonly involves spawning identical processes, each with a subset of the problem data. This style is called single-program/multiple-data (SPMD). SPMD programs could execute in SIMD fashion but may not be efficient in that context due to the aforementioned load-balancing problems introduced by the excessively fine, instruction-level synchronization. On the other hand, there is no guarantee that a program with load-balancing problems on a SIMD machine will be more efficient than on a MIMD machine, all other factors being equal, unless the processors that would be idle in the SIMD approach actually have some other useful work they can do before the next synchronization point.

Parallel machines are also commonly classified by the physical layout of the processors and memory(s), into *distributed-memory* and *shared-memory* machines. The examples of SIMD- and MIMD-style computers given above are all distributed-memory machines: each processor is associated with local memory that is physically separate from the memory of every other processor. One can include in the distributed-memory category machines such as the Kendall Square Research KSR-1 and the Cray T3D, which are logically shared-memory machines. Truly shared-memory machines include the Cray C90 and SGI Power Challenge machines, for example. In a shared-memory machine, a single high-speed bus connects all the processors to a single memory. *Memory contention*, the problem of two processors trying to access the same address at the same time, strongly limits the number of processors which can be used. The performance can be greatly improved in many cases (the effect is algorithm-dependent) by using local cache memories for each processor, but additional cost is incurred by the need to maintain cache coherence. The key to the performance of distributed-memory machines lies in the interconnection network. Examples of interconnection networks are rings, meshes, hypercubes, trees, and switches. Algorithm-independent comparisons between networks are made in terms of connectivity and diameter, but the actual performance of applications depends strongly on the algorithm, specifically on the frequency and pattern of communication.

From a programming standpoint, the distinction between *multicomputers* and *multiprocessors* is more important than the distinction of distributed and shared memory. Multicomputers are those distributed-memory machines such as the Paragon and the CM-5 that use as the fundamental unit the processor-memory pair, as found in serial computers. Multiprocessors are shared-memory machines as well as those distributed-memory machines whose memory modules are separated from the processors but connected to them by a processor-memory interconnection network. The multiprocessor/multicomputer distinction reflects the differences in programming techniques used to affect synchronization and communication. Multiprocessors use *global addressing* while multicomputers use *message passing*. Multiprocessor machines can emulate message passing, though. In the global-addressing approach, the memory associated with each processor is made distinct from all the other processor's memories by the hardware of the interconnection network. Thus, remote memory references in the program will transparently access the remote datum across the interconnection network. This is the strategy used in the Cray T3D and the KSR-1. Full switches, which contain $\mathcal{O}(N^2)$ elemental switches to fully connect N processors to N memories, are the ideal processor-memory network, but they quickly become too expensive and difficult to build as the number of processors grows. Butterfly and shuffle-exchange switches contain $\mathcal{O}(N \log 2N)$ elemental switches, and thus are easier to scale up to many processors, but network congestion is then a problem for some applications.

In the message-passing approach, the memory associated with each processor is locally addressed. Remote accesses have to be recognized as such by the programmer or compiler and replaced by explicit send and receive operations. Examples of message-passing computers are the Intel Paragon and workstation clusters. In general, message passing is slower than global addressing, but the actual message-passing performance depends strongly on the communication interface. At the highest level, all the processing of messages, to check for errors and determine the routing, is done in software, and therefore it generates significant latencies per message. Relatively large messages are necessary to amortize this processing cost. At the lowest level, the communication interface itself does the work and the communication delay is effectively proportional to the message size. In addition, the message-passing cost depends on the routing algorithms used and the number of messages simultaneously in the network.

Disregarding architectural differences in favor of a performance-based categorization, one can consider the *communication–computation granularity* of the machine, which is the ratio of the time required for a basic remote memory access in a distributed-memory machine to the time required for a floating-point operation. If this ratio is small in relative terms, the machine is called *fine grained*, otherwise it is *coarse grained*. In terms of efficiency, fine-grained machines are preferable. However, if the relatively fast communication is simply a reflection of slow processors, more processors will be needed to obtain the same actual runtime.

The number of processors is known as the *processor granularity*. Massively parallel systems are fine grained; systems with only a few processors are coarse grained. At present, overall runtime and direct cost considerations seem to be favoring machines that use fewer, more powerful processors. However, with regard to parallel efficiency (indirect costs), fine-grained machines are better, assuming the problem has a sufficient amount of parallelism to make use of many processors. Specific models of the direct and indirect costs of parallel computation can be found in the recent literature (see, for example, Fischer and Patera 1994).

3.2.3   Parallel Algorithms: Implementation

From the programming perspective, the fundamental issue of parallel as opposed to serial computing is that the design and performance of algorithms depends strongly on the parallel architecture. Serial computation can be considered to be a degenerate case of parallel computation in which interprocessor communication is free. This viewpoint mostly explains why computing issues have received less attention in CFD in the past than have physical and numerical issues.

On serial computers it is sometimes possible to identify a "best" algorithm for a particular problem. For example, to solve a system of tridiagonal equations, the $O(N)$ tridiagonal matrix algorithm is optimal. However, when algorithm and architecture interact, one can no longer devise parallel algorithms that are optimal for all possible configurations. There is considerable difference among the best parallel algorithms for different parallel architectures. There are also many issues that are specific to classes of algorithms. Some examples of classes of algorithms are those dealing with dynamic programming, with solving sparse systems of linear equations, with solving dense systems of linear equations, and with optimization. For sparse systems of linear equations, the matrix storage scheme and the type of method, direct or iterative, are major considerations in the design and implementation of parallel algorithms for a given machine.

Complicating the matter is the fact that the question of implementation does not always enter the algorithm/code development procedure at the same level. In a fine-grained implementation, such as that used in conjunction with vector supercomputers (vectorization is a form of parallel computation which we neglect to cover here), SIMD machines, and automatic parallelizing compilers, the ratio of the time spent by concurrently executing processes to the process creation time is small. The programming challenge is to set up loop constructs and to order operations in such a way as to facilitate the parallel execution of these low-level tasks. Use of a FORALL-type statement is typical of a fine-grained implementation of either task- or data-parallel algorithms. Coarse-grained implementations generally use some sort of an explicit fork-join operation by which subprocesses are created and merged. The cost of creating multiple processes can be significant if too little computation is done by each subtask. However, such implementations may, in the data partitioning case, consist of only a single fork or "spawn" operation at the beginning of the program, and therefore be of no consequence to the overall runtime.

There are no universally agreed-upon categories for parallel algorithms and certainly no unique way to parallelize existing sequential algorithms. However, many descriptive terms are in use. We have already used the term "task-parallel," which refers to a parallel algorithm in which multiple tasks execute concurrently. Generally these independent tasks depend on shared data and so are well matched to shared-memory computers. The implementation can be further classified as fine- or coarse-grained, though. In the example of searching a graph, there are initially a relatively small number of independent paths down which to go, but as new paths are discovered as part of the solution, these can be assigned to the next avialable processor, the so-called pool-of-tasks paradigm. Such problems are sometimes said to exploit fine-grained task parallelism because the lifetime of each of the concurrent processes is relatively short compared to the time spent enforcing synchronization and the overall runtime. In contrast, a coarse-grained task parallelization would be one for which the same algorithm is applied to analyze a data set multiple times, changing only a parameter or an initial condition. In this case the multiple instances can be computed concurrently.

Data parallelism refers to the situation where the data are replicated or when there exist multiple sets of data to be processed. Then the concurrency comes from the data. There is no cooperation among processes with replicated data or multiple data sets, so no communication of any kind will be required – such problems are called "embarrassingly" or "trivially" parallel. The more interesting case, however, deals with the solution of a single problem, and in this context "data parallel" refers to "data partitioning," in which the entirety of the problem data is subdivided into as many parts as there are processors, and each part is mapped to a processor. The assignment of data subsets to processors is known as the *mapping* and is a key issue in the present context, so we discuss it in more detail below.

3.2.3.1 PARTITIONING, MAPPING, AND LOAD BALANCING

The assignment of the data subsets in a data-partitioned problem can be critical to the overall program performance. In a distributed-memory environment, interprocessor communication is necessary to access remotely held data. Typically, the cost of accessing a remote memory location using the network is several orders of magnitude slower than accessing local memory. Thus, algorithms and/or data structures should be designed to increase the *locality* of the processor work and to minimize communication.

Data partitioning in computational fluid dynamics is typically grid-based – the grid variables are considered to be the problem data. Thus, in a 2D problem, the grid variables of individual rows, columns, or blocks consisting of a subset of rows and columns of the grid can be considered independently. The extension to 3D is straightforward. Grid-based partitionings are popular on both distributed- and shared-memory machines. For the latter, any local cache memory can be utilized to a great extent with the grid-based mapping to reduce memory contention.

To assign each of the independent data subsets to a specific processor, one must first imagine the interconnection network to be mapped into a logical interconnection network that matches the grid-based partitioning. This logical network is called the processor layout. The motivation behind matching the data partitioning to the processor layout is to service the communication requirements of the algorithm. Ideally, all communication can be accomplished using physically existing links between processors. Only in that case can the algorithm achieve scalability for the architecture. A grid-based processor layout is frequently adequate for the algorithms that arise in finite-difference types of computations. To give a contrasting example, though, consider that a grid-based processor layout is not adequate for the parallel fast Fourier transform, which actually requires hypercube connections to support all communication with direct communication links (Averbuch et al. 1990, Swartztrauber 1987).

The logical interconnection network is *embedded* into the actual network. A ring can be embedded in a 2D mesh, which in turn can be embedded in a 3D mesh, which can in turn be embedded in a hypercube, which can in turn be embedded in a fully connected network such as a switch. Thus, a switch can accommodate all algorithms, but a ring network, such as that sometimes found in a local area network of workstations, cannot provide independent paths even for a 2D mesh virtual topology.

The core numerical tasks of computational fluid dynamics algorithms frequently include the solution of sparse systems of linear equations. With reference to the matrix form of a system of equations, grid-based partitionings of the unknowns are equivalent to row-based partitionings of the matrix because the coefficients in each equation (matrix row) are located at the same grid point and hence are on the same processor. The ordering of equations then determines the communication requirements of numerical algorithms applied to solve the system of equations. The optimal situation with respect to communication occurs using structured grids because the regularity of the equation numbering allows a diagonal storage format for the matrix coefficients, which allows the basic matrix operations such as inner products and matrix-vector multiplications to be accomplished with the minimum possible amount of communication.

For unstructured grids, compressed row storage or compressed column storage schemes are traditional for serial computation. However, neither of these data structures is suitable for parallel computation because they cannot avoid all-to-all broadcasts in parallel matrix-vector multiplications. Thus, novel partitionings are required, which generally requires partitioning the matrix along rows and columns. This topic is an active area of research. Sparse matrix storage schemes suitable for parallel computations have been discussed in Duff et al. (1990) and Ferng et al. (1993), and various aspects of sparse parallel numerical linear algebra are discussed in Bertsekas and Tsitsiklis (1989), Bodin et al. (1993), Gupta et al. (1993), and Health and Raghavan (1993). The partitioning problem which arises for unstructured grid problems has been discussed by Hammond (1992), Nicol and Saltz (1990), and Simon (1991). In many cases, it is still possible to devise parallel iterative or direct algorithms that are quite efficient, but considerable effort must be expended in computing an optimal partitioning, one which minimizes communication.

In addition to communication, the data partitioning also clearly influences the *load balance* of the computation. If one processor has to handle more data than another, it will typically require longer to do so. At synchronization points, the processor with the most work to do may cause the others to wait. Thus the data partitioning should try to give equal computational workload to each processor.

However, in practice, perfect load balance is rarely achieved, for several reasons. First, the processors may not all run at the same speed, and in fact they may run at effective speeds that are unknown a priori. For example, in a workstation cluster, the processors may be different models and may be more or less loaded by other users. In such cases, a dynamic load-balancing strategy may be required to repartition the problem as the computation proceeds.

3.2.3.2 PROGRAMMING

The preceding subsection stressed the importance of matching the algorithm to the architecture in the context of the data-partitioning and mapping problem. An equally important consideration is the manner in which the parallel algorithm will be expressed in a programming language. Bal et al. (1989) give an indication of the breadth of possibilities here. The implementation provides the concrete details of process creation, communication, and synchronization and therefore can strongly affect the performance of the resulting code both in terms of the effective computational speed as well as scalability.

Parallel programming techniques differ depending on the machine and the type of parallelism being exploited. Thus, there are few bases for comparison. Most approaches, though, can be described either as working with existing sequential programming languages as they are, with explicit message passing or automatic parallelization, or as extending or replacing a sequential language by adding new language primitives or new variable types.

Explicit message passing is quite popular because it allows total flexibility. In the message-passing approach, the sequential programming language (Fortran, C, C++, Lisp) is supplemented by libraries designed to carry out message passing, process creation and destruction, synchronization, and so on. The libraries either are supplied by the vendor (CM-5, Paragon, KSR-1) or may be designed to be portable (PVM, MPI, Parasoft EXPRESS, PICL). The programmer has both the burden and the advantage of having all responsibility for interprocessor communication. The library approach is sometimes difficult to use because the programmer needs to be familiar with both message-passing programming techniques and the application in order to arrive at an efficient implementation. Some of the more frequently occurring message-passing programming idioms are (1) local synchronization, (2) overlapping zones, and (3) the host–worker/hostless control model. Hatcher and Quinn (1991) discuss some of these message-passing programming issues in the context of data-parallel applications.

Local synchronization refers to a commonly used shortcut for grid-based partitionings. Processors synchronize only with their neighbors instead of with all the other processors. Explicit integrations of partial differential equations and iterative methods for sparse linear systems of equations can make use of local synchronization because the local computations depend only on values obtained from nearest-neighbor processors. Global synchronization requires all-to-one communication patterns, whose cost depends on the number of processors. In contrast, the time to complete a local synchronization is constant. To do local synchronization between two processors in a message-passing environment, each processor simply sends a message to the other, then executes a (blocking) receive operation.

Overlapping zones is another simplification that arises in grid-based partitionings. In this technique, the local subgrid is augmented or padded by additional rows and columns that contain the "ghost" elements needed in the local computations but that are officially held by another processor. The ghost rows and columns are updated all at once instead of having individual data elements updated separately, which minimizes the number of communications. Fewer long messages are more efficient than many small messages because of the overhead cost of each message. Furthermore, the overlapping zones model can greatly facilitate programming because the additional rows/columns can be integrated into the local data arrays.

In an SPMD program, the host–worker and hostless models are alternative styles for program control. In the host–worker model, the host processor spawns the workers and detects when they are finished, takes care of input and output, and may be the target for global reduction computations such as the computation of a global residual by summing the local residuals. Also, the host may provide synchronization services.

In the hostless model, all such operations must be accomplished by designating one processor to do the job of the host. One advantage to the host–worker model is that it is conceptually easier. Separating the computational kernel from the control and I/O operations requires two programs to be written, but since the host code may be quite system-dependent, this separation makes the overall program easier to modify and maintain. The hostless model may be better suited when parallel input or output are involved. In terms of performance there is little difference. If the program contains a lot of sequential code, then the host may cause the workers to wait in the host–worker model; the same problem shows up in the hostless style in the form of a load imbalance since one processor will have more work to do than others.

The important point to note is that in message-passing programming, such programming idioms as illustrated above are critical to the performance of applications. Unfortunately, the "tricks" are highly application-dependent. As a result, parallel programming tools and environments which could simplify the programming task are difficult to develop and not always efficient.

Automatic parallelization of existing sequential programs is at the other end of the spectrum from explicit message-passing but is similar in the sense that the program is written in a sequential programming language. The compiler bears the entire responsibility for generating communication, and the programmer has no control. Development in this area has been restricted by the difficulty of the challenge – there are often a number of ways to code the same algorithm on a serial computer, and some programs are much easier than others to analyze. Also, given the broad range of parallel architectures, the portability aspect of parallelizing compilers is a serious concern. Practically, automatic parallelization has not extended much beyond the analysis of loop constructs. Thus parallel computation is generated at a relatively low level and is best matched to task-parallel problems.

In between automatic parallelization and explicit message passing are languages based on a variety of paradigms. Data-parallel languages add new parallel variables. Data-parallel Fortrans use arrays as first-class objects. The partitioning of arrays is automatic but can be guided by compiler directives. The basic program is sequential, and parallelism is exploited at a low level, which somewhat restricts the possible sources of efficient parallelism. Examples of data-parallel languages are CM-Fortran, MP-Fortran, Fortran D, and High-performance Fortran. Fortran D (Hiranandani et al. 1992) is similar to CM-Fortran and MP-Fortran in that it supports both the block and cyclic mappings (to be defined below) with BLOCK and CYCLIC compiler directives. Also, there are DISTRIBUTE and ALIGN directives to inform the compiler about the data partitioning. However, even with this information, the compiler cannot produce the most efficient code possible, even for a simple matrix multiplication. Dependence analysis is complicated, and high-level constructs will rarely be adequate to indicate all potential optimizations. High-performance Fortran (HPF) is a superset of CM-Fortran/MP-Fortran with the goal of portability. Recently, Applied Parallel Research released a compiler that can compile the NAS parallel benchmarks on the Cray T3D, IBM SP-2, and Intel Paragon parallel computers (SC-NET 1995). Data-parallel C-language variations include Dataparallel C and C*; they also treat arrays as distributed objects.

Portability is widely recognized as an essential characteristic of parallel programs, languages, and tools. Typically, a parallel program requires extensive changes in moving from one machine to another. The message-passing community has recently created what they hope will be a portable standard for message-passing programs, called MPI (Gropp et al. 1994). Automatic parallelizing compilers also can achieve portability from the viewpoint of the programmer who does not express parallelism explicitly. From the viewpoint of the compiler designer, however, portability will be restricted to specific forms of parallelism that can be detected and accommodated in a general parallel architectural model.

For global-addressing parallel computers, the parallel languages in use build upon fork-join operations, local/shared variables, mechanisms for synchronization, and atomic operations, such as locks and channels. In the present context of data-partitioned problems, though, better performance may be obtained in many cases by emulating message passing. The basic technique is to logically set up communication buffers as shared variables and to declare all other variables local (private, depending on the language primitives for this purpose). Then interprocess communication is accomplished only by writing to and reading from these buffers. This strategy generally decreases memory contention by minimizing the use of shared variables.

3.2.4   Parallel Algorithms: Performance

Due to the complexity of the machines on which they are running, the effective computational speed of a program depends strongly on the problem size and the number of processors, in addition to external factors such as the number of other users. As a result, the performance of parallel programs is difficult to predict. Several measures and systematic experiments are in common use for assessing performance.

First and foremost is simply runtime. For serial computations, the runtime is modeled at a very low level for many purposes in computer science. Multiplication, addition, assignment, control, and logical operations all execute at different rates. Furthermore, the time to access operands depends on their location in the memory hierarchy (registers, cache, etc.). The performance of a real serial machine for a particular algorithm will depend on the problem size as the machine's main memory becomes exhausted. Thus, at this level of detail, modeling the performance of algorithms is a complex task. However, a useful simplification is commonly employed in algorithm development and in comparisons with parallel computations, namely that the machine performance is assumed to be independent of problem size (i.e., infinite memory). Furthermore, all floating-point operations are assumed to take the same amount of time and to dominate the cost of other operations. Thus, with such assumptions, one can take the serial runtime to be proportional to the computational complexity of the algorithm.

By analogy, the parallel runtime depends on the *communication complexity* in addition to the computational complexity. The communication cost can be modeled at a low level, including the latency, the transmission time, and models of communication delay (some nondeterministic). For performance assessment, though, a similar coarse modeling of the cost is appropriate. Unfortunately, one cannot assume that all the basic communication operations (one-to-all, all-to-all, shift) take the same amount of time. The time for these operations depends on the specific interconnection network of the parallel machine and the amount of data to be transmitted, the latter of which depending on the algorithm and its implementation.

Consequently, there are several performance measures, each with a different interpretation and utility. The most common measure is *speedup*, S, which may be defined as

$$S = \frac{T_1}{T_p} \tag{3.1}$$

where T_p is the measured runtime using n_p processors, and T_1 is the modeled runtime of the parallel algorithm on one processor. Parallel efficiency is the ratio of the actual speedup to the ideal (n_p) and reflects the overhead costs of doing the computation in parallel,

$$E = \frac{S_{\text{actual}}}{S_{\text{ideal}}} = \frac{T_1}{T_p n_p}. \tag{3.2}$$

There are many models of the parallel workload which are more or less approximate models of a low-level communication/computation model. At the crudest level of approximation is Amdahl's law, which considers individual operations in a parallel algorithm as either totally parallel or sequential and assumes that the execution time of each operation is the same. The fraction of operations that are sequential is f_s, and the fraction of operations that are parallel is $f_p = 1 - f_s$. The latter, because they are executed concurrently, require a time to execute that is proportional to f_p/n_p, where n_p is the number of processors. Thus, the speedup by Amdahl's law is

$$S_{\text{Amdahl}} = \frac{1}{f_s + \frac{f_p}{n_p}}. \tag{3.3}$$

Amdahl's law, though inaccurate, is useful because it provides an upper bound on the speedup that can be obtained, since communication costs are ignored and the parallel operations are executed with perfect parallelism (no processors are left idle). Amdahl's law also reveals a fundamental idea regarding *scalability*, namely that for a given problem size (fixed f_s and f_p), the maximum speedup is bounded by $1/f_s$ no matter how many processors are used. Thus the benefits of parallel computing, for the purpose of speeding up a given instance of a problem, are limited.

Another useful model divides a parallel algorithm's runtime into contributions from serial computation, parallel computation, and communication. The time spent on serial operations, T_{serial}, cannot be reduced by adding processors. The time spent in parallel computation is T_{comp}, and the model assumes that all n_p processors are busy doing useful work. That is, if the parallel work were done on one processor, the required time would be $n_p T_{\text{comp}}$. T_{comm} is the time spent by the processors doing interprocessor communication for a distributed-memory machine, or, more generally, one could interpret it as any overhead cost of parallelization, for example, load imbalance.

Assuming the serial and parallel work do not overlap, the parallel runtime is approximately $T_{comp} + T_{comm} + T_{serial}$, and on a single processor the runtime is $n_p T_{comp} + T_{serial}$. Thus, using Eq. (3.2) the efficiency E can be written in the following form

$$E = \frac{1 + \frac{T_{serial}}{n_p T_{comp}}}{1 + \frac{T_{serial} + T_{comm}}{T_{comp}}}. \qquad (3.4)$$

Generally an algorithm is not suitable for parallelization unless $T_{serial} \ll T_{comp}$, because of Amdahl's law. Thus, assuming that T_{serial} is negligible, one can approximate the parallel efficiency further as

$$E = \frac{1}{1 + \frac{T_{comm}}{T_{comp}}}. \qquad (3.5)$$

Since time is work divided by speed (Gustafson 1990) in computational science, one can see that parallel efficiency depends on both machine-related factors (the speed at which communication and computation are carried out) and the algorithm's communication requirements (the amount of communication and computation work to be done). High parallel efficiency is not necessarily a product of fast processors or fast communications considered alone; instead the ratio of speeds and the amount of work are what is important.

Another performance measure and key concept, one which appears in more than one context, is *scalability*. A scalable parallel architecture is one whose interprocessor communication network provides the same bandwidth as the number of processors grows. This means that the time required for a basic communication operation, such as an all-to-all broadcast, depends linearly on the number of processors. Increasing the number of processors does not diminish the network performance.

Scalable parallel algorithms are those whose computational and communication complexity both depend only on the problem size per processor. Storage is also an issue. Global data structures that are replicated in each processor cause the per-processor memory requirement to be a function of the number of processors, a situation that can cause trouble if using many processors. Lack of scalability in any of these contexts can be interpreted as an indirect cost of parallel computing (Fischer and Patera 1994). To be able to make efficient use of many processors, scalable algorithms on scalable architectures are needed. Thus, scalability should be considered a very important measure of the performance of a parallel algorithm, not an afterthought to speedup or speed measured in MFlops.

A more intuitive alternate definition of a scalable algorithm is one which obtains linear speedup in the scaled-size experiment (Gustafson et al. 1988, Kumar et al. 1994). In the scaled-size experiment, the problem size is increased along with the number of processors to maintain a constant local problem size for each of the parallel processors. In the sections to follow, we denote the problem size per processor as N, which is approximately the number of grid points N_{gp} divided by the number of processors n_p. The local problem size is sometimes called the *subgrid size*.

The aim in the scaled-size experiment is to maintain a high level of parallel efficiency, which indicates that the added additional processors increased the speedup in a one-for-one trade. Unfortunately, the introduction of the idea of speedup to assess scalability is somewhat cumbersome, since it depends on how one defines speedup (Gustafson 1990). Speedup can be a ratio of speeds or a ratio of times. Furthermore, the one-processor time can refer to an optimal serial implementation or to a parallel implementation using only one processor.

A more current development in the assessment of scalability is the isoefficiency function approach (Kumar and Singh 1991, Kumar et al. 1994). This method uses the runtime of the serial algorithm, instead of the storage requirements, as a measure of the problem size. In the context of iterative algorithms, including the present algorithm, the overall runtime is the product of the number of iterations multiplied by the cost per iteration. Thus, the viability of large-scale parallel computations, that is, solving large problems using many processors, depends on two factors: the scalability of the cost on a per-iteration basis, and the scalability of the convergence rate as the problem size increases. The isoefficiency approach is useful in assessing scalability in this broader sense of the cost-per-iteration/convergence-rate product.

3.3 Multigrid Method for Convergence Acceleration

The convergence rate of single-grid algorithms for the Navier-Stokes equations depends on many factors, including the coupling between pressure and velocity in the particular flow problem of interest, the dimensionless parameters such as Reynolds number, the boundary and initial conditions, and the grid characteristics. In the context of pressure-correction algorithms, there are additional algorithmic factors, namely the values of the relaxation factors, the type of inner iterative solution method, and the number of inner iterations. Thus, in general, the convergence rate dependence on problem size is not simply a reflection of the asymptotic convergence rate of the inner iterative methods used (Shyy 1994).

Nevertheless, if the coupling between unknowns in the inner iterative technique is local within iterations, as is typical, then the convergence rate of the outer iterations will likely deteriorate rapidly as the single-grid problem size (number of grid points) increases. Multigrid acceleration techniques are of interest because they can maintain good convergence rates with respect to increases in the problem size. For parallel computers, such convergence-rate scalability is critical, since the target problem size is quite large.

A solid theoretical understanding for the success of multigrid methods has been established for linear equations. For sufficiently smooth elliptic problems, for example, Poisson's equation, multigrid methods can achieve problem size–independent convergence rates (Hackbusch 1980a, Nicolaides 1979). The recent book by Briggs (1987) introduces the major concepts of multigrid methods in the context of Poisson equations. Briggs and McCormick (1987), Hackbush (1980b), and Luchini and Dalascio (1994) consider multigrid convergence properties for more general linear equations.

For the incompressible Navier-Stokes equations, there is less analytical background available but considerable practical experience. Brandt gave the initial description of the multigrid procedure, emphasizing the practical techniques and special considerations for nonlinear equations (Brandt 1977, 1984). Pressure-correction multigrid methods of the type used here have been considered in Shaw and Sivalo-ganathan (1988a,b), Shyy and Sun (1993), and Sockol (1993), and other multigrid results for incompressible Navier-Stokes equations can be found in Bruneau and Jouron (1990), Ghia et al. (1982), Lien and Leschziner (1991), and Michelson (1991).

Experience has indicated that multigrid accelerations work well for the pressure-based Navier-Stokes formulations for both compressible and incompressible flows (Shyy 1994, Shyy et al. 1992b), provided that suitable multigrid components – the smoother, restriction and prolongation procedures – and multigrid techniques are employed. Speedups of two orders of magnitude over the convergence rate of single-grid methods are typical. However, there are still many unsettled issues. The physical and numerical considerations that affect the convergence rate of single-grid computations carry over to the multigrid framework and are compounded there by the coupling between the evolving solutions on multiple grid levels, and by the particular grid scheduling used.

The purpose of this section is to illustrate some of the important issues in the context of parallel computing. In particular, we consider the impact of the multilevel convection scheme on the stability and convergence rate of V-cycle convergence rates, the role of the initial guess and the full multigrid starting procedure by which the initial fine-grid solution is generated, and the restriction procedure for transferring residuals and solutions from fine to coarse grid levels. A heuristic technique relating the convergence tolerances for the coarse grids to the truncation error of the discretization is developed. In numerical experiments the technique is found to be effective and robust. For example, with second-order upwinding on all grid levels, a five-level 320×80 step flow solution was obtained in 20-V cycles, which corresponds to a smoothing rate of 0.7 and required 25 s on a 32-node CM-5. Overall, the convergence rates obtained in the following illustrations are comparable to the most competitive findings reported in the literature.

The multigrid approach also appears to be a promising way to exploit the increased computational capabilities that parallel computers offer. The standard V-cycle full multigrid (FMG) algorithm has an almost optimal operation count, $\mathcal{O}(\log^2 N)$ for Poisson equations, on parallel computers. However, there are still many issues to be resolved in this area, some of which are discussed in the next section. Assuming that scalable implementations can be obtained, though, multigrid methods appear to provide a nearly optimal path to large-scale parallel computations.

3.3.1   **Background**

The goal of multigrid iteration is to effectively damp all solution error components, smooth and oscillatory, at the same rate. Standard iterative techniques such as point-Jacobi iteration have the smoothing property – eigenvectors of the solution error are damped at a rate that is proportional to their corresponding eigenvalues. Consequently, oscillatory errors are damped quickly but smooth errors disappear very slowly. Thus the asymptotic convergence rate reflects the rate of error reduction for smooth components. To achieve the multigrid goal, smooth error components, with respect to the fine-grid scale, are eliminated on coarse grids. On coarse grids the smooth errors are effectively more oscillatory, so iterations can quickly eliminate them. Since the coarse grid typically has only 1/4 as many grid points as the fine grid, in two-dimensional problems the smoothing iterations on the coarse grid are cheaper as well as more effective at reducing the smooth error components than the smoothing iterations on the fine grid.

Because of this uniform (in terms of frequency) treatment of solution errors, multigrid methods are highly efficient for physical problems with multiple, disparate physical scales. Traditionally, multigrid iteration is based on a uniformly fine/coarse grids. However, multigrid iterations are similar and complementary to local grid refinement, whose goal is to obtain a uniformly accurate discretization by varying the grid resolution according to some error criteria. The latter causes substantial modifications to the discrete equation operator, though, and therefore can be viewed as an additional complication.

3.3.1.1 CORRECTION AND FAS MULTIGRID SCHEMES

The components of a multigrid method are the smoothing procedure by which the equations on each grid level are solved and the restriction and prolongation procedures by which the equation residuals and correction quantities are transferred between grid levels. The *multigrid scheme* specifies how these components interact, that is, how coarse-grid problems are generated from the fine-grid problem and in what order the multiple grid levels are visited. The latter is known as the *multigrid cycle* type, and common choices include V, W, and F cycles.

For linear equations the "correction" multigrid scheme is sufficient. In this scheme, coarse-grid problems are developed for the solution error instead of the solution itself. This approach saves some storage and slightly reduces the cost of the multigrid iteration, compared to a "full-approximation scheme," since one can take the initial guess on coarse grids to be zero instead of approximating it through restriction of a finer-grid solution.

To more formally present the multigrid iteration, we can write the discrete problem to be solved on the fine grid as $A^h u^h = S^h$, where superscript h denotes the grid spacing on the fine grid. This matrix form of a system of linear algebraic equations is presumed to have been derived from some linear differential equation $L[u] = S$. The set of values u^h for a two-dimensional problem is defined by

$$\{u_{i,j}\} = u(ih, jh), \quad (i, j) \in ([0:N], [0:N]) \approx \Omega^h \qquad (3.6)$$

where Ω^h is the domain of interest and where for ease of presentation we assume the same grid spacing in both directions. Similarly, u^{2h} can be defined on the coarser grid Ω^{2h} with grid spacing $2h$. For generalization to a system of linear differential equations, one can interpret the variable u as either a scalar or a vector. The same basic notation is also extensible to the nonlinear problem by letting the operator A be a function of u.

With this notation, a two-level multigrid cycle using the correction scheme based on a smoother whose iteration matrix is P consists of the following steps:

Do ν fine-grid iterations $\quad v^h \leftarrow P^\nu v^h$

Compute residual on $\Omega^h \quad r^h = A^h v^h - S^h$

Restrict r^h to $\Omega^{2h} \quad r^{2h} = I_h^{2h} r^h$

Solve exactly for $e^{2h} \quad e^{2h} = (A^{2h})^{-1} r^{2h}$

Correct v^h on $\Omega^h \quad (v^h)^{\text{new}} = (v^h)^{\text{old}} + I_{2h}^h e^{2h}$

I_h^{2h} and I_{2h}^h symbolize the restriction and prolongation procedures (Shyy 1994).

The quantity v^h is the current approximation to the discrete solution u^h. The *algebraic* error is the difference between them, $e^h = u^h - v^h$. The *discretization* error is the difference between the exact solutions of the continuous and discrete problems, $e_{\text{discr}} = u - u^h$. The *truncation error* is obtained by substituting the exact solution into the discrete equation,

$$\tau^h = A^h u - S^h = A^h u - A^h u^h. \tag{3.7}$$

The two-level algorithm has the following steps: pre-smoothing iterations, restriction to a coarse grid, solving the coarse-grid problem for coarse-grid corrections, prolongation of corrections to the fine grid, and post-smoothing iterations. The procedure begins on the fine grid with ν iterations of the smoother, which produces a new approximation to the solution, v^h, with a certain algebraic error. The correction scheme calls for an approximation to the fine-grid residual equation for the algebraic error, namely

$$A^{2h} e^{2h} = r^{2h}. \tag{3.8}$$

To obtain the coarse-grid source term r^{2h}, the restriction procedure I_h^{2h} is applied to the fine-grid residual r^h,

$$r^{2h} = I_h^{2h} r^h. \tag{3.9}$$

Equation (3.9) is an averaging type of operation. Two common restriction procedures are straight injection of fine-grid values to their corresponding coarse-grid grid points and averaging r^h over a few fine-grid grid points that are near the corresponding coarse-grid grid point. The initial error on the coarse grid is taken as zero.

For analytical purposes, in the two-level algorithm the solution for e^{2h} is assumed to be obtained accurately, by either smoothing or by direct methods. In practice, though, the solution for e^{2h} is obtained by recursion on the two-level cycle – $(A^{2h})^{-1}$ is not explicitly computed except perhaps on the coarsest grid if direct solution is cheap. Otherwise a few smoothing iterations generally suffice to obtain the coarsest-grid solution for e^{2h}.

The next step is to interpolate e^{2h} to the fine grid and to correct the fine-grid solution,

$$v^h \leftarrow v^h + I_{2h}^h e^{2h}. \tag{3.10}$$

For I_{2h}^h, common choices are bilinear or biquadratic interpolation.

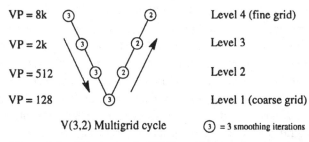

VP = 8k ③ ② Level 4 (fine grid)

VP = 2k ③ ② Level 3

VP = 512 ③ ② Level 2

VP = 128 ③ Level 1 (coarse grid)

V(3,2) Multigrid cycle ③ = 3 smoothing iterations

Figure 3.1 Schematic of a V(3,2) multigrid cycle.

The use of recursion to solve for e^{2h} in the two-level algorithm leads to a V cycle, as shown in Fig. 3.1. In the figure the number of pre-smoothing and post-smoothing iterations are shown inside the circles; it is a fixed V(3,2) cycle. Three smoothing iterations are taken before restricting to the next coarser grid, and two iterations are taken after the solution has been corrected. The purpose of the post-smoothing is to damp any high-frequency errors introduced by the prolongation. Other cycles can be envisioned. In particular the W cycle is popular (Brandt 1984) in which two corrections are applied before post-smoothing, on each level.

The most important consideration for the correction scheme has been saved for last, namely the definition of the coarse-grid discrete equation A^{2h}. One possibility is to discretize the original differential equation directly on the coarse grid. However this choice is not always the best one. The convergence-rate benefit from the multigrid strategy is derived from the particular coarse-grid approximation to the fine-grid *discrete* problem, not the continuous problem. Because the coarse-grid solutions and residuals are obtained by particular averaging procedures, there is an implied averaging procedure for the fine-grid discrete operator A^h that should be honored to ensure a useful homogenization of the fine-grid residual equation. This issue is critical when the coefficients and/or dependent variables of the governing equations are not smooth (Dendy 1982).

For the Poisson equation, the Galerkin approximation $A^{2h} = I_h^{2h} A^h I_{2h}^h$ is the right choice. The discretized equation coefficients on the coarse grid are obtained by applying suitable averaging and interpolation operations to the fine-grid coefficients, instead of by discretizing the governing equation on a grid with a coarser mesh spacing. Briggs has shown, by exploiting the algebraic relationship between bilinear interpolation and full-weighting restriction operators, that initially smooth errors begin in the range of interpolation and finish, after the smoothing correction cycle is applied, in the null space of the restriction operator (Briggs 1987). Thus, if the fine-grid smoothing eliminates all the high-frequency error components in the solution, the correction scheme is a direct solver for the Poisson equation. In general, the convergence rate of multigrid methods using the Galerkin approximation is more difficult to analyze if the governing equations are more complicated than Poisson equations, but significant theoretical advantages for application to general linear problems have been indicated (Wesseling 1987). There are several good analyses of the functioning of multilevel algorithms in other contexts as well (Briggs 1993, Douglas and Douglas 1993, Hwang and Parsons 1994).

In contrast to the correction scheme, the full-approximation storage (FAS) scheme, first described by Brandt (1977), forms the coarse-grid equations in a manner that can be successful for nonlinear problems. In the FAS scheme, one approximates the actual fine-grid equation on a coarse grid instead of the fine-grid residual equation for the solution error. Since the solution is not smooth on the scale of the coarse grid (the error is, after the pre-smoothing iterations), appropriately averaged source terms must be added to the coarse-grid equations. Ideally, the coarse-grid solution corresponds to the fine-grid solution at the same spatial locations. The coarse-grid operator A^{2h} is generally derived in the same manner as the fine-grid one, namely by discretizing the original equations, on the coarse grid.

Starting again with $A^h u^h = S^h$ on the fine grid, with nonlinear A^h, pre-smoothing with an appropriate technique produces an approximate solution v^h that defines the residual on the fine grid,

$$A^h v^h = S^h + r^h. \tag{3.11}$$

A correction, the algebraic error $e^h_{alg} = u^h - v^h$, is sought that satisfies

$$A^h \left(v^h + e^h_{alg} \right) = S^h. \tag{3.12}$$

The fine-grid residual equation is formed by subtracting Eq. (3.11) from Eq. (3.12), cancelling S^h,

$$A^h (v^h + e^h) - A^h (v^h) = -r^h \tag{3.13}$$

where the subscript indicating algebraic error is dropped for convenience. For linear equations the $A^h v^h$ terms cancel, leaving Eq. (3.8). Equation (3.13) does not simplify for nonlinear equations. Assuming that the smoother has done its job, r^h will be smooth and the solution to Eq. (3.13) will be approximately the solution to the coarse-grid residual equation

$$A^{2h} (v^{2h} + e^{2h}) - A^{2h} (v^{2h}) = -r^{2h}. \tag{3.14}$$

The error e^{2h}, once found, is interpolated back to Ω^h according to $e^h = I^h_{2h} e^{2h}$ and added to v^h so that Eq. (3.12) is satisfied. In Eq. (3.14), the known quantities are v^{2h}, which is a "suitable" restriction of v^h, and r^{2h}, likewise a restriction of r^h. Different restrictions can be used for residuals and solutions. Thus, Eq. (3.14) can be written

$$A^{2h} \left(I^{2h}_h v^h + e^{2h} \right) = A^{2h} \left(I^{2h}_h v^h \right) - I^{2h}_h r^h. \tag{3.15}$$

Since Eq. (3.15) is not an equation for e^{2h}, one solves instead for the sum $I^{2h}_h v^h + e^{2h}$. Expanding r^h and regrouping terms, Eq. (3.15) can be written

$$A^{2h} (u^{2h}) = A^{2h} \left(I^{2h}_h v^h \right) - I^{2h}_h r^h \tag{3.16}$$

$$= \left[A^{2h} \left(I^{2h}_h v^h \right) - I^{2h}_h (A^h v^h) + I^{2h}_h S^h - S^{2h} \right] + S^{2h} \tag{3.17}$$

$$= \left[S^{2h}_{numerical} \right] + S^{2h}. \tag{3.18}$$

Equation (3.18) is basically in the same form $A^{2h}u^{2h} = S^{2h}$ as the original equation when it is discretized directly on the coarse grid, except for the extra numerically derived source term. Once $I_h^{2h}v^h + e^{2h}$ is obtained, the coarse-grid approximation to the fine-grid error, e^{2h}, is computed by first subtracting the initial coarse-grid solution $I_h^{2h}v^h$,

$$e^{2h} = u^{2h} - I_h^{2h}v^h, \tag{3.19}$$

and then interpolating back to the fine grid and combining with the current solution,

$$v^h \leftarrow v^h + I_{2h}^h(e^{2h}). \tag{3.20}$$

3.3.1.2 ISSUES IN THE EXTENSION TO THE NAVIER–STOKES EQUATIONS

The incompressible Navier-Stokes equations are a system of coupled, nonlinear equations. Consequently the FAS scheme given above for single nonlinear equations needs to be modified. If we denote by the variables u_1^h, u_2^h, and u_3^h the Cartesian velocity components and pressure, respectively, and use corresponding subscripts to identify each equation's source term, residual, and discrete operator, then the three equations for momentum and mass conservation can be written in matrix form for a grid level whose spacing is represented by superscript h,

$$\begin{bmatrix} A_1^h & 0 & G_x^h \\ 0 & A_2^h & G_y^h \\ G_x^h & G_y^h & 0 \end{bmatrix} \begin{bmatrix} u_1^h \\ u_2^h \\ u_3^h \end{bmatrix} = \begin{bmatrix} S_1^h \\ S_2^h \\ S_3^h \end{bmatrix}. \tag{3.21}$$

Thus, for the u_1-momentum equation, Eq. (3.21) is modified to account for the pressure gradient, $G_x^h u_3^h$, which also is an unknown. The continuity equation source term S_3^h is zero on the finest grid, but for coarser grid levels it may not be zero and so it is retained in Eq. (3.21).

The FAS scheme for the system of equations is straightforward to adapt to the system of equations. The approximate solutions are v_1^h, v_2^h, and v_3^h corresponding to u_1^h, u_2^h, and u_3^h. For the u_1-momentum equation, for example, the approximate solution satisfies

$$A_1^h v_1^h + G_x^h v_3^h = S_1^h + r_1^h. \tag{3.22}$$

The fine-grid residual equation, corresponding to Eq. (3.13) for the single equation case, is modified to

$$A_1^h \left(v_1^h + e_1^h\right) - A_1^h \left(v_1^h\right) + G_x^h \left(v_3^h + e_3^h\right) - G_x^h \left(v_3^h\right) = -r_1^h \tag{3.23}$$

which is approximated on the coarse grid by the corresponding coarse-grid residual equation,

$$A_1^{2h} \left(v_1^{2h} + e_1^{2h}\right) - A_1^{2h} \left(v_1^{2h}\right) + G_x^{2h} \left(v_3^{2h} + e_3^{2h}\right) - G_x^{2h} \left(v_3^{2h}\right) = -r_1^{2h}. \tag{3.24}$$

The known terms are $v_1^{2h} = I_h^{2h} v_1^h$, $v_3^{2h} = I_h^{2h} v_3^h$, and $r_1^{2h} = I_h^{2h} r_1^h$. Expanding r_1^h and regrouping terms, Eq. (3.23) can be written

$$
\begin{aligned}
A_1^{2h} &\left(u_1^{2h}\right) + G_x^{2h}\left(u_3^{2h}\right) \\
&= A_1^{2h}\left(I_h^{2h} v_1^h\right) + G_x^{2h}\left(I_h^{2h} v_3^h\right) - I_h^{2h}\left(A_1^h v_1^h + G_x^h v_3^h\right) + I_h^{2h} S_1^h \\
&= \left[A_1^{2h}\left(I_h^{2h} v_1^h\right) + G_x^{2h}\left(I_h^{2h} v_3^h\right) - I_h^{2h}\left(A_1^h v_1^h + G_x^h v_3^h\right)\right. \\
&\quad \left. + I_h^{2h} S_1^h - S_1^{2h}\right] + S_1^{2h} = \left[S_{1,\text{numerical}}^{2h}\right] + S_1^{2h}.
\end{aligned}
\tag{3.25}
$$

Equation (3.26) is in the same form as the first equation of Eq. (3.21), except that numerically derived source terms supplement the physical source terms. Thus, the resulting coarse-grid system of equations can be solved by the same pressure-correction technique used on the fine grid. The coarse-grid variables are not the same as would be obtained from a discretization of the original continuous governing equations on the coarse grid. Instead the coarse-grid solution corresponds to the fine-grid solution at corresponding spatial locations.

The u_2-momentum equation is adapted similarly, and the coarse-grid continuity equation is

$$
G_x^{2h} u_1^{2h} + G_y^{2h} u_2^{2h} = G_x^{2h}\left(I_h^{2h} u_1^h\right) + G_y^{2h}\left(I_h^{2h} u_2^h\right) - I_h^{2h} r_3^h.
\tag{3.26}
$$

In the finite-volume staggered grid formulation, one can make the right side of Eq. (3.26) identically zero on all grid levels by setting the coarse-grid velocity components equal to the average of the two fine-grid velocity components on the corresponding coarse-grid control volume face. We discuss this point in some detail below.

While the FAS scheme is appropriate for our solution techniques, correction schemes are also appropriate in some cases. The role of multigrid iteration depends on the solution technique. If the governing equations are fully coupled and solved by iteration, then the solution of the linearized equations can employ the correction scheme. Usually the coarse-grid equations for the errors are derived directly from the discretized fine-grid equations by the Galerkin approximation (Zeng and Wesseling 1994), which is computationally cumbersome and expensive.

Projection methods, a popular class of sequential integration techniques for unsteady flow problems, also frequently employ multigrid correction schemes to solve the pressure Poisson equation (Chorin 1967, Minion 1994). As a result, the cost per timestep scales linearly with the problem size. The timestep size is still limited by the viscous stability constraint on the fine grid, however. Consequently, in application to steady-state problems, the cost of projection methods still scales with the number of grid points because more timesteps are needed to reach the steady state.

Similarly, in sequential solution procedures such as the present one, correction schemes may be used for the pressure equation (Braaten and Shyy 1987, Hutchinson and Raithby 1986, Rhie 1989, Sathyamurthy and Patankar 1994). However, the convergence rate deterioration of the outer iterations with increasing problem size persists because the velocity-pressure coupling is not addressed except on the fine grid. In contrast, the FAS scheme allows the multigrid iterations to be used as the outermost computational shell. The equations on each grid level are the nonlinear, implicitly discretized, incompressible Navier-Stokes equations, and these are smoothed by the pressure correction technique. Consequently, FAS addressed the pressure–velocity coupling on multiple scales and can therefore accelerate the convergence rate when seeking the steady-state solution directly. To summarize, correction schemes are useful for methods that iterate on the linearized fully coupled system of Navier-Stokes equations and for projection methods for unsteady flow. For pressure correction methods, though, FAS is appropriate. The discrete problems at each grid level are solved as in the single-grid case but coupled to each other by the FAS prescription in order to obtain improved convergence of the fine-grid iterations.

Due to the nonlinearity of the governing equations, errors propagate through the multigrid cycle in unpredictable manners. Consequently, there has been much discussion about which numerical methods for the steady-state incompressible Navier-Stokes equations perform best in the FAS multigrid setting (Yavneh 1993). Efficient single-grid methods such as pressure correction methods are not necessarily good smoothers. The distinction between the terms "smoother" and "solver" emphasizes the role of the method in the multigrid context, which is to eliminate only those errors that are oscillatory on the scale of the grid level under consideration.

Vanka (1986) has developed a locally coupled explicit method for the steady-state incompressible Navier-Stokes equations. As a single-grid solver it is slowly converging, but in the multigrid context it is evidently a good smoother. A staggered-grid finite-volume discretization is employed as in the pressure correction method. In Vanka's smoother, though, the velocity components and pressure of each control volume are updated simultaneously, but the coupling between control volumes is not taken into account, so the calculation of new velocities and pressures is explicit. The control volumes are visited in lexicographic order in the original method which is therefore aptly called BGS (block Gauss-Seidel). Line variants have been developed to couple the flow variables in neighboring control volumes along lines (Sockol 1993, Thompson and Ferziger 1989).

Linden et al. (1990) reviewed multigrid methods for the steady-state incompressible Navier-Stokes equations and stated a preference for Vanka's method over sequential pressure correction methods, but no direct comparisons were made. The argument given was that the local coupling of variables is better suited to produce local smoothing of residuals and therefore leads to a more accurate coarse-grid approximation of the fine-grid problem. Similar reasoning appears to have influenced Ferziger and Peric (1993), Ghia et al. (1982), and Vanka (1986). A simplified Fourier analysis of locally coupled smoothing for the Stokes equations confirmed good smoothing properties for the locally coupled explicit method. Shaw and Sivaloganathan (1988b) have found that SIMPLE (with the SLUR solver) also has good smoothing properties for the Stokes equations, assuming that the pressure correction equation is solved completely during each iteration. Thus there is some analytical evidence that both pressure correction methods and the locally coupled explicit technique are suitable as multigrid smoothers. However, the analytical work is oversimplified – numerical comparisons are needed on a problem-by-problem basis.

Sockol (1993) has compared the performance of BGS, two line-updating variations on BGS, and the SIMPLE method with successive line underrelaxation for the inner iterations on three model flow problems with different characteristics, varying grid aspect ratios, and a range of Reynolds numbers. Each method was best, in terms of work units, in some range of the parameter space. Brandt and Yavneh (1992) have studied the error smoothing properties of a multigrid method which uses a line-interative sequential smoother for the incompressible Navier-Stokes equations. Good convergence rates were observed for "entering-type" flow problems in which the flow has a dominant direction and is aligned with grid lines. Evidently, line relaxation has the effect of providing nonisotropic error smoothing properties to match the physics of the problem. Alternating line-Jacobi relaxation appears to be better for multigrid smoothing than alternating successive line-underrelaxtion (SLUR) (Wesseling 1991). Other experiences with steady-state incompressible flows via multigrid methods include those of Luchini and Dalascio (1994) and Shyy (1994). Cumulatively these researches indicate that sequential pressure-based methods are viable multigrid smoothers, and that the competitiveness with Vanka's smoother is a problem-dependent consideration.

The following results are obtained with the SIMPLE pressure correction method as smoother, using for the inner iterations damped point-Jacobi relaxation. In most cases the single-grid SIMPLE method with point-Jacobi inner iterations requires small relaxation factors and consequently converges slowly. However, with multigrid good convergence rates are attainable as our results show. The multigrid procedure also adds robustness in comparison to the single-grid SIMPLE method (Shyy 1994, Shyy et al. 1992b). Overall the approach presented below is very promising.

3.3.2   Issues Relating to the Multigrid Convergence Rate

To illustrate some of the factors that are important in the present context, we discuss numerical experiments using two model flow problems with different physical characteristics: lid-driven cavity flow at Reynolds number of 5000 and flow past a symmetric backward-facing step at Reynolds number 300 based on a second-order upwind scheme for convection treatment (Shyy et al. 1992a). Streamlines, velocity, vorticity, and pressure contours are shown in Figs. 3.2 and 3.3. In the lid-driven cavity flow, convection and cross-stream diffusion balance each other in most of the domain, and the pressure gradient is insignificant except in the corners. The flow is not generally aligned with the grid lines. In the symmetric backward-facing step flow, the pressure gradient is in balance with viscous diffusion and, except in the (weak) recirculation region, is strongly aligned with the grid lines. Thus, the two model problems pose different challenges for our smoothing procedure, and at the same time they are representative of broader classes of flow problems.

3.3.2.1   STABILIZATION STRATEGY

With FAS, the discrete problem on each grid level is obtained by discretizing the differential equations. However, it is not always possible to use the same discretization on each grid level. For example, central-differencing the convection terms is likely to be unstable on coarse grids in most problems, depending on the boundary conditions (Shyy 1994). The multigrid iterations will diverge if the smoothing iterations diverge on the coarse grid(s). However, it is also problematic to use central differencing on

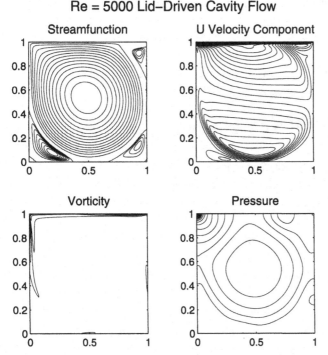

Re = 5000 Lid–Driven Cavity Flow

Figure 3.2 Streamfunction, vorticity, and pressure contours for $Re = 5000$ lid-driven cavity flow, using the second-order upwind convection scheme. The streamfunction contours are evenly spaced within the recirculation bubbles and in the interior of the flows, but these spacings are not the same. The actual velocities within the recirculation regions are relatively weak compared to the core flows.

the fine grid for accuracy and first-order upwinding on the coarse grids, unless special restriction and prolongation procedures are employed, or unless the convection effect is not dominant.

We have experimented with two strategies for stabilizing the multigrid iterations in the presence of the convection–diffusion stability problem. One is the defect correction approach, as described in Altas and Burrage (1994), Blosch (1994), Sockol (1993), Thompson and Ferziger (1989), and Vanka (1986). The other is the use of a robust second-order convection scheme on each grid level, namely second-order upwinding.

In the defect correction approach, the convection terms on all the grid levels are discretized by first-order upwinding, but on the finest grid, a source-term correction is applied which allows the second-order-accurate central-difference solution (provided it is stable) to be recovered when the multigrid iterations converge. The fine-grid discretized equations for one of the velocity components ϕ at iteration $n + 1$, for a control volume, are set up in the following form:

$$\left[a_P^{u1}\phi_P - a_E^{u1}\phi_E - a_W^{u1}\phi_W - a_N^{u1}\phi_N - a_S^{u1}\phi_S - b_P^{u1} \right]^{n+1}$$
$$= [r^{u1}]^{n+1} = [r^{u1} - r^{ce}]^n \tag{3.27}$$

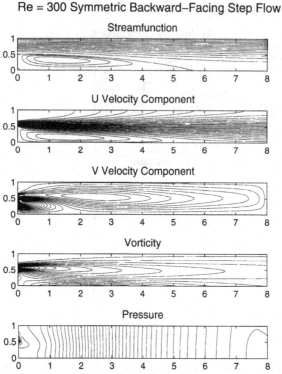

Figure 3.3 Streamfunction, vorticity, pressure, and velocity component contours for $Re = 300$ symmetric backward-facing step flow, using the second-order upwind convection scheme. The streamfunction contours are evenly spaced within the recirculation bubbles and in the interior of the flows, but these spacings are not the same. The actual velocities within the recirculation regions are relatively weak compared to the core flows.

where the superscripts $u1$ and ce denote first-order upwinding and central differencing, n denotes the current values, r denotes equation residuals, b denotes source terms, and the coefficient subscripts denote the spatial coupling between the east, west, north, and south ϕs. The second-order central-difference solution, $r^{ce} \to 0$, is recovered when $[r^{u1}]^{n+1}$ is approximately equal to $[r^{u1}]^n$.

In the second-order upwinding approach, the equation coefficients are computed as in Shyy et al. (1992a). It is interesting to note that the cost of the momentum equation coefficient computations on the CM-5 using either second-order upwind or defect correction is about the same and approximately 30% more than central differencing. While the second-order upwind stencil involves two upwind neighbors and consequently has more communication, the defect correction scheme has more computational work, due to the source terms.

Shyy and Sun (1993) compared the second-order upwinding approach with the approach of using first-order upwinding and central differencing on all grid levels for those cases in which the latter approach was stable. Comparable multigrid convergence rates were obtained for all three convection schemes in $Re = 100$ and

Table 3.1 *Number of single-grid SIMPLE iterations to converge to* $\|r_u\| \le 10^{-5}$, *for the lid-driven cavity flow on an* 81×81 *grid. The* L_1 *norm is used, normalized by the number of grid points*

Lid-driven cavity	Inner iterative method	Convection scheme			
		First-order upwinding	Defect correction	Central differencing	Second-order upwinding
$Re = 1000$	Point-Jacobi	2745	3947	1769	4419
$Re = 1000$	Line-Jacobi	2442	3497	1543	3610
$Re = 1000$	SLUR	2433	3482	1534	3568
$Re = 3200$	Point-Jacobi	16526	>20000	12302	>20000
$Re = 3200$	Line-Jacobi	16462	>20000	12032	>20000
$Re = 3200$	SLUR	16458	>20000	11985	>20000

$Re = 1000$ lid-driven cavity flows, whereas for single-grid computations there were relatively large differences in the convergence rates.

Table 3.1 and Fig. 3.4 help to illustrate some of the issues with regard to the convection scheme, in the single-grid context, for a lid-driven cavity flow problem. Comparison between the alternative stabilization strategies in the multigrid context are made in connection with the discussion of Figs. 3.5 and 3.6 later.

The table records the number of iterations to converge both of the momentum equations to the level $\|r\| \le -0.5$, where $\|r\|$ is the L_1 norm of the equation residual

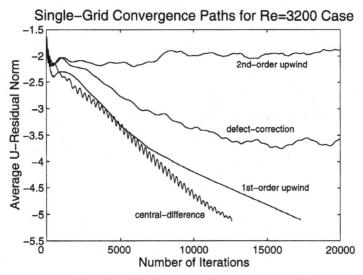

Figure 3.4 Decrease in the norm of the u-momentum equation residual as a function of the number of SIMPLE iterations, for different convection schemes. The results are for a single-grid simulation of $Re = 3200$ lid-driven cavity flow on an 81×81 grid. The alternating line-Jacobi method is used for the inner iterations. The results do not change significantlly with the point-Jacobi or the SLUR solver.

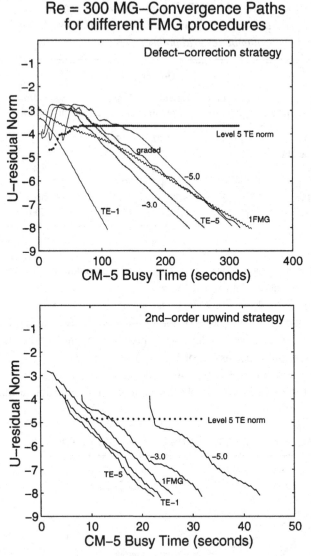

Figure 3.5   The convergence path of the u-residual L_1 norm on the finest grid level in the 5-level $Re = 300$ symmetric backward-facing step flow. The relaxation factors used were $\omega_{uv} = 0.6$ for momentum equations and $\omega_c = 0.4$ for pressure correction.

divided by the number of grid points. The continuity equation is also monitored, but for the cavity flow it converges at a faster rate than the momentum equations.

For the $Re = 1000$ case, the numbers of inner iterations used were $\nu_u = \nu_v = 3$ and $\nu_c = 9$, and the corresponding relaxation factors were 0.4 for u and v and 0.7 for pressure. For the $Re = 3200$ case, the numbers of inner iterations were increased to $\nu_u = \nu_v = 5$ and $\nu_c = 10$. The relaxation factors were reduced until a converged solution was obtainable using central differencing. Then these relaxation factors, 0.1 for the momentum equations and 0.3 for pressure, were used for all the other

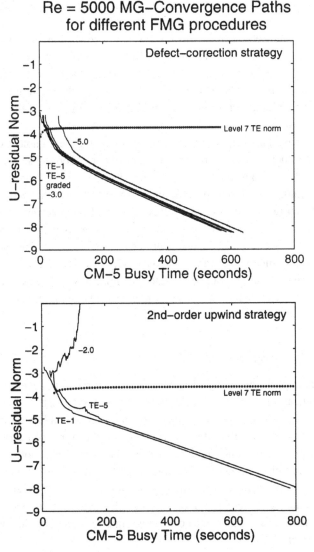

Figure 3.6 The convergence path of the u-residual L_1 norm on the finest grid level in the 7-level $Re = 5000$ lid-driven cavity flow. The relaxation factors used were $\omega_{uv} = 0.5$ for momentum equations and $\omega_c = 0.4$ for pressure correction.

convection schemes as well. Actually, in the lid-driven cavity flow, the pressure plays a minor role in comparison with the balance between convection and diffusion. Consequently, the pressure relaxation factor can be varied between 0.1 and 0.5 with negligible impact on the convergence rate. The convergence rate is very sensitive to the momentum relaxation factor, however.

For the $Re = 1000$ cases, the convergence rate of defect-correction iterations is not quite as good as central differencing or first-order upwinding, but it is slightly better than second-order upwinding. This result makes intuitive sense for those cases where central differencing does not have stability problems, because the defect-correction discretization can be viewed as a less implicit version of central differencing. Similarly, the convergence rate of SIMPLE with defect correction should be expected to be slightly lower than with first-order upwinding due to the presence of source terms, which vary with the iterations. As the table shows, the inner-iterative method used (line-Jacobi, point-Jacobi, SLUR) has no influence on the convergence rate for either Reynolds number tested. From experience it appears that the lid-driven cavity flow is unusual in this regard.

For the $Re = 3200$ cases, convergence is difficult to obtain for these choices of relaxation factors and numbers of inner iterations. The defect correction and second-order upwind schemes do not converge, but they do not diverge either. It appears that the amount of smoothing is insufficient to handle the source terms in the second-order upwind and defect correction schemes at this Reynolds number. In the multigrid context, however, the $Re = 3200$ case with either convection scheme converges, using much larger relaxation factors as well. Second-order central differencing does not normally look as good in comparison. The lid-driven cavity flow is a special case for which central difference solutions can be obtained for relatively high Reynolds numbers due to the shear-driven nature of the flow and the relative unimportance of the pressure gradient.

3.3.2.2   FMG CYCLING TECHNIQUE

Another factor that impacts the multigrid convergence rate is the initial guess for the velocity components and pressure on the fine grid. Thus, a systematic, nested iteration approach for generating the starting fine-grid solution, also called a "full multigrid" (FMG) strategy, is generally effective at improving the convergence rate compared to a random or zero initial solution. We have observed that both the convergence rate and the stability of multigrid iterations can be affected by the initial fine-grid guess, depending on the stabilization strategy and flow problem. With regard to cost, the FMG procedure generally contributes to a small fraction of the overall runtime. However, on parallel computers the cost can be significant, a point that we discuss in more detail later.

In the present context the following FMG technique is used. Starting with a zero initial guess on the coarsest grid, a few pressure-correction iterations are made to obtain an approximate solution. This solution is prolongated to the next grid level, and two-level multigrid V cycles are initiated, continuing until some convergence criterion is met. The solution is prolongated to the next finer grid, multigrid cycling (3-level this time) resumes again, and so on, until the finest grid level is reached. The converged solution on level $k_{max} - 1$, where k denotes the grid level, interpolated to the fine grid, is then the starting solution for the multigrid V cycles using the full complement of grid levels. Figure 3.7 shows a "1-FMG" V(3,2) cycle for a 4-level computation. Only one V cycle is taken for each coarse grid in the FMG procedure. In our approach, V cycles are continued on each level until the prescribed coarse-grid convergence tolerances are met. The coarse-grid tolerances are determined as part of the solution procedure, as described below.

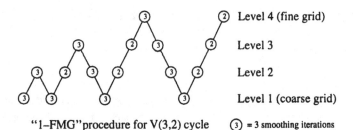

"1–FMG" procedure for V(3,2) cycle   ③ = 3 smoothing iterations

Figure 3.7   Schematic of an FMG V(3,2) multigrid cycle.

Based on discussions in Brandt and Ta'asan (1984), Press et al. (1992), and Thompson and Ferziger (1989), an estimate of the solution truncation error can be developed, and it is this quantity that we monitor during the FMG cycling to assess convergence. Ultimately the FAS scheme tries to obtain the same solution at corresponding locations on every grid level by adding on source terms related to the difference in truncation errors between adjacent levels in the grid hierarchy. Thus the solutions on all the coarse grids in the FMG procedure should be made close to the differential solution,

$$\| A^h u - A^h v^h \| \le \varepsilon, \tag{3.28}$$

where A^h is the discrete equation operator on some coarse grid whose grid spacing is denoted by h, u is the exact differential solution, v^h is an approximate solution to the discrete problem whose exact solution is u^h, and ε is a small number. The development for the coupled system of momentum and continuity equations is given in Blosch (1994), but here we discuss a single equation for brevity.

Since u is unknown, one can only assess the equation residual $r^h = \| A^h u^h - A^h v^h \|$. To relate the residual to the level of truncation, use the triangle inequality and the definition of truncation error,

$$\| A^h u^h - A^h v^h \| \le \| A^h u^h - A^h u \| + \| A^h u - A^h v^h \| \le \| \tau^h \| + \varepsilon \tag{3.29}$$

where τ^h denotes the solution truncation error on the coarse grid. Thus for small ε,

$$\| r^h \| \le \| \tau^h \| \tag{3.30}$$

should be the convergence criterion. The truncation error is in turn related to the discretization of two adjacent grid levels as follows, assuming a second-order scheme,

$$\tau^h = \frac{[A^{2h} u - S^{2h}] - [A^h u - S^h]}{3} \tag{3.31}$$

where S^{2h} just indicates the source term on the grid with spacing $2h$; by the problem definition, $S^{2h} = A^{2h} u^{2h}$, and likewise for S^h. Equation (3.31) is evaluated on the grid whose spacing is $2h$ to give the truncation error estimate for the grid whose spacing is h. Substituting the most current approximations for u, v^h, and v^{2h}, Eq. (3.31) becomes

$$\tau^h \simeq \frac{[A^{2h}v^{2h} - S^{2h}] - [A^h v^h - S^h]}{3}. \tag{3.32}$$

The second term in brackets is just the residual r^h, which as discussed in the background is approximated on the grid with spacing $2h$ by summing over the appropriate control volumes for the finite-volume methodology. Thus it can be observed that the truncation error is just the numerically derived part of the source term in the discretized equation on the grid whose spacing is

$$\tau^h \simeq \frac{S^{2h}_{\text{numerical}}}{3}. \tag{3.33}$$

Thus Eq. (3.30), the convergence criterion, is

$$\|r^h\| \le \left\| \frac{S^{2h}_{\text{numerical}}}{3} \right\|. \tag{3.34}$$

Two levels are always available to make this estimate for the coarse-grid cycling in the FMG procedure. The L_1 norm is used, divided by the appropriate number of control volumes; for a gridvector v on an $N \times N$ mesh,

$$\|v\| = \sum_{\text{all } i,j} \frac{v_{i,j}}{N^2}. \tag{3.35}$$

Equation (3.34) is evaluated on the fly and without additional cost since the source term $S^{2h}_{\text{numerical}}$ is already evaluated as part of the coefficient computations for the previous multigrid cycle's post-smoothing iterations on the grid whose spacing is $2h$.

Consider the step flow first (Fig. 3.5). For this simulation, a 321×81 fine grid was used, with 5 multigrid levels. V(3,2) cycles were used, and for the smoothing iterations, 3, 3, and 9 inner iterations of the point-Jacobi type were taken. The relaxation factors were $\omega_{uv} = 0.6$ and $\omega_c = 0.4$. Figure 3.5 compares the convergence paths for different coarse-grid convergence tolerances in the FMG procedure (i.e., for different initial solutions on the finest grid). The second-order upwind and defect-correction stabilization strategies are compared, also. The curves shown have the following meanings. "TE-1" and "TE-5" refer to the truncation error criterion described above, with the denominator set to 1 and 5. We have treated Eq. (3.30) as a heuristic to observe the sensitivity of this approach to the approximations that have been made. "1FMG" refers to the FMG cycle taking only one V cycle on each coarse-grid level as shown in Fig. 3.7. The numbers "–3.0" and "–5.0" refer to constant coarse-grid convergence tolerances on each level (e.g., $\log_{10} \|r^h\| = -3.0$).

The "graded" tolerances, shown for the defect-correction scheme only, refer to our attempt to pick reasonable values for each level a priori. Specifying graded tolerances is equivalent to specifying a fixed tolerance using a residual that is normalized by a momentum flux instead of the number of control volumes, which is the approach used by Shyy and Sun (1993). Since the equation residuals physically represent integrated quantities in the finite-volume formulation, the net residual regardless of the grid level should be roughly the same, and hence graded tolerances are appropriate. However, the truncation error will not in general be spaced evenly (i.e., according to a factor of 4) because it depends on the solution, and so it becomes very difficult to choose a priori a good set of graded tolerances. Note also in the figure that CM-5 busy time is used since equivalent fine-grid iterations (work units) are not an accurate approximation for parallel computations. Two observations regarding Fig. 3.5 (the step flow) are:

(1) Second-order upwinding performs better than defect correction. The u-residual norm reaches -8.0 in slightly more than 20 seconds on the CM-5, which equates to 140 work units, 20 fine-grid V(3,2) cycles. This corresponds to an amplification factor of 0.7 per cycle. Since the convergence is fast, the contribution of the FMG startup procedure to the overall cost is a significant fraction of the overall parallel runtime, whereas it would not be on a serial computer.

(2) The convergence rate for defect correction depends strongly on the initial fine-grid guess. The best case is "TE-1," whose rate of error reduction per cycle corresponds to a smoothing rate of 0.95. As discussed earlier, in the defect-correction multigrid cycle we use first-order upwinding on all the coarse-grid levels, applying the source term corrections only on the finest grid. So the present results are consistent with the truncation error derivation sketched above, which suggests a denominator 1 in the case of first-order discretizations.

It should be stressed that for each of the curves, identical procedures are used after the FMG procedure (i.e., once the fine-grid cycles are initiated), which is the point where the curves begin. Thus the fact that the asymptotic convergence rates differ reflects differences in the initial fine-grid guess. Brandt and Ta'asan (1985) have shown that there can exist certain error modes in the initial fine-grid solution that are damped very slowly by the smoothing-correction multigrid combination when the flow is aligned with the grid and the convection terms are first-order upwinded, as in the present case; this may explain the observed convergence behavior.

For the lid-driven cavity simulation, a 321×321 fine grid was used, with 7 multigrid levels. Again, V(3,2) cycles were used, and for the smoothing iterations, 3, 3, and 9 inner iterations of the point-Jacobi type were taken. The relaxation factors were $\omega_{uv} = 0.5$ for momentum equations and $\omega_c = 0.5$ for pressure correction. Some observations regarding Fig. 3.6 (the lid-driven cavity flow) are:

(1) The convergence rate for this recirculating-type flow, which corresponds to an error smoothing rate of about 0.99, is not as good as for the entering-type flow. However, the results appear to be consistent with the results obtained by Sockol (1993). For the $Re = 5000$ lid-driven cavity flow, using SIMPLE with $v_u = v_v = 1$ and $v_c = 4$ inner line iterations and a W(1,1) multigrid cycle, Sockol found that 86 work units were needed to reach convergence (800 seconds on an Amdahl 5980). To reach a similar convergence tolerance, the present computation needed 30 cycles on the

fine grid, which is 200 work units (64 seconds on the CM-5). Additional experiments have shown that for the cavity flow, V(2,1) cycles are sufficient to obtain the same convergence rate as with V(3,2) cycles, and so our work units are effectively overstated by about 2.7 work units per cycle, yielding an equivalent of 119 work units. Experience has shown that for this unique flow problem, the inner iterative method (line relaxation or point-Jacobi) has no impact on the convergence rate of the outer iterations, and so this procedural difference between our work and Sockol's is irrelevant here. The slightly faster convergence observed by Sockol may be attributable to the use of a W cycle instead of a V cycle. However, for this 7-level problem size, W(1,1) cycles take twice as long as V(2,1) cycles on the CM-5, so the total runtime is less using the latter.

(2) For either defect correction or second-order upwinding, the convergence rate does not depend on the criterion used in the FMG procedure, that is, on the initial fine-grid guess. In experiments with the defect correction strategy, we have found that no matter how stringently the intermediate grid-level solutions are converged, the error norm begins at approximately the same value, as shown in Fig. 3.8. The figure compares the convergence paths of the u-residual norm during the FMG procedure for the "TE-5" and "−3.0" curves, that is, for the time preceding the starting time for the corresponding curves in the top plot of Fig. 3.6. These results indicate that the number of defect-correction iterations used on a given outer grid level during the FMG procedure does not seem to matter, an observation that is consistent with a convection-dominated flowfield: if the true velocity field has a strong upwind effect, the defect-correction source terms are small.

In the second-order upwind cases of Fig. 3.6, convergence was obtained only for those cases that used the truncation error criterion. Experience with single-grid computations suggests that for high-Reynolds-number lid-driven cavity flows, the second-order upwind scheme is harder to converge than either the defect correction, first-order upwind, or central-difference schemes for a given set of relaxation factors. Thus the second-order upwind cases that diverged can be made to converge by increasing the amount of smoothing and/or the relaxation factors. It is also possible that the bilinear interpolation prolongation procedure used here may be less compatable with a second-order upwinded solution than a central-differenced one.

3.3.2.3 RESTRICTION TECHNIQUE

The restriction technique also affects the convergence rate. For finite-volume discretizations, *conservation* is the restriction procedure that is consistent with the finite-volume discretization. The reason is that the terms in the discrete equations represent integrals over an area (in two dimensions). The method of integration for source terms determines the actual restriction procedure; we use piecewise constant. In either a staggered grid or cell-centered finite-volume formulation, the mass residual in a coarse-grid control volume is the sum of the mass residuals in the four fine-grid

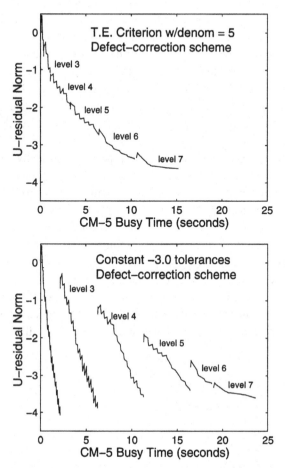

Figure 3.8 The convergence path of the u-residual L_1 norm during the FMG procedure for the 7-level $Re = 5000$ cavity flow, using the defect correction strategy, contrasting two criteria for controlling the coarse-grid cycling.

control volumes that comprise the coarse-grid control volume. In addition, the u-momentum equation residuals on the fine grid are treated as piecewise constant and summed over the region corresponding to the coarse-grid u control volume under consideration. Due to the staggered grid, this involves summation over six fine-grid u control volumes, taking only half the contribution from four of the control volumes. The same procedure applies to the v-momentum equation residuals. For the prolongation step, we use a linear interpolation adapted to the staggered grid variables.

Typically the mass residuals are treated in this manner (conservation as the restriction procedure), but the summation of momentum residuals has proven problematic in other work (Shyy 1994, Shyy and Sun 1993, Sockol 1993) even though it is physically consistent. By summing the mass residuals and restricting u^h and v^h by cell-face averaging, satisfaction of the continuity equation can be identically maintained on the coarse grids at all times. This is not strictly necessary except at convergence, and it generates source terms in the coarse-grid momentum equations, since the cell face–averaged solutions will not in general satisfy the momentum equations. Thus, due to the source terms, the smoothing iterations may diverge when summation of residuals is used in conjunction with cell-face averaging of the velocity variables. In that case, cell-face averaging of the residuals may be better because, by effectively reducing (by 1/4) the magnitude of the numerically derived source terms, convergence may be easier to obtain, albeit with a much slower convergence rate. Clearly, summation of both mass and momentum residuals is desirable, but what about the solution variables?

In the FAS formulation, the coarse-grid equations incorporate numerically derived source terms which are the difference between the restricted fine-grid equation residuals and the coarse-grid equation residuals based on an "initial" coarse-grid solution, which may be obtained by restricting the fine-grid variables (Shyy and Sun 1993). In the original description of FAS, both solution variables and residuals were restricted (Brandt 1977). However, the initial coarse-grid solution may also be taken from the previous multigrid cycle's "upstroke," that is, the most recent solution, and we have observed that in some cases this strategy is preferable.

Figure 3.9 compares the convergence rates of V(3,2) cycles in two cases: curve 1 indicates cell-face averaging of the velocity variables and bilinear interpolation of pressure, and curve 2 indicates taking the most recent coarse-grid solution. All other

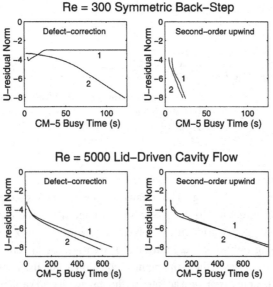

Figure 3.9   The convergence path of the u-residual L_1 norm on the finest grid level in the 5-level $Re = 300$ symmetric backward-facing step flow and 7-level $Re = 5000$ lid-driven cavity flow. Contrasted are the two alternative treatments for restriction of solution variables: (1) cell-face averaging for velocities with bilinear interpolation for pressure; and (2) use of the most recent coarse-grid solution. In all cases the residuals are restricted by summation.

parameters/procedures are identical and are described below. The two approaches give comparable performance except in the step flow using defect corrections, for which the solution does not converge when both solutions and residuals are restricted.

The problematic case indicates a competitive balance instead of a complementary effect, from the fine grid's perspective, between smoothing and coarse-grid correction. Consider the situation when the fine-grid mass and momentum residuals are everywhere zero, that is, convergence. To maintain convergence in the case where the initial coarse-grid solution is taken from the previous cycle, the coarse-grid correction quantities Δu^{2h} as in

$$u^{2h}_{\text{to be prolongated}} = u^{2h}_{\text{smoothed, upstroke}} - u^{2h}_{\text{initial, downstroke}}, \tag{3.36}$$

where u refers to either velocity components or pressure, and where the subscripts are self-explanatory, must be zero. Since we are summing the fine-grid residuals, that contribution to the coarse-grid source term that comes from the fine grid is zero. However, the initial coarse-grid solution does not satisfy the governing equations discretized on the coarse grid; rather there *must be*, in the general case, nonzero source terms, since the coarse-grid solutions are an approximation to the fine-grid solution at corresponding locations. In the case of cell-face averaging, however, the coarse-grid continuity equation has no artificial source terms, and so the pressure field is not able to adjust to reestablish the satisfaction of the momentum equations destroyed by the uncoordinated restrictions of u and v. If the coarse-grid source terms are important, as they generally are for the defect-correction cases since one is effectively switching between a central-difference and a first-order upwind convection scheme, convergence may be impossible, as we observe in the step flow of Fig. 3.9. The key point is that for a system of coupled equations, the restriction of the solution variables must be coordinated with the restriction of the residuals. Our experience indicates that it is a more robust approach to let the initial coarse-grid values float, effectively letting the coarse-grid equations respond only to the approximation of the fine-grid residuals, which are obtained in a manner consistent with the finite-volume discretization of the governing equations. However, this technique should be investigated further.

3.4 Data-Parallel Pressure-Correction Methods

The sequential solution procedure that is characteristic of pressure-correction and projection methods is a natural match for SIMD platforms because of its frequent synchronization requirements. Every update of the velocity components and pressure in both the inner and outer iterations requires synchronization. Also, local equation residuals are accumulated frequently in order to monitor convergence. Since SIMD machines automatically enforce synchronization and incorporate accumulation and broadcast operations into the system software, our initial work in this area has been on SIMD machines.

We have compared the performance of the data-partitioned algorithm on three machines, Thinking Machines' CM-2 and CM-5, and MasPar's MP-1. As follows from the brief review of parallel programming given previously, the programming for SIMD machines requires attention to parallelization at the level of individual instructions. The languages in which we have programmed are the data-parallel array-based languages, CM-Fortran and MP-Fortran; these are very similar to each other. It is also possible to use the CM-5 as a MIMD machine using explicit message passing. Architecturally the CM-5 is a MIMD machine with a special control network to provide automatic synchronization to allow SIMD execution.

Our experience indicates that the pressure-correction algorithm is dominated by operations that act on all elements of the data arrays. Thus, the array-based languages are sufficiently expressive for the present application. Furthermore, the task of programming in CM-Fortran or MP-Fortran is simplified because accumulation, broadcast, and interprocessor communication operations are generated by the compiler in the form of calls to a runtime library. We are further motivated to choose the array-based Fortrans over serial Fortran or C combined with explicit message passing by recent developments in portable parallel compilers for high-performance Fortran (SC-NET 1995). CM-Fortran and MP-Fortran are essentially subsets of HPF. Thus, a research code will be easily portable to different architectures.

Although SIMD *architectures* have lost ground to MIMD and shared memory, the data-partitioning parallelization and the host-worker SPMD computational model, which are implicit in SIMD execution, are ubiquitous and probably will increase in popularity as portable data-parallel compilers mature.

Qualitatively the following results will be the same using the host-worker SPMD model on a MIMD machine. The key factor in the present context is the parallelization by a block partitioning of the grid variables. In terms of the runtime, SIMD computation differs from MIMD by the additional per-instruction overhead cost of broadcasting instructions and host-located data from the host to the workers. This cost is a constant with respect to the number of processors and thus does not affect scalability. The amount of interprocessor communication, assuming a MIMD implementation that uses overlapping arrays, is the same as the SIMD implementation. Thus, the variation of speedup and efficiency with problem size and number of processors is qualitatively the same. However, since the relative speed of communication compared to computation is typically much faster on SIMD than on MIMD machines, the parallel efficiencies achieved can be higher.

3.4.1 Single-Grid Computational Issues

The cost of each pressure-correction iteration on a single grid depends on the choice of the relaxation method (solver) for the systems of linear algebraic equations, the number of inner iterations (ν_u, ν_v, and ν_c), the computation of coefficients for each system of equations, the correction step, and the convergence checking and the serial work done in controlling execution. The pressure-correction equation, since it is not underrelaxed, typically needs to be given more iterations than the momentum equations. Typically ν_u and ν_v are the same and are ≤ 3 and $\nu_c \leq 5\nu_u$.

The majority of effort is spent on solving the discretized momentum and pressure-correction equations. It is a feature of the numerical method that the systems of equations arising during the course of iterations are not solved to strict tolerances but instead are solved approximately by applying a few iterations of an inexpensive iterative technique. The successive line underrelaxation technique (SLUR) has been found to be a robust and inexpensive inner iterative solver for serial computations (Braaten and Shyy 1987).

The successive line underrelaxation procedure is effective for serial computation because it uses the tridiagonal matrix algorithm (TDMA, whose operation count is optimal, $\mathcal{O}(N)$), and because it allows long-distance coupling between flow variables (along lines), which promotes convergence in the outer iterations. The TDMA is intrinsically serial. Thus, for parallel computations, a parallel tridiagonal solver must be developed. In the present work, the method of parallel cyclic reduction is used. The time per iteration comes from computation, $\mathcal{O}(N \log_2 N)$, where N is the number of unknowns per line, and also from communication. The communication cost depends on the implementation and the performance of the data network.

For extension of the single-grid method to multigrid we have also considered point-Jacobi inner iterations. Generally, point-Jacobi iteration is not sufficiently effective for complex flow problems. However, as part of a multigrid strategy, good convergence rates have been obtained as illustrated in the previous section, even for flow problems with strong alignment to grid lines. Furthermore, because it only involves regular grid-type interprocessor communication, which is fast, point-Jacobi iteration provides an upper bound for parallel efficiency against which other solvers can be compared.

Below we consider the comparative performance of these two different inner iterative solvers in terms of efficiency. Variations on the implementation have also been studied. These shed some light on the communication cost in the algorithm. The focus is on general trends that are common to all grid-based partitionings of the data arrays, independent of the machine architecture.

3.4.1.1 IMPLEMENTATION ISSUES

Figure 3.10, based on timings made on a 32-node CM-5, plots parallel efficiency E against the local problem size N. E was computed via Eq. (3.5) by directly timing the communication and computation times T_{comm} and T_{comp}, using profiling tools available from Thinking Machines. Some explanation of the plotted data is necessary.

The implementation labeled "NEWS," with the symbol "o" in the figure, uses the virtual processor concept so that interprocessor communication in a certain sense occurs for all array elements, as explained below. "NEWS" refers to the regular north-east-west-south communication pattern which is detected by the compiler; this is the jargon of CM-Fortran.

The virtual processor concept is a compiler model of SIMD machines. In this model the machine is viewed as having as many processors as array elements. Typically, the number of array elements exceeds the number of physical processors available, and thus the model requires the assignment of several virtual processors to each physical processor (the mapping problem in the SIMD context). The term "virtual

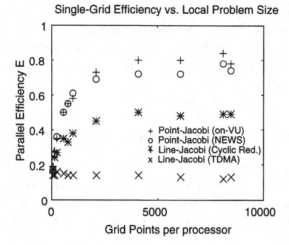

Figure 3.10   Parallel efficiency, E, as a function of problem size and solver, for the CM-5 single-grid cases. The number of grid points is the virtual processor ratio, N, multiplied by the number of processors, 128. E is computed from Eq. (3.5). It reflects the relative amount of communication, compared to computation, in the algorithm.

processor ratio," VP, is in common use in this context; VP is the same as the local problem size N. The virtual processor model is also useful in algorithm design (Kumar et al. 1994).

To map the virtual processors, which are organized with the same dimension extents as the given data array, to the physical processors, the latter must be logically laid out with the same number of dimensions. The logical processor layout we use is a two-dimensional mesh. A two-dimensional mesh can be fully embedded in the networks used by the CM-2 (hypercube) and CM-5 (fat tree). Thus we know that all grid-type communication can occur independently for each physical processor. The MP-1 has two data networks, one specifically for grid-type communication (X-Net) and one for global communication patterns (a three-stage crossbar switch). Then the compiler assigns virtual processors to physical processors by either a block or cyclic mapping. The difference between the two is clarified by Fig. 3.11.

In the cyclic mapping, nearest-neighbor virtual processors end up on different physical processors. Thus, the cost of a nearest-neighbor communication of distance one will be proportional to N. In the block mapping, nearest-neighbor virtual processors are likely to be on the same physical processor, and the number of array elements to be physically moved in a nearest-neighbor communication is only proportional to $N^{\frac{1}{2}}$. This is called the area-perimeter effect of the block data partitioning.

Thus, using block data partitioning, a nearest-neighbor type of communication (of distance 1) only needs to exchange boundary array elements with boundary array elements from the neighboring physical processor. Interior elements, although they are the property of different virtual processors, could be accessed within the same physical processor by a normal memory read operation.

In the "NEWS" implementation, though, the area-perimeter effect is only half of the picture. Because of the virtual processor machine model used by the compiler,

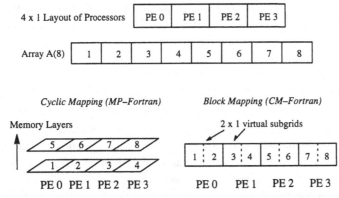

Figure 3.11 Mapping an 8-element array A onto 4 processors. For the cyclic mapping, nearest-neighbor array elements are mapped to nearest-neighbor physical processors. For the block mapping, nearest-neighbor array elements are mapped to nearest-neighbor virtual processors, which may be on the same physical processor.

some cost must be incurred for *every* array element during a communication operation, because every virtual processor must do the same thing in SIMD execution. Specifically, the cost incurred is a call to a runtime communication library which tells whether or not a given pair of virtual processors actually reside on the same physical processor. If they do, then the corresponding array elements are fetched from memory instead of through the data network, which is much quicker. However, the overhead cost of calling the library is still present. This overhead cost is counted as part of the interprocessor communication cost. Thus, the total interprocessor communication cost is proportional to the area of the subgrid, N, as well as the perimeter. Asymptotically the area cost dominates, but the perimeter elements may significantly affect the efficiency for practical problem sizes, if the network communication speed is much slower than the memory bandwidth.

The implementation labeled "on-VU," with the symbol "+" in the figure, is free from the aforementioned overhead cost of the virtual processor model, in the inner iterations. In CM-Fortran, and in an analogous fashion in high-performance Fortran, compiler directives for the data layout can be used to indicate explicitly an array's mapping onto the logical processor layout. By using four-dimensional arrays, with two indices for the area or "on-processor" data elements and two indices for the perimeter or "off-processor" elements, it is possible for the programmer to isolate the operations that generate real communication between physical processors from those that generate communication between virtual processors on the same physical processor. This avoids the unnecessary overhead cost of communication in the virtual processor model. More details are available in Blosch and Shyy (1994). The notation "on-VU" is used because in the CM-5 the independent processing elements are vector units. (A general comparison of the CM-5, CM-2, and MP-1 is discussed in conjunction with Fig. 3.13 below.)

Returning now to the discussion of the figure, we observe that for all cases, the efficiency is initially low but increases gradually to an asymptotic constant value as the problem size increases. This trend is a consequence of the increasing amount of parallel computation done by the program as the problem size increases, compared to the fixed amount of serial work (mostly control statements). Also, the trend of increasing efficiency with N reflects the area-perimeter effect.

For both implementations, the peak value of parallel efficiency is quite high, 0.8 for the "NEWS" version and 0.85 for the "on-VU" version. The timings indicate that for $N > 2k$, ($k = 1000$), in the "NEWS" implementation of the point-Jacobi solver, computation is taking about 3/4 of the time with the remainder split evenly between host-worker communication and interprocessor communication. In the "on-VU" version, the overhead cost of communication between interior virtual processors is eliminated. However, this benefit comes at the expense of more code blocks, that is, more host-worker communication. Consequently it takes $N > 4k$ to reach peak efficiency instead of 2k with the "NEWS" version. For $N > 4k$, however, E is about 5–10% higher than for the "NEWS" version, because the total communication cost has been reduced.

The efficiencies in Fig. 3.10 reflect the entire cost of the pressure-correction iterations, not just the inner iterations. The results shown were based on typical numbers of inner iterations, 3 each for the u- and v-momentum equations and 9 for the pressure-correction equation. In the present case, approximately half of the total runtime is spent in inner iterations. Obviously, if more inner iterations are taken, the difference between the efficiencies of the "NEWS" and "on-VU" cases would be greater.

Red–black analogues to the "NEWS" and "on-VU" versions of point-Jacobi iteration have also been implemented and timed. Red–black iterations done in the "on-VU" manner do not generate any more host-worker communication than the point-Jacobi "on-VU" version. Consequently, the runtime is almost identical to point-Jacobi, and this approach is therefore the preferred one due to the improved convergence rate of red–black iteration. However, with the "NEWS" implementation, red–black iterations require two code blocks instead of one, and also halve the amount of computation per code block. This results in a substantial (~35% for the $N = 8k$ case) increase in runtime. Thus, because of the virtual processor model inherent in SIMD computation, red–black iterations are probably not cost effective.

We have also implemented Vanka's method (Vanka 1986) (block Gauss-Seidel, BGS) with a red–black updating scheme (BRB) to make it suitable for parallelization. On the CM-5, the runtime per iteration of the pressure-correction method using 3, 3, and 9 inner point-Jacobi iterations is virtually the same as that of BRB. The line variants of Vanka's scheme cannot be implemented in CM-Fortran without generating excessive interprocessor communication.

Figure 3.10 also shows two implementations of line-Jacobi iteration. In both implementations, one iteration consists of forming a tridiagonal system of equations for the unknown in each vertical line of control volumes (treating the east/west terms as source terms), solving the multiple systems of tridiagonal equations simultaneously by a direct method, and repeating this procedure for the horizontal lines.

The first implementation uses parallel cyclic reduction to solve the multiple tridiagonal systems of equations (see Jespersen and Levit (1989) for a clear presentation). The basic technique for a single tridiagonal system of equations is a recursive combining of each equation with its neighboring equations in the tridiagonal system. This decouples the original system into two sets of equations, each half the size of the original. Repeating the reduction step $\log_2 N$ times, where N is the number of equations, leaves N independent systems of equations of size 1. At each step, N equations are active; thus, the total computational operation count is $\mathcal{O}(N \log_2 N)$. On a parallel machine, interprocessor communication is necessary for every unknown at every step. Thus, the communication complexity is also $\mathcal{O}(N \log_2 N)$, but in fact for most machines the communication time will dominate the time spent on floating-point operations.

Furthermore, the distance between communicating (virtual) processors increases each step of the reduction by a factor of 2. Each system of equations obtained for each line of unknowns in the 2D problem is contained in a one-dimensional row or column of processors. Thus, in the first step, nearest-neighbor communication occurs – only edge values on the processors' "virtual subgrids" are involved. In the second step, two edge values must be sent/received, and in the third step, four edge values, and so on. In the present implementation we terminate the reduction process when the ratio of the size of the off-diagonal elements to the size of the diagonal falls below a certain threshold, in this case 10^{-4}. This eliminates the last few steps of the reduction, which involve communication over the longest distances, and reduces the effect of the logarithmic factor in the communication complexity. However, on average, the communication in each step of the reduction takes longer than just a single nearest-neighbor communication. Thus while the line-Jacobi method using cyclic reduction with early cutoff is nearly scalable, the runtime per (inner) iteration is longer than the point-Jacobi method. Because a greater proportion of the runtime comes from communication, the line-Jacobi iterations are less efficient than the point-Jacobi iterations as well.

Accordingly, we observe from Fig. 3.10 that the parallel efficiency peaks at about 0.5 as compared to 0.8 for point-Jacobi iteration ("NEWS" implementation). The timings indicate the interprocessor communication is taking as much time as computation for $N > 4k$, $(k = 1000)$. In terms of runtime, each line-Jacobi iteration takes approximately 3 times longer than a corresponding point-Jacobi iteration. Whether or not the additional cost is offset by improvement in the convergence rate depends on the flow characteristics, that is, the pressure–velocity coupling, and on the relaxation factors and the number of inner iterations used on each equation. In many problems the performance of the inner iterative method is reflected in the outer iterative convergence rate; cavity flow is a notable exception.

The second implementation uses the standard TDMA algorithm along the lines by remapping the necessary coefficient, solution, and source arrays from a two-dimensional mapping to a one-dimensional mapping for each step. No communication is generated in the solution procedure itself since entire lines are within the same processor. However, the remapping steps are communication-intensive and generate calls to slower, more general routing algorithms because the communication pattern is more complicated.

The resulting performance in Fig. 3.10 is poor. E is approximately constant at 0.14 except for very small N. Constant E implies from Eq. (3.5) that T_{comm} and

T_{comp} both scale in the same way with problem size, which in this case is proportional to N since TDMA is $\mathcal{O}(N)$. This is an indication that the full bandwidth of the CM-5's fat tree is being utilized in these remapping operations. One would expect that as the compiler matures, better optimizations will be obtained for computations on "SERIAL" dimensions of parallel arrays. However, the degree of improvement that can obtained is limited because computation is not the problem here.

The disappointing performance of the standard line-iterative approach using the TDMA is due to the global communication (the remapping) within the inner iterations. There is simply not enough computation to amortize slow communication in the solver for any problem size. With parallel cyclic reduction, where the regularity of the data movement allows faster communication, the efficiency is much higher, although still significantly lower than for point-Jacobi or red–black iterations.

The remapping which occurs in the TDMA-based line-Jacobi method is just a reordering of the equations in a row- or column-major order followed by a linear mapping of blocks of rows to processors. Such mappings are sometimes used in packaged parallel sparse matrix solvers. Consequently such solvers are unlikely to be useful in the context of the pressure-correction algorithm. Projection methods, which are time-stepping predictor corrector methods for the incompressible Navier-Stokes equations, do require an accurate solution to a system of equations for the pressure, in contrast to pressure-correction methods. Thus for projection methods the remapping cost may be amortized by the solution cost. Still better efficiencies could be achieved, though, using methods that work with the data layout imposed by the grid partitioning of the problem, for example, parallel multigrid solvers.

To further explore the area-perimeter effect on communication cost, we next examine the secondary effect of the aspect ratio of the virtual processor subgrid on the parallel efficiency. The major influence on E is N (i.e., the subgrid size), but the subgrid *shape* matters, too. Higher-aspect-ratio subgrids have higher area-to-perimeter ratios and thus relatively more off-processor communication than a square subgrid of equal N.

Figure 3.12 gives some idea of the relative importance of the subgrid aspect ratio effect on the CM-5. Along each curve the number of grid points is fixed, but the grid dimensions vary, which for a given processor layout causes the subgrid aspect ratio to vary. For example, on the 32-node CM-5 with an 8×16 processor (vector unit) layout, the following grids were used, corresponding to the $N = 1024$ CM-5 curve: 256×512, 512×256, 680×192, and 1024×128. These cases give subgrid aspect ratios of 1, 4, 7, and 16. T_{news} is the time spent in "NEWS" communication, and T_{comp} is the time spent doing computation. The times are based on 100 pressure-correction iterations using point-Jacobi inner iterations.

For the $N = 1024$ CM-5 case, increasing the aspect ratio from 1 to 16 causes T_{news}/T_{comp} to increase from 0.3 to 0.5. This increase in T_{news}/T_{comp} increases the runtime for 100 iterations from 15 s to 20 s and decreases the efficiency from 0.61 to 0.54. For the $N = 8192$ CM-5 case, increasing the aspect ratio from 1 to 16 causes T_{news}/T_{comp} to increase from 0.19 to 0.27. This increase in T_{news}/T_{comp} increases the runtime for 100 iterations from 118 s to 126 s and decreases the efficiency from 0.74 to 0.72. Thus, the aspect ratio effect diminishes as N increases due to the increasing

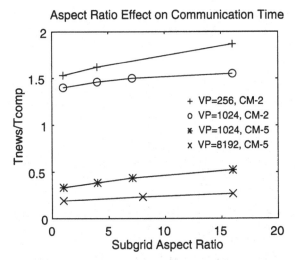

Figure 3.12   Effect of subgrid aspect ratio on interprocessor communication time, T_{news}, for the block data mapping (CM-2 and CM-5). T_{news} is normalized by T_{comp} in order to show how the aspect ratio effect varies with problem size, without the complication of the fact that T_{comp} varies also.

area of the subgrid, as one would expect. The CM-2 results are similar to the CM-5 results. However, on the CM-2 the on-processor type of communication is slower than on the CM-5, relative to the computational speed. Thus, $T_{\text{news}}/T_{\text{comp}}$ ratios are higher on the CM-2.

3.4.1.2   RELATIVE SPEED OF COMPUTATION AND COMMUNICATION

Now that the effect of problem size on efficiency has been discussed, we turn to the machine-related factors that are important, specifically the relative speed of computation and communication. Figure 3.13 compares the performance for the pressure-correction algorithm using the point-Jacobi solver and three SIMD machines, the CM-5, a CM-2, and the MP-1. The MP-1 is a smaller machine, so results are shown only in the range of problem sizes that can be handled on the MP-1. Also, because the computers have different numbers of processors, the number of grid points is used instead of N to define the problem size.

A brief comparison of these SIMD machines is helpful. The results in Fig. 3.13 are based on timings obtained on a 32-node CM-5 with vector units, a 16k processor CM-2, and a 1k processor MP-1. The CM-5 has 4 Gbytes total memory, while the CM-2 has 512 Mbytes and the MP-1 has 64 Mbytes. The peak speeds of these computers are 4, 3.5, and 0.034 Gflops, respectively, in double precision. Per processor, the peak speeds are 32, 7, and 0.033 Mflops, with memory bandwidths of 128, 25, and 0.67 Mbytes/s (Schreiber and Simon 1992, Thinking Machines Corporation 1992). Clearly these are computers with very different capabilities, even taking into account the fact that peak speeds, which are based only on the processor speed under ideal conditions, are not an accurate basis for comparison.

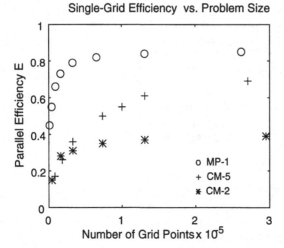

Figure 3.13   Comparison between the CM-2, CM-5 and MP-1. The variation of parallel efficiency with problem size is shown for the model problem, using point-Jacobi relaxation as the solver.

The host processors (front ends) are Sun-4 workstations in the CM-2 and CM-5, while in the MP-1 a Decstation 5000 is used. In the CM-2 and MP-1, there is an intermediate processor, called either a sequencer or an array control unit, which simultaneously broadcasts the program instruction to all processors. In the 32-node CM-5, there are 32 SPARC microprocessors, each of which plays the role of a sequencer for its four vector units. Thus a 32-node CM-5 has 128 independent processors. In the CM-2 each processor actually consists of a floating-point unit coupled with 32 bit-serial processors. Each bit-serial processor is the memory manager for a single bit of a 32-bit word. Thus, the 16k processor CM-2 actually has only 512 independent processing elements. This strange CM-2 processor design came about basically as a workaround introduced to improve the memory bandwidth for floating-point calculations (Schreiber 1990). Compared to the CM-5 VUs, the CM-2 processors are about one-fourth as fast, with larger overhead costs associated with memory access and computation. The MP-1 has 1024 4-bit processors; compared to either the CM-5 or CM-2 processors, the MP-1 processors are very slow. The generic term "processing element" (PE) is used to refer to either one of the VUs, one of the 512 CM-2 processors, or one of the MP-1 processors, whichever is appropriate.

For the present study, the processing elements are logically arranged in a 2D mesh, which is a layout that is well supported by the data networks of each machine. The data network of the 32-node CM-5 is a fat tree of height 3, which is similar to a binary tree except that the bandwidth stays constant upwards from height 2 at 160 Mbytes/s (details in (Thinking Machines Corporation 1992)). One can expect approximately 480 Mbytes/s for regular grid communication patterns (i.e., between nearest-neighbor SPARC nodes) and 128 Mbytes/s for random (global) communications. The randomly directed messages have to go farther up the tree, so they are slower. The CM-2 network, a hypercube, is completely different from the fat tree network, and its performance for regular grid communication between nearest-neighbor processors is roughly 350 Mbytes/s (Schreiber and Simon 1992). The grid network on the CM-2, NEWS, is a subset of the hypercube connections selected at runtime. The MP-1 has two networks: regular communications use the X-Net (1.25 Gbytes/s, peak), which connects each processor to its eight nearest neighbors, and random communications use a 3-stage crossbar (80 Mbytes/s, peak). To summarize the relative speeds of these three SIMD computers it is sufficient for the present study to observe that the MP-1 has very fast nearest-neighbor communication compared to its computational speed, whereas the exact opposite is true for the CM-2. The ratio of nearest-neighbor communication speed to computation speed is smaller still for the CM-5 than the CM-2.

Returning to Fig. 3.13, we observe that the MP-1 reaches peak efficiency at a much smaller problem size than either the CM-2 or CM-5. The MP-1 uses a cyclic mapping of array elements to processors instead of a block mapping, so all communication is of the "off-processor" type. Thus there is no area-perimeter effect on the MP-1. The increase in efficiency with problem size is entirely due to the amortization of the host–worker type of communication by parallel computation. Peak E is reached for approximately $N > 32$. The reason that parallel computation amortizes the host–worker overhead so quickly is because the MP-1's processors are relatively slow. This is also reflected in the peak E achieved, 0.85. On the CM-2, the peak E is only 0.4, and this efficiency is reached for approximately $N > 128$. On the CM-5, the peak E is 0.8, but this efficiency is not reached until $N > 2k$. If computation is fast, as is the case with the CM-5, then the rate of increase of E with N depends on the relative speed of on-processor, off-processor, and host–worker communication. The CM-5 memory bandwidth is better balanced to the processor speed than the CM-2 for this application, which results in a higher peak efficiency. However, the network speeds are so much slower than the computation rate that the peak efficiency is not reached until very large problem sizes, $N > 2k$, 64 times larger than on the MP-1. With regard to efficiency, nothing is gained by speeding up computation.

Table 3.2 summarizes the relative performance of SIMPLE on the CM-2, CM-5, and MP-1 computers, using the point- and line-iterative solvers. In the first three cases the "NEWS" implementation of point-Jacobi relaxation is the solver, while the last two cases are for the line-Jacobi solver using cyclic reduction.

In Table 3.2, the speeds reported are obtained by comparing the timings with the identical code timed on a Cray C90, using the Cray hardware performance monitor to determine Mflops. In terms of Mflops, the CM-2 version of the SIMPLE algorithm's performance appears to be consistent with other CFD algorithms on the CM-2. Jesperson and Levit (1989) report 117 Mflops for a scalar implicit version of an approximate factorization Navier-Stokes algorithm using parallel cyclic reduction to solve the tridiagonal systems of equations. This result was obtained for a 512×512 simulation of two-dimensional flow over a cylinder using a 16k CM-2 as in the present study (a different execution model was used, see Blosch and Shyy (1994) and Levit (1989) for details. The measured time per timestep per grid point was 1.6×10^{-5} seconds. By comparison, the performance of the SIMPLE algorithm for the 512×1024

Table 3.2 *Performance results for the SIMPLE algorithm for 100 iterations of the model problem. The solvers are the point-Jacobi ("NEWS") and line-Jacobi (cyclic reduction) implementations. 3, 3, and 9 inner iterations are used for the u, v, and p equations, respectively.*

Machine	Solver	Problem size	VP	T_p	Time/Iter./Pt. (s)	Speed* (Mflops)	% Peak speed
512 PE CM-2	Point-Jacobi	512× 1024	1024	188 s	2.6×10^{-6}	147	4
128 VU CM-5	Point-Jacobi	736× 1472	8192	137 s	1.3×10^{-6}	417	10
1024 PE MP-1	Point-Jacobi	512× 512	256	316 s	1.2×10^{-5}	44	59
512 PE CM-2	Line-Jacobi	512× 1024	1024	409 s	7.8×10^{-6}	133	3
128 VU CM-5	Line-Jacobi	736× 1472	8192	453 s	4.2×10^{-6}	247	6

*The speeds are for double-precision calculations, except on the MP-1.

problem size using the line-Jacobi solver is 133 Mflops and 7.8×10^{-6} seconds per iteration per grid point. Egolf (1992) reports that the TEACH Navier-Stokes combustor code based on a sequential pressure-based method with a solver that is comparable to point-Jacobi relaxation obtains a performance which is 3.67 times better than a vectorized Cray X-MP version of the code, for a model problem with 3.2×10^4 nodes. The present program runs 1.6 times faster than a single Cray C90 processor for a 128×256 problem (32k grid points). One Cray C-90 processor is about 2–4 times faster than a Cray X-MP. Thus, the present code runs comparably fast.

3.4.1.3 TREATMENT OF BOUNDARY COMPUTATIONS

Parallelizing an algorithm by partitioning the data is a relatively straightforward approach. Nonetheless, modification to the algorithm or special attention to the implementation is often required to obtain good performance. The reason is that algorithms contain operations that can be categorized as inherently serial, partially parallelizable, and fully parallelizable, as discussed in the development of Eq. (3.5). To obtain good parallel efficiency and to ensure scalability, it is important to eliminate as much of the serial and partially parallelized operations as possible.

For example, when parallelizing the standard serial multigrid algorithm, one is potentially faced with a load-balancing problem when mapping the coarse grid onto the processor array, since the number of processors may exceed the number of coarse-grid control volumes. In this case the severity of the problem with regard to efficiency and scalability will be impacted by the granularity of the parallel machine. One option is to reduce the number of multigrid levels so that all processors are still active on the coarsest grid, but for a massively-parallel machine this restriction would hinder the effectiveness of the multigrid technique. Another approach would be to allow the idle processors to proceed asynchronously to do more smoothing iterations on a different grid level, using fixed boundary values from those processors which are busy working on the coarse grid. Similar load-balancing problems can arise when there is local grid refinement. The point is that the serial computational procedure is modified to avoid the decrease in efficiency and loss of scalability associated with idle processors. In this example, the algorithm is modified, changing the results in exchange for scalability and higher parallel efficiencies.

In the present context of single-grid computation on SIMD machines, a potential load-balancing problem derives from the need for special formulas for the coefficients and source terms for boundary control volumes. Since the boundary control volumes are contained in $\mathcal{O}(n_p^{1/2})$ processors using a block mapping, these operations leave interior processors idle. Furthermore, for SIMD computation, each of the four boundaries (in two dimensions) and interior control volumes are treated sequentially. The resulting effect on parallel efficiency can be significant.

One can generally employ the technique used below to achieve full parallelization on SIMD machines. It is less applicable to MIMD computations because the focus on parallelization at a very fine granularity (i.e., at the level of individual instructions) is not as important as for SIMD computation. The reason is that a MIMD machine, with autonomous processing elements, always develops load imbalances even when the application is SPMD, due to slight differences in processor speeds or configurations or operating system factors. Relative to these "natural" imbalances, extra computations for boundary work *may not* be a major source of load imbalance – obviously the importance is problem dependent. However, in the ideal case of an otherwise fully parallelized computation, some processors would have to wait at synchronization points in a MIMD computation unless the boundary work could be folded in with the interior work.

We simultaneously compute boundary and interior equation coefficients using the interior control volume formulas and mask arrays. Oran et al. (1990) have called this technique the uniform boundary condition approach. For example, consider the source terms for the north boundary u control volumes, which are computed by the formula (for index notation, see Fig. 2.1)

$$b = a_N u_N + (p_w - p_e)\Delta y. \tag{3.37}$$

Recall that a_N represents the discretized convective and diffusive flux terms, u_N is the boundary value, and in the pressure gradient term, Δy is the vertical dimension of the u control volume; p_w/p_e are the west/east u-control-volume face pressures on the staggered grid. Similar modifications show up in the south, east, and west boundary u control volume source terms. To compute the boundary and interior source terms simultaneously, the following implementation is used:

$$b = a_{\text{boundary}} u_{\text{boundary}} + (p_w - p_e)\Delta y \tag{3.38}$$

where

$$u_{\text{boundary}} = u_N I_N + u_S I_S + u_E I_E + u_W I_W \tag{3.39}$$

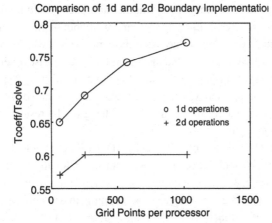

Figure 3.14   Normalized coefficient computation time as a function of problem size. In the timings marked "1d," the boundary coefficients are computed by a sequences of four 1D array operations (one for each boundary). In the timings marked "2d," the boundary coefficients are computed simultaneously with the interior coefficients.

and

$$a_{\text{boundary}} = a_N I_N + a_S I_S + a_E I_E + a_W I_W. \tag{3.40}$$

I_N, I_S, I_E, and I_W are the mask arrays, which are set to 1 in the appropriate boundary control volumes and are set to 0 everywhere else.

Figure 3.14 shows the importance of fully parallelizing all the operations in the pressure-correction algorithm. The figure compares the time spent on computation of coefficients per pressure-correction iteration, T_{coeff}, using the uniform approach (denoted "2d") and the direct parallelization of the serial algorithm, in which the boundaries are treated separately from the interior (denoted "1d"). The times are normalized by the time per iteration spent in the inner iterative solver, T_{solve}, in this case a total of 15 point-Jacobi relaxations. This gives some idea as to the relative impact of the coefficient computations compared to the total cost per outer iteration.

The ratio remains constant at 0.6 for $N > 256$, reflecting the fact that both T_{coeff} and $T_{\text{solve}} \sim N$ using the uniform approach. With the 1D implementation, though, the coefficient cost demonstrates a strong square-root dependency until we reach the point where the cost of the interior coefficient computations (asymptotically $\mathcal{O}(N)$) dominates the boundary work (asymptotically $\mathcal{O}(N^{1/2})$). Thus, beyond this problem size, even the naive implementation is scalable. However, the overall time per iteration is greater than the uniform approach because the source terms computations are not fully parallelized, requiring five steps instead of one. Our experiences indicate that it is important to fully parallelize these boundary operations because more complex problems with many dependent variables can easily exhaust the memory for $N \sim 1024$. In such problems, a 1D treatment of boundary computations would be inefficient for the entire range of possible problem sizes.

3.4.1.4 SCALABILITY

Figure 3.15, which is based on the point-Jacobi MP-1 timings, incorporates the above information into one plot, which has been called an isoefficiency plot by Kumar and Singh (1991). Along isoefficiency lines, the parallel efficiency E remains constant as the problem size and the number of processors n_p are varied. In the present case, we use N as the measure of the problem size instead of T_1, because we know in advance that $T_1 \sim N$. In general, though, the scalability is assessed by plotting isoefficiency curves on a plot of n_p against T_1. Lines of constant efficiency can be determined by curve fitting the timings over a range of n_p and T_1; frequently, T_1 has to be estimated because the problem is too large for a single processor. The curves can be written in a form $y \sim x^\alpha$ to quantify scalability. Each isoefficiency curve may have a different scalability. Ideally, $\alpha = 1.0$. In most algorithms, though, the communication and computation costs do not scale with the local problem size N but instead depend on N and n_p separately. For these problems, the isoefficiency curves have $\alpha > 1$.

In the figure, the isoefficiency lines are straight ($\alpha = 1.0$). This reflects the fact that the runtime does not depend on the number of processors; all the operations in the pressure-correction algorithm are fully parallel and the communication is dominated by regular grid-type nearest-neighbor communications. The efficiency is only a function of the local problem size N. Thus, along isoefficiency lines, the runtime per iteration is constant.

With more dependent variables, the maximum problem size is smaller than the present case. On the MP-1, though, the asymptotic parallel efficiency may still be obtained with smaller N, as shown in Fig. 3.13. Thus, the more complex problem can still be efficiently scaled up by adding relatively few processors, for example, by following the $E = 0.80$ isoefficiency curve.

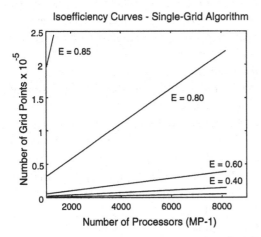

Figure 3.15 Isoefficiency curves based on the MP-1 cases and SIMPLE method with the point-Jacobi solver. Efficiency E is computed from Eq. (3.5). Along lines of constant E the cost per SIMPLE iteration is constant with the point-Jacobi solver and the uniform boundary condition implementation.

3.4.1.5 REMARKS

The present section has illustrated some of the important issues for data-partition-ed parallel computation. Implementation makes a difference in performance. The parallelization strategy and the development of suitable parallel algorithms are critical preliminaries, but in programming, one must still take great care to use programming idioms and techniques that keep all processors busy as much of the time as possible and that minimize the amount of communication.

In the SIMD context, this goal requires making all operations fully parallel as we have done with the uniform boundary condition approach. The communication factor splits into two parts in the SIMD model: host–worker communication and interprocessor communication. Actual interprocessor communication (between vir-tual processors on different physical processors) is minimized by using a block data partitioning. However, the effective cost is always asymptotically proportional to N in SIMD execution because the compiler must perform a check for every virtual pro-cessor, even in regular grid-type communication. Thus, in the SIMD model, cyclic mappings are not necessarily worse than block mappings; the MP-1 results prove that the key factor is the relative speed of computation and communication. In the SPMD context, the block mapping should be used because the interprocessor communication cost then varies with $N^{1/2}$ asymptotically. Consequently, for large enough problem sizes, efficiencies approaching 1 (linear speedup) are theoretically possible.

The most important machine-related factor is the relative speed of computation and communication. The algorithm and problem size determine the relative amount of communication, parallel computation, and serial computation. Together the speed and amount determine the time spent in communication and computation, which is strongly connected to efficiency by Eq. (3.5). In the SIMD context, the host–worker communication can have a strong effect on the variation of efficiency with problem size. This form of communication is a more-or-less constant overhead, and therefore it becomes amortized as the problem size (amount of parallel work) increases. Consequently we see the characteristic trend of increasing efficiency with problem size for fixed n_p. The interprocessor communication cost is also significant. It varies with the local problem size asymptotically, but for the CM-2 and CM-5, which use the block data–partitioning strategy, there is also a strong contribution from the subgrid perimeter which is $\mathcal{O}(N^{1/2})$. The perimeter contribution was assessed by studying the aspect ratio effect in conjunction with Fig. 3.12.

In the SPMD context, host–worker communication is less important than inter-processor communication assuming one takes advantage of local synchronization. However, the relative disparity of the speeds of communication and computation are typically much greater than on SIMD machines. Consequently, the parallel effi-ciency shows the same trend with problem size, corresponding to the amortization of the communication $\mathcal{O}(N^{1/2})$ by the computation $\mathcal{O}(N)$.

The CM-5 results illustrate the performance problem posed by disparate com-munication/computation speeds. On the CM-5, $N > 2k$ is required to reach peak E. This problem size is about one-fourth the maximum size that can be accommodated in the present program (8k with about 20 arrays). In programs designed to solve more complex physical problems, more data arrays are needed. Thus, it is probable that the maximum N which can be accommodated for problems of practical interest may still be too small to reach the peak efficiency of the machine for the given algorithm. Competition with other users for memory exacerbates the problem.

Lower efficiency means longer runtimes or, in other words, lower effective computational speeds. Thus, the speed we have estimated here for our application on a 32-node CM-5 using CM-Fortran, 420 Mflops, may be an upper figure for practical purposes. Since this is still in the same range as traditional vector supercomputers, it appears that parallel computers of this size, with disparate communication/computation speeds, can handle much bigger problems than supercomputers, but cannot do so significantly faster. Communication is a bottleneck. By extension to the SPMD case, where the communication/computation speed disparity typically is even greater than for the CM-5, one can expect that on machines with $n_p \sim \mathcal{O}(10^2)$ or larger, the speed of the nodes will be virtually irrelevant to the effective computational speeds obtained. This observation means that the value of large-scale parallel computation lies primarily in the ability to handle larger problems instead of the ability to speed up a calculation of fixed size. On the other hand, if the communication speed is relatively fast, it then becomes possible to solve a fixed-size problem significantly faster by adding processors, up to the point at which the parallelism is exhausted.

Scalability is another important issue. To achieve scalability, all of the operations in the algorithm need to be scalable on the particular architecture in use. The present algorithm uses only nearest-neighbor communication and involves all processors in every operation. Thus, it will be scalable on any architecture which supports a 2D mesh interconnection scalably.

Of course, one must also consider the bigger picture in the present context: Scalability of the cost per iteration is desirable, but the algorithm must be scalable in terms of convergence rate as well. Otherwise, as the problem size is increased to match the number of processors in use, the overall cost of solution will still be untenable because of the decrease in convergence rate, even though the cost per iteration remains small. The convergence rate of the pressure-correction method does not directly correspond to the convergence rate of the inner iterative solver, point-Jacobi iterations in many of our examples. The coupling between pressure and velocity is more important in many flow problems; in at least one problem, cavity flow, the outer convergence rate is independent of the inner solver used. Nevertheless, acceleration of the outer convergence rate is needed for very large problems, and our answer has been to utilize the multigrid techniques described in the previous section. In this way we benefit from the computational efficiency and scalability of point-Jacobi iterations but still obtain a method with robust convergence properties.

3.4.2 Multigrid Computational Issues

Parallel multigrid algorithms have been specifically addressed in several recent references (Alef 1994, Gupta et al. 1993, Linden et al. 1994, McBryan et al. 1991, Michielse 1993). These techniques are receiving attention because of their almost ideal scalability ($\mathcal{O}(\log_2 N)$ for problem size N on $n_p = N$ processors, in theory) on

Poisson equations. However, the performance of the parallelized multigrid algorithm depends strongly on the communication costs of a particular parallel platform and the parameters of the multigrid iteration. Specifically, the type of multigrid cycle and the starting procedure are important considerations for parallel computation.

In our research (Blosch and Shyy 1994, 1996) the FAS multigrid algorithm, using the sequential pressure-based solution procedure described previously to solve the governing equations on each grid level, has been parallelized as described previously with the data-parallel array-based language, CM-Fortran. The results obtained on the CM-5 are representative of what would be obtained for block data-partitioning multigrid implementations on MIMD distributed-memory machines.

With regard to implementation, the main concern beyond the single-grid case is the multigrid storage problem. Dendy et al. (1992) have recently described a multigrid method on the CM-2. However, to accommodate the data-parallel programming model they had to dimension their array data on every grid level to the dimension extents of the fine-grid arrays. This approach is very wasteful of storage, but it is difficult to avoid with distributed-memory parallel computers. Consequently the size of problems which can be solved is greatly reduced. Recently the CM-Fortran compiler has incorporated a feature which is the analog of EQUIVALENCE for serial Fortran. Using this feature, we have been able to develop an implementation which circumvents the storage problem. Our implementation has the same storage requirements as serial multigrid algorithms. Consequently, we have been able to consider problems as large as 1024×1024 on a 32-node CM-5 (4 Gbytes of memory). For discussion of the programming technique, see Blosch and Shyy (1996). The high-performance Fortran compilers currently becoming available will allow the same approach to be employed for multigrid algorithms on a variety of MIMD distributed-memory machines.

With regard to performance, the critical step(s) of the parallelized multigrid algorithm are the smoothing iterations on the coarse grid levels. The single-grid results showed that small problem sizes were inefficient, and so the natural concern is that the coarse grids may be a bottleneck to efficient performance of the overall multigrid cycle.

Figure 3.16 displays the cost of the smoothing iterations in terms of run time, for the coarse grid levels in a 9-level multigrid calculation on the 32-node CM-5. The finest grid level, level 9, corresponds to a 1024×1024 grid, and the coarsest grid level corresponds to 4×4. The times given are for five SIMPLE iterations, that is, they correspond to one V(3,2) cycle, using 3, 3, and 9 inner point-Jacobi iterations on the u, v, and pressure-correction equations, respectively. The smoothing cost would be translated up or down depending on the number of pre- and post-smoothing iterations per cycle.

From the figure, one can see that for the CM-5 the coarse-grid levels' smoothing cost, while still small in comparison with the cost of fine-grid pressure-correction iterations, is no longer negligible. In fact, there is basically no decrease in the smoothing times beneath level 5, which corresponds to a 128×128 grid. In Fig. 3.16, the bar on the left is the CM-5 busy time, and the bar on the right is the corresponding elapsed time. Busy time is the time spent doing parallel computation

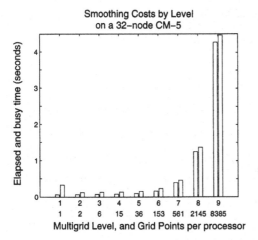

Figure 3.16 Smoothing cost, in terms of elapsed and busy time on a 32-node CM-5, as a function of the multigrid level for a case with a 1024×1024 fine grid. The elapsed time is the one on the right (always greater than the busy time). The times correspond to five SIMPLE iterations.

and interprocessor communication operations. The elapsed time is the busy time plus the contribution from host–worker communication, that is, the passing of code blocks and host data to the workers. Beneath level 5, this overhead dominates, and consequently we do not see the coarse-grid smoothing cost (elapsed time) go to zero as N goes to zero, as is the case for serial computation. The busy times do not scale linearly either, but this trend comes from the effect of problem size on the efficiency of interprocessor communication and vectorized computation for the CM-5.

3.4.2.1 CHOICE OF CYCLE TYPE

The effect of the coarse grids on the efficiency can be seen from Fig. 3.17. Figure 3.17 plots the elapsed and busy times for 7 level V(3,2) and W(3,2) cycles, as measured on the 32-node CM-5, against the virtual processor ratio N of the finest grid level. The number of levels is kept fixed as the finest grid dimensions increase.

The cost of one V cycle using n_{level} levels and $(n_{\text{pre}}, n_{\text{post}})$ pre- and post-smoothing iterations can be modeled as

$$\frac{\text{Time(s)}}{\text{V cycle}} = \sum_{k=1}^{n_{\text{level}}} s_k (n_{\text{pre}} + n_{\text{post}}) + \sum_{k=2}^{n_{\text{level}}} (r_k + p_k) \tag{3.41}$$

where s_k, r_k, and p_k are the times for one pressure-correction iteration on level k, for restriction from level k to level $k - 1$, and for prolongation to level k from level $k - 1$. For W cycles, the runtime can be modeled as

$$\frac{\text{Time(s)}}{\text{W cycle}} = \sum_{k=1}^{n_{\text{level}}} s_k (n_{\text{pre}} + n_{\text{post}}) 2^{n_{\text{level}}-k} + \sum_{k=2}^{n_{\text{level}}} (r_k + p_k) 2^{n_{\text{level}}-k}. \tag{3.42}$$

Thus the number of smoothing iterations on the coarsest grid ($k = 1$) increases geometrically with the number of levels for a W cycle. A given grid level k is visited

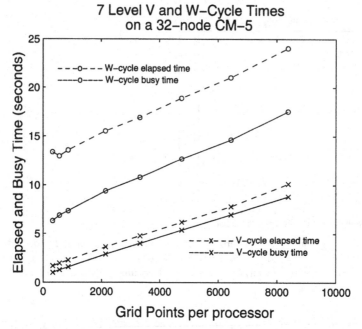

Figure 3.17   Elapsed and busy time per cycle on a 32-node CM-5, as a function of the problem size, given in terms of N. The number of levels is fixed at seven as the dimensions of the finest grid are increased.

2 to the power $n_{level} - k$ times, as compared to twice for a V cycle. On a serial computer it is generally reasonable to neglect the restriction and prolongation cost in comparison with smoothing, and to assume that the smoothing cost is proportional to the number of unknowns, that is, the cost on level $k - 1$ is 1/4 of the cost for level k in two-dimensional cases. Furthermore, this model allows one to measure the multigrid cycle cost in terms of work units, equivalent fine-grid iterations. The cost on parallel computers must be measured directly in seconds because the assumption of equal smoothing efficiency regardless of problem size is no longer valid.

From the figure, we observe that the runtime per cycle is approximately three times that of the V cycle. Efficiency is roughly the ratio of busy to elapsed time. Thus one can deduce that the W cycle efficiency is much less than the V cycle efficiency. Furthermore, since the frequency of visitation of coarse grids scales geometrically with the number of grid levels, the efficiency of W cycles decreases as the problem size increases, if additional levels are added. In contrast, V cycles are almost scalable, as discussed below.

As one may expect, similar comparisons using fewer levels show less difference between the efficiency of V and W cycles. However, using a truncated number of levels will deteriorate the multigrid convergence rate. Generally the tradeoff between V and W cycles, and the choice of the number of levels, is problem-dependent, but the tradeoff is shifted in favor of V cycles on the CM-5 in comparison to the situation for serial computations. Satisfactory convergence rates can be obtained using V cycles and the dynamic full-multigrid strategy described in the previous chapter.

3.4.2.2 RELATIVE COST OF RESTRICTION AND PROLONGATION

As in serial multigrid computations, the contribution to the cost per cycle from the restriction and prolongation operation is small, as can be seen from Fig. 3.18. The restriction cost is not shown for clarity but is slightly less than that of prolongation and shows the same trend. The ratio of the times for restriction, prolongation, and smoothing tends toward 1:2:13 on the 32-node CM-5, as the problem size increases. Obviously, the relative contribution from restriction and prolongation will depend on the amount of smoothing.

The cost of the intergrid transfer steps scales linearly with problem size, as was the case for smoothing (for large N), which reflects the fact that restriction/prolongation are just local averaging/interpolation operations applied to every residual. However, unlike the smoothing iterations, the time for restriction and prolongation depends on the number of processors as well. In the example above, we find that the prolongation to the grid level with $N = 4k$, which is roughly 360×360 on 32 nodes and 1440×1440 on 512 nodes, takes 0.4 s on 32 nodes but 0.5 s on 512 nodes. Thus the compiler does not generate scalable code for this operation.

The prolongation and restriction operations generate random communication patterns as far as the compiler is concerned because the array expressions for these

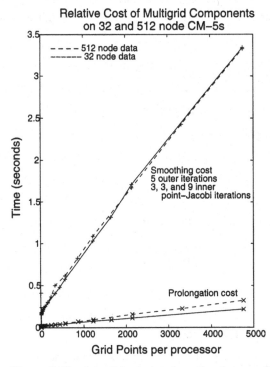

Figure 3.18 Smoothing and prolongation times per V cycle as a function of the problem size for 32- and 512-node CM-5 computers (128 and 2048 processing elements, respectively). The times are elapsed times, for V(3,2) cycles: 5 smoothing iterations, 1 restriction, and 1 prolongation, at each grid level. The restriction cost is not shown for clarity. The trend is the same as for prolongation and the time is roughly the same.

operations involve arrays that are declared to have different dimensions, that is, one defined on the coarse grid and one defined on the fine grid. The mapping of arrays to processors is done at runtime; it must be, because the number of processors is unknown a priori. Thus, the compiler has no knowledge that could enable it to recognize the regularity and locality of the communication patterns in the restriction and prolongation operations. Consequently, the CM-5 fat tree network is used with a general routing algorithm despite the fact that the communication pattern is actually quite regular.

According to the Thinking Machine Corporation document (1992), the global communication network of a 32-node CM-5 is a fat tree of height 3. The fat tree is similar to a binary tree except the bandwidth stays constant upwards from height 2 at 160 Mbytes/s. Thus, it is strictly true to say only that the performance of the network is scalable beyond height 2. Based on our results, it is evident that the restriction and prolongation operations generate communications that travel farther up the tree on 512 nodes than on 32 nodes, reducing the effective communication bandwidth slightly.

3.4.2.3 SCALABILITY

Figures 3.19 and 3.20 integrate the information contained in the preceding figures. In Fig. 3.10, the variation of parallel efficiency of 7-level V(3,2) cycles with problem

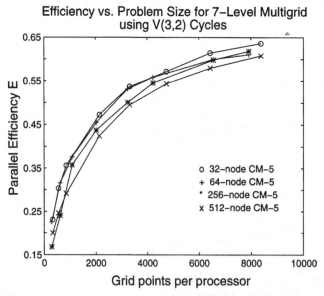

Figure 3.19 Parallel efficiency of the 7-level multigrid algorithm on the CM-5, as a function of the problem size. Efficiency is determined from Eq. (3.5), where T_p is the elapsed time for a fixed number of V(3,2) cycles and T_1 is the parallel computation time (T_{node}–cpu) multiplied by the number of processors. The trend is the same as for the single-grid algorithm, indicating the dominant contribution of the smoother to the overall multigrid cost.

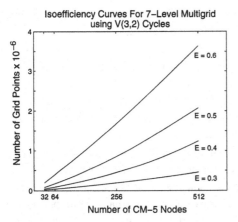

Figure 3.20   Isoefficiency curves for the 7-level pressure-correction multigrid method, based on timings of a fixed number of V(3,2) cycles, using point-Jacobi inner iterations. The isoefficiency curves have the general form $N - N_0 = c(n_p - 32)\alpha$, where $\alpha = 1.1$ for the effficiencies shown.

size is summarized. The problem size is that of the finest grid level, normalized by the number of processors. Figure 3.19 is similar to Fig. 3.10 obtained using the single-grid pressure-correction algorithm. For small N the useful work (parallel computation in this context) is dominated by the interprocessor and host–worker communication, resulting in low parallel efficiencies. The efficiency rises as the time spent in parallel computation increases relative to the overhead costs. The highest efficiency obtained is almost 0.65. This figure is somewhat smaller than the 0.8 efficiency obtained for the single-grid method on the CM-5. We attribute the majority of this decrease in efficiency to the increased contribution from the coarse grids during the multigrid V cycles.

Unlike the single-grid case, however, the efficiency does not reach an asymptotic peak value at the upper limit of problem size considered here, $N = 8192$. The parallel efficiency of the multigrid computations reaches an asymptotic value only when the coarsest grid is large enough, approximately $N \sim 1024$ for CM-Fortran on the CM-5. The coarsest grid level for the 7-level problem has $N = 128$.

Furthermore, note that the range of N for the levels of a 7-level multigrid cycle (a realistic cycle) span three orders of magnitude. Thus the best performance of multigrid methods will be from fine-grained machines like the MP-1, for which the single-grid pressure-correction iterations are efficient over a broad range of problem sizes ($32 < N < 1k$ in this case).

Figure 3.19 also shows that the impact on efficiency due to the nonscalable restriction/prolongation operations is minor. Consequently we would expect the isoefficiency curves to be almost linear, and indeed they are, as shown in Fig. 3.20. Each of the isoefficiency curves can be accommodated by an expression of the form

$$N - N_0 = c(n_p - 32)^\alpha \tag{3.43}$$

with $\alpha \simeq 1.1$ and c a constant. The symbol N_0 is the initial problem size on 32 nodes that corresponds to the particular value of E for the curve.

For straight isoefficiency curves, "scaled-speedup" (Gustafson 1990, Gustafson et al. 1988, Kumar et al. 1994) is achieved. In the present program, the parallel runtime T_p at some initial problem size and number of processors increases slightly as the problem size and the number of processors are increased in proportion. In other words, along the isoefficiency curves T_p remains the same, but n_p increases at a rate which is slightly greater than the rate of increase in problem size.

Note also that although we have used point-Jacobi inner iterations, similar scalability characteristics (although longer run times for a particular problem size) are obtained using line-Jacobi inner iterations. The line iterative method is $\mathcal{O}(N \log_2 N)$, but the effect of the logarithmic factor is generally quite small due to the early cutoff tolerance in the cyclic reduction kernel used in the line solver, which effectively reduces the amount of long-distance communication.

3.4.2.4   FULL MULTIGRID EFFICIENCY

Another issue which is important in the context of parallel multigrid computation is the cost of the full-multigrid starting procedure. As discussed previously, the full-multigrid procedure greatly improves the convergence rate by generating a good initial guess. The additional work that precedes the beginning of multigrid cycling on the fine grid is usually negligible on serial computers. For parallel computation, however, the cost of the FMG procedure is more of a concern, because coarse-grid cycles cost relatively more than they do on serial computers.

Figures 3.15 and 3.16 of the previous chapter illustrate this issue. In the top plot of each figure, the convergence path in the initial part of the computation, the FMG procedure, is plotted in terms of work units (serial computer). The bottom plot uses actual runtime on the CM-5. By comparing the two it is clear that the parallel cost of the FMG procedure is between 5 and 50 times as much as on a serial computer. Whether or not the additional cost makes FMG worthwhile depends on the problem. With reference to the aforementioned figures, we can observe that the FMG cost is still largely insignificant in comparison with the total solution cost. If one defines convergence according to the level of truncation instead of an algebraic criterion, though, the FMG cost for the step flow using second-order upwinding is 25% of the entire solution cost. As was discussed in the previous chapter, it has been argued that the solution accuracy with respect to the actual physical solution does not improve beyond the level of convergence to the truncation error. Thus it appears that there may be instances for which the parallel FMG cost may be significant, and in that case it is important to minimize the coarse-grid cycling. The techniques presented in the previous section should be helpful in this regard.

Tuminaro and Womble (1993) have recently modeled the parallel runtime of the FMG cycle on a distributed-memory MIMD computer, a 1024-node nCUBE2. They developed a grid-switching criterion to account for the inefficiencies of smoothing on coarse grids. The grid-switching criterion effectively reduces the number of coarse-grid cycles taken during the FMG procedure. They do not report numerical tests of their model, but the theoretical results indicate that the cost/convergence rate tradeoff can still favor FMG cycles for multigrid methods on parallel computers, in agreement with the present findings.

3.4.2.5 REMARKS

By counting floating-point operations we can estimate the effective speed of the multigrid code running on a 32-node CM-5 for the 7-level V(3,2) cycle case to be roughly 333 Mflops. The cost per V(3,2) cycle is about 1.5 s on a 128-vector unit CM-5 for a 7-level problem with a 321×321 fine grid. The efficiency and speed improve slightly with fewer multigrid levels, but in the general case this comes at the expense of convergence rate. The single-grid SIMPLE method using the point-Jacobi solver ran at 420 Mflops. Figure 3.19, which is based on detailed timing information and Eq. (3.5), shows that the parallel efficiency was 0.65, compared to 0.8 for the single-grid program (about a 20% decrease). The decrease in effective speed simply reflects less efficient execution. The decrease in efficiency occurs because the multigrid cycle cost is composed of a range of problem sizes. The smaller problems (the coarse grid smoothing iterations) are less efficient. There is additional work from restriction and prolongation but the impact of the efficiency of these operations is relatively small, the same as in serial computations.

Several ideas are being explored to improve the efficiency with which the coarse-grid level work is carried out. One idea is to use multiple coarse grid to increase the efficiency of the computations and communications (see, for example, Griebel 1993, Overman and Van Rosendale 1993, Smith and Weiser 1992, and the references therein for recent research along this line). In absolute terms, more computation and communication are done per cycle, but the overall efficiency is higher if the multiple coarse grids are treated in parallel. The obvious problem is making the extra work pay off in terms of convergence rate. Another idea is to alter the grid schedule to visit the coarsest grid levels only every couple of cycles. This approach can lead to nearly scalable implementations (Gupta et al. 1992, Linden et al. 1994) but may sacrifice the convergence rate. "Agglomeration," an efficiency-increasing technique used in MIMD multigrid programs, refers to the technique of duplicating the coarse-grid problem in each processor so that computation proceeds independently (and redundantly). Such an approach can also be scalable (Lonsdale and Shuller 1993). The degree to which the aforementioned ideas improve the fundamental efficiency/convergence rate tradeoff is of course problem-dependent. There have also been efforts to devise novel multilevel algorithms with more parallelism for SIMD computation (Frederickson and McBryan 1981, Gannon and Van Rosendale 1986, Griebel 1993). These efforts and others have been reviewed in Chan and Tuminaro (1988), McBryan et al. (1991), and Womble and Young (1990).

Overall the results of our work have been very encouraging regarding the potential of the pressure-based multigrid method for large-scale parallel computing. Using an efficient storage scheme, very large problem sizes can be accommodated, up to $4096 \times$

4096 on a 512-node machine (64 Gytes memory). Also, the results of numerical experiments indicate that the parallelized multigrid method is nearly scalable.

Several points should be addressed regarding the results just presented to clarify the breadth of their validity/utility. First, the numerical experiments were conducted using two-dimensional Cartesian grids. Storage, or rather lack of storage, was the major reason why two- instead of three-dimensional calculations were used to illustrate the effect of the coarse grids on parallel efficiency and scalability. In the 4096×4096 case one can use 11 multigrid levels, but for the equivalent three-dimensional case, $256 \times 256 \times 256$, one can use only 4 or 5 multigrid levels, on 512 processors, fewer if the investigations are carried out on 32 processors. Thus, three-dimensional computations show a lesser effect from the coarse grids in the timings. For the purpose of assessing the performance of V/W cycles and the parallel cost of the FMG procedure, two-dimensional computations are preferrable. In three-dimensions, the number of control volumes increases by a factor of 4 with each finer grid, and so relatively more time is spent smoothing the fine grids in comparison with two-dimensional, calculations. Thus, in three dimensions, the overall efficiency of both V and W cycles would be increased. Qualitatively the trends observed remain unchanged.

Second, the results presented above were obtained on a particular machine, operating in the SIMD style of parallel execution. However, it has been established that the key factor affecting efficiency and scalability is the amount and relative speed of interprocessor communication compared to computation. The amount of communication depends on the mapping of data to processors. Using a block data-partitioning strategy, actual communication is needed only for the edge regions of the processor subgrids, the area-perimeter effect. Thus, the present results are representative of the results that would be obtained on MIMD computers using the SPMD approach.

Third, the multigrid techniques in use for incompressible Navier-Stokes equations vary widely. Consequently, there are many issues left uncovered in this review that are the subjects of current research. One of the most important of these is load balancing, which is an inherent problem for multilevel computations that derive from local refinement. In such methods (see, for example, Howell 1994, McCormick 1989, Minion 1994) there are a base coarse grid that covers the whole domain and local patches of refinement which are created dynamically by monitoring the solution truncation error. Thus, for parallel computation, the partitioning of the grid must be dynamic to accommodate the varying grid-point density of the composite grids.

3.5 Concluding Remarks

Parallel computing is advantageous because large problem sizes can be accommodated. The obtainable computational speeds (e.g., Mflops), however, are strongly algorithm-dependent. The effective computation rate depends on the relative amounts and speeds of computation and interprocessor communication. Raw communication speeds are typically orders of magnitude slower than floating-point operations. Thus, more often than not, the communication steps in the algorithm – and the network performance for these steps – strongly influence the parallel runtime. The convergence rate of the multigrid method for the Navier-Stokes equations depends on many factors, including the restriction/prolongation procedures, the amount of pre- and post-smoothing, the initial fine grid guess, the stabilizing strategy, and the flow problems. Our experience with sequential pressure-based smoothers has been that none of these factors is unimportant. Overall, V cycling in conjunction with the truncation-error-controlled nested method seems to be a reasonable approach for the problems assessed.

4 Multiblock Methods

4.1 Introduction

There are many physical problems that exhibit multiple length and time scales in the form of, for example, high velocity and temperature gradients, recirculating zones, and phase change fronts, as well as geometric complexities due to the irregular shapes of the flow domain. To handle these characteristics, a multiblock grid method is a very useful technique, allowing different numbers and densities of grid points to be distributed in different regions without forcing the grid lines to be continuous in the entire domain. As reviewed in Chapter 1, there are alternative techniques proposed in the literature to allow gridding flexibility, including structured and unstructured meshes. In the present text, we will concentrate on the structured, multiblock method. The multiblock structured grid is a useful approach because (i) it can reduce the topological complexity of a single structured grid system by employing several grid blocks, permitting each individual grid block to be generated independently so that both geometry and resolution in the boundary region can be treated more satisfactorily; (ii) grid lines need not be continuous across grid interfaces, and local grid refinement and adaptive redistribution can be conducted more easily to accommodate different physical length scales present in different regions; (iii) the multiblock method also provides a natural route for parallel computations.

A detailed account will be given here to address the important issues involved in multiblock methods. Specifically, the interface treatment in the region containing discontinuous grid lines, and the satisfaction of the physical conservation laws consistent with the numerical resolution, will be emphasized. Because the grid lines may not be continuous across grid block interfaces, information between the blocks needs to be transferred; the effect of such an information transfer procedure on the solution accuracy, computational efficiency, and robustness of the overall algorithm is of great concern. It is preferable that the information transfer method be easy to implement, while maintaining consistent accuracy and robust performance. In the following discussion, only structured grid layout will be considered to maintain the necessary focus.

4.2 Overview of Multiblock Method

Multiblock structured grids can be broadly classified as either patched grids (Rai 1984) or overlapping grids (Steger 1991). Patched grids are individual grid blocks of which any two neighboring blocks are joined together at a common grid line without overlap. With overlapping grids, the grid blocks can be arbitrarily superimposed on each other to cover the domain of interest. Compared to patched grids, overlapping grids are more flexible to generate and can be more easily fitted to complex geometry, but information transfer between blocks is more complicated and flux conservation is more difficult to maintain. For both grid arrangements, the issues of interface treatment regarding both conservation laws and spatial accuracy need to be addressed. For flow problems involving discontinuities or sharp gradients of flow variables, it is well known that it is often advantageous to use a numerical scheme in conservation form. The need for accurate conservative grid interfaces is illustrated by Benek et al. (1983). Rai (1985, 1986) has developed a conservative interface treatment for patched grids and has conducted calculations demonstrating the shock capturing capability with the discontinuous grid interface. Berger (1987) has given a discussion of conservative interpolation on overlapping grids. Chesshire and Henshaw (1990, 1994) have conducted analyses on grid interface treatment and developed data structures for conservative interpolation on general overlapping grids. They solved slowly and fast-moving shock problems with both conservative and nonconservative interpolation and found that, with the appropriate form of artificial viscosity, both conservative and nonconservative interface schemes can give good results. Part-Enander and Sjogreen (1994) also compared the effects of both conservative and nonconservative interpolation on slowly moving shock problems. They found that the nonconservative interpolation could lead to large error; in their approach, the conservative interpolation alone was not stable, and the conservative flux interpolation with a characteristic decomposition and nonlinear filter were necessary to produce a satisfactory solution. Meakin (1994) investigated the spatial and temporal accuracy of an overlapping method for moving body problems and suggested that the issue with interface treatment was not necessarily one of conservative versus nonconservative, but one of grid resolution. The issue of accurate and conservative treatment of discontinuous grids has also been investigated by Kallinderis (1992). A related work on conservative interface treatment for overlapping grids can also be found in Moon and Liou (1989) and Wang et al. (1995). From the literature cited above, it can be seen that several factors can affect the solution quality for flow problems involving sharp flow gradients and discontinuous grid interfaces, such as the order of interpolation accuracy, conservative or nonconservative interpolation, convection schemes (or forms of artificial viscosity), and grid resolution. Which factor (or combination of several different factors) has the most critical effect is still not clear. Ideally, the physical conservation laws should be satisfied in the entire domain, but when discontinuous grids are encountered, to what extent flux should be conserved is still an open question. For some problems, compromise has to be made between maintaining flux conservation and interpolation accuracy consistently in both grid interface and interior regions. All these problems need to be investigated further.

For solving incompressible flow problems, because of the decoupling of thermodynamic pressure and velocity fields, the way to maintain mass conservation is crucial for designing an effective solution algorithm. Several algorithms have been developed to reach this goal, such as the projection method (Chorin 1967), the SIMPLE method (Patankar 1980), and the PISO method (Issa 1985), to name a few. A common point in these methods is that, instead of solving the continuity equation directly, a pressure (or pressure-correction) equation (formed by manipulating the continuity and momentum equations) is solved in conjunction with the momentum equations. One of the key points for the success of these algorithms lies in whether the pressure (or pressure-correction) equation and the associated boundary conditions truly represent the original equation – the continuity equation and the corresponding boundary conditions – in other words, whether the conservation of mass flux over the whole flow domain is strictly satisfied or not. For multiblock computations of incompressible flow, the maintenance of mass conservation across discontinuous grid interfaces is also an important issue. Some efforts have been made with different interface treatments for mass flux. For example, two-dimensional incompressible flow problems have been computed by employing patched grids (Lai et al. 1993, Perng and Street 1991, Thakur et al. 1996). In these methods, the grid interface arrangements are either continuous or consist of one coarse grid including an integer number of fine grids so that the maintenance of mass conservation across the grid interface is relatively easy. Several authors (Henshaw 1994, Hinatsu and Ferziger 1991, Meakin and Street 1988, Strikwerda and Scarbnick 1993) solved the flow problems using the overlapping grids and just employed direct interpolation of the dependent variables across the grid interface. The extent of satisfaction of mass conservation there depends solely on the order of the interpolation methods used. Yung et al. (1989) and Tu and Fuchs (1992) have used different mass correction methods to achieve global mass conservation across the grid interface, but the effects of those methods on solution accuracy have not been assessed. Wright and Shyy (1993) have developed a pressure-based multiblock method based on the discontinuous overlapping grid technique for solving the incompressible Navier-Stokes equations in Cartesian grid systems. A locally conservative interface scheme, with first-order accuracy, is devised to ensure that local and global conservation of mass and momentum fluxes are maintained. This overlapping grid methodology has been extended to nonorthogonal curvilinear coordinates (Shyy et al. 1994). A related work on domain decomposition and parallel computation methods can be found in Gropp and Keyes (1992a,b). Although much progress has been made in this field, some fundamental issues remain unsolved. For mass flux treatment, an interface scheme with global mass conservation may not be sufficient to yield accurate solution. On the other hand, a locally conservative interface scheme may have difficulties maintaining equal accuracy between the interface and the interior regions. Furthermore, for incompressible flow computations, the pressure is generally known up to an arbitrary constant. When the flow solver is applied to multiblock grids, how to couple pressure fields in different blocks, and at the same time maintain mass flux conservation across interface, is an important issue to be addressed.

4.3 Analysis of Model Equations on Multiblock Grids

In this section, the notation of multiblock grids is introduced in order to lay the foundation for the development of a numerical method for multiblock computations in a later section. Suppose an entire computational domain is denoted as D. The whole domain D can be partitioned into a series of subdomains (or blocks) D_i, which either can overlap or patch one another, as indicated in Fig. 4.1. Then the whole domain D can be considered as a union of subdomains D_i,

$$D = D_1 \cup D_2 \cup \cdots \cup D_N. \tag{4.1}$$

Similarly, suppose Φ is the solution of a particular equation or a system of equations over the entire domain D. The solution Φ can be composed of solutions on each subdomain,

$$\Phi = \Phi_1 \cup \Phi_2 \cup \cdots \cup \Phi_N. \tag{4.2}$$

4.3.1 1-D Poisson Equation

Let us first consider a 1-D Poisson equation as an example to illustrate possible interface treatment methods. The Poisson equation is a second-order elliptic equation which can represent many different physical processes and can be considered as the degenerate equation of the more general convection–diffusion equation. Suppose the governing equation and the associated boundary conditions are as follows:

$$\begin{cases} \Phi_{xx} = f, & x \in (0, 1) \\ \Phi(0) = a, & \Phi(1) = b \end{cases} \tag{4.3}$$

where a and b are constants, and f is a function of x. Here, the boundary conditions

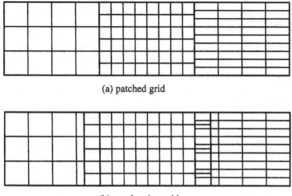

(a) patched grid

(b) overlapping grid

Figure 4.1   Multiblock grid layouts.

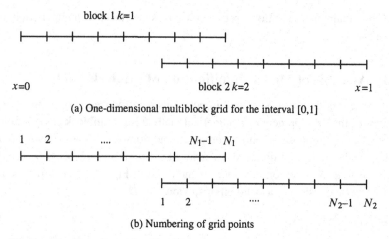

(a) One-dimensional multiblock grid for the interval [0,1]

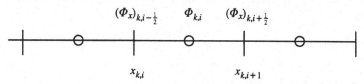

(b) Numbering of grid points

Figure 4.2   One-dimensional multiblock grid organization.

$$(\Phi_x)_{k,i-\frac{1}{2}} \qquad \Phi_{k,i} \qquad (\Phi_x)_{k,i+\frac{1}{2}}$$

$$x_{k,i} \qquad\qquad x_{k,i+1}$$

Figure 4.3   Location of discrete variables.

at the two ends of the domain are both Dirichlet conditions. For simplicity, only two grid blocks are considered. The governing equation is discretized on a 2-block overlapping grid as illustrated in the following expressions and Fig. 4.2:

$$D_1 = \{x_1(i) \mid x_1(i) = x_1(1) + i h_1, \quad i = 1, 2, \ldots, N_1\} \tag{4.4}$$

$$D_2 = \{x_2(i) \mid x_2(i) = x_2(1) + i h_2, \quad i = 1, 2, \ldots, N_2\} \tag{4.5}$$

$$d = x_1(N_1) - x_2(1) \geq 0. \tag{4.6}$$

Let $\Phi_{k,i}$ denote the discrete approximation to Φ on the component grid $k = 1, 2$ for $i = 1, 2, \ldots, N_k$. Here $\Phi_{k,i}$ is defined at the midpoints of the grid cells, as shown in Fig. 4.3. The grid points will be denoted by $x_{k,i}$ with grid spacings $h_{k,i} = x_{k,i+1} - x_{k,i}$. In the present example, for simplicity, the uniform grid spacing is used on each grid block $k = 1, 2$, so that $h_k = h_{k,i}$, for $i = 1, 2, \ldots, N_k$. Therefore, the governing equation can be discretized as:

$$(\Phi_x)_{k,i+\frac{1}{2}} - (\Phi_x)_{k,i-\frac{1}{2}} = h_k f_{k,i} \tag{4.7}$$

$$\Phi_{k,i+1} - 2\Phi_{k,i} + \Phi_{k,i-1} = (h_k)^2 f_{k,i}. \tag{4.8}$$

The discretized equation on each block can be solved with either direct methods or iterative methods. However, the boundary conditions at the interfaces have to be provided to connect the solutions on the two blocks together so that a global solution can be obtained. In the following, different interface treatments will be discussed.

4.3.1.1 DIRICHLET INTERFACE CONDITION

On block grid 1, Φ_{1,N_1} is unknown, and on block grid 2, $\Phi_{2,1}$ is unknown. They can be interpolated from the other grid in the following form:

$$\Phi_{1,N_1} = \sum_{j=L_1}^{L_2} \gamma_j \Phi_{2,j}, \quad 1 \le L_1 \le L_2 \le N_2 \tag{4.9}$$

$$\Phi_{2,1} = \sum_{j=M_1}^{M_2} \beta_j \Phi_{1,j}, \quad 1 \le M_1 \le M_2 \le N_1 \tag{4.10}$$

where γ_j and β_j are interpolation coefficients. How many terms are involved in the interpolation is determined by the order of interpolation formula chosen. With this treatment, the unknown value of the dependent variable at the interface is interpolated from the other block and is supposed to be known, so that it is called the Dirichlet-type interface treatment or, simply, the D-type treatment. If both blocks use the D-type treatment at the interfaces, this combination is called the D-D-type treatment. With the above interface treatment, an iteration process between two blocks can be conducted to solve the equation on the whole domain.

4.3.1.2 NEUMANN INTERFACE CONDITION

For the second-order equation, Eq. (4.7), an alternative interface treatment can be used. Instead of Φ, the first-order derivative Φ_x is interpolated at the interface.

$$(\Phi_x)_{1,N_1-\frac{1}{2}} = \sum_{j=L_1}^{L_2} \gamma_j (\Phi_x)_{2,j+\frac{1}{2}}, \quad 0 \le L_1 \le L_2 \le N_2 - 1, \tag{4.11}$$

$$(\Phi_x)_{2,\frac{1}{2}} = \sum_{j=M_1}^{M_2} \beta_j (\Phi_x)_{1,j+\frac{1}{2}}, \quad 0 \le M_1 \le M_2 \le N_1 - 1. \tag{4.12}$$

With this treatment, the derivative of Φ is interpolated from the other block and is supposed to be known, so that it is called the Neumann-type interface treatment, or the N-type treatment. If both blocks use the N-type treatment, then it is called the N-N-type interface treatment. It is noted that for a two-block 1-D Poisson equation, if the N-N-type interface treatment is used, a physically unrealistic solution may be obtained, since the N-N-type treatment only requires the first-order derivatives to be equal at the interface, that is,

$$(\Phi_x)_1 = (\Phi_x)_2. \tag{4.13}$$

Across the interface, a constant may exist between the solutions from the two blocks in the overlapping region, namely,

$$\Phi_1 = \Phi_2 + c \tag{4.14}$$

because Eq. (4.13) satisfies Eq. (4.14) for any value of c. A remedy to this problem is to use the N-type treatment at one side and the D-type treatment at the other side of the

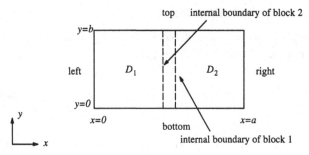

Figure 4.4   Two-dimensional two-block grid layout.

interface, which is called the D-N-type interface treatment. The D-N-type treatment can eliminate the possible nonuniqueness problem with the N-N-type treatment. The different interface treatments considered here may have different effects on the convergence rate when solving the equation on multiblock grids (Rodrigue and Shah 1989).

4.3.2   2-D Poisson Equation

Here, the 2-D Poisson equation is considered as the following:

$$\Phi_{xx} + \Phi_{yy} = f. \tag{4.15}$$

The computational domain is a rectangular domain, as illustrated in Fig. 4.4. It is decomposed into two overlapping domains D_1 and D_2 with a overlapping distance $d > 0$ in the x direction. With this configuration, the type of interface treatment that can be applied also depends on the kind of physical boundary condition of the problem. For example, if the physical boundary conditions at all four domain boundaries are Dirichlet type, then all the interface treatments discussed above can yield unique solutions (Rodrigue and Shah 1989). On the other hand, if at the top and bottom boundaries the physical boundary conditions are Neumann-type, that is,

$$\Phi_y = c \tag{4.16}$$

where c is a constant, the N-N-type interface treatment can not guarantee a unique solution (Wright and Shyy 1996). In addition, if at the top and bottom boundaries the physical boundary conditions are periodic, namely,

$$(\Phi)_{\text{top}} = (\Phi)_{\text{bottom}} \tag{4.17}$$

the application of the N-N-type interface treatment can not guarantee a unique solution either, because there is no constraint at the top and bottom boundaries to fix the solution in the overlapping region.

4.3.3   1-D Convection–Diffusion Equation

Here we consider different interface treatments for a steady one-dimensional convection–diffusion equation discretized on a two-block overlapping grid. The

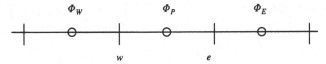

Figure 4.5 Illustration of the control volume.

governing equation is

$$(\varrho u \Phi)_x = (\Gamma \Phi_x)_x. \tag{4.18}$$

Integration of Eq. (4.18) over the control volume as indicated in Fig. 4.5 gives

$$(\varrho u \Phi)_e - (\varrho u \Phi)_w = (\Gamma \Phi_x)_e - (\Gamma \Phi_x)_w. \tag{4.19}$$

If we take

$$\Phi_e = \frac{1}{2}(\Phi_E + \Phi_P) \quad \text{and} \quad \Phi_w = \frac{1}{2}(\Phi_P + \Phi_W) \tag{4.20}$$

and use the central difference scheme for the diffusion terms, Eq. (4.19) can be written as

$$\frac{1}{2}(\varrho u)_e(\Phi_E + \Phi_P) - \frac{1}{2}(\varrho u)_w(\Phi_P + \Phi_W)$$
$$= \frac{\Gamma(\Phi_E - \Phi_P)}{\Delta x} - \frac{\Gamma(\Phi_P - \Phi_W)}{\Delta x}. \tag{4.21}$$

If we define

$$F = \varrho u, \qquad D = \frac{\Gamma}{\Delta x} \tag{4.22}$$

the discretized equation becomes

$$a_P \Phi_P = a_E \Phi_E + a_W \Phi_W \tag{4.23}$$

where

$$a_E = D_e - \frac{F_e}{2}, \qquad a_W = D_w - \frac{F_w}{2} \tag{4.24}$$

$$a_P = D_e + \frac{F_e}{2} + D_w - \frac{F_w}{2}. \tag{4.25}$$

The first interface treatment to be considered is simply of Dirichlet type. If the two-block domain is the same as that shown in Fig. 4.2, at the right boundary of block 1, the boundary value of $\Phi_{1,N}$ can be interpolated from block 2. In order to evaluate the associated coefficient a_W at the interface (i.e., $a_{1,N}$), we have to obtain the coefficients D_w and $F_w/2$ using the values from block 1 and the interpolated values from block 2. The same method is applied to the left boundary of block 2.

An alternative interface condition can be derived directly from Eq. (4.19), that is, at the right boundary of block 1, the total flux

$$-(\varrho u \Phi)_w + (\Gamma \Phi_x)_w \tag{4.26}$$

can be directly interpolated from block 2:

$$(-\varrho u \Phi + \Gamma \Phi_x)_{1,N_1-\frac{1}{2}} = \sum_{j=L_1}^{L_2} \gamma_j (-\varrho u \Phi + \Gamma \Phi_x)_{2,j+\frac{1}{2}},$$

$$0 \leq L_1 \leq L_2 \leq N_2 - 1. \tag{4.27}$$

This interface treatment is of Rubin type, or simply, the R type. Similar to the N-N-type interface treatment, if the R-R-type treatment is applied, the nonuniqueness problem may appear. The D-R-type interface treatment may fix this problem, or an extra constraint can be used to match the solutions in two blocks together (Chesshire and Henshaw 1994).

4.3.4　Analysis of the Navier-Stokes Equations

The interface treatments for solving the model equations on two overlapping grids have been discussed in the last section. It can be extended to the Navier-Stokes equations. However, there are some special issues associated with the Navier-Stokes computations in multiblock configurations that need to be addressed.

We begin by considering two domains Ω_1 and Ω_2 which can overlap or patch together. We denote the Navier-Stokes equations, Eqs. (2.2)–(2.3), by the following:

$$\mathcal{N}(\vec{u}, p) = 0. \tag{4.28}$$

The continuity equation, Eq. (2.1), is replaced by the pressure-correction equation, Eq. (2.32), which is denoted as

$$P(p') = f(\vec{u}^*) \tag{4.29}$$

where $\vec{u}^*$ is the intermediate velocity. We will solve Eqs. (4.28)–(4.29) on each separate domain. Boundary conditions for each domain are specified only on that part of the boundary that is not interior to the other domain. For the purpose of discussion, we assume that the velocity $\vec{u}$ is specified by the boundary value $\vec{b}$ on the boundary of $\Omega_1 \cup \Omega_2$. For that part of $\partial\Omega_1$ (the boundary of Ω_1) that lies within or patches with Ω_2, we require that the velocity and pressure be equal to those obtained from the solution on Ω_2, and similarly for that portion of $\partial\Omega_2$ within or patching with Ω_1. When grid lines of the two domains in the overlapping region or at the patched boundary are not exactly matched, flow variables have to be interpolated between the two domains. The interpolation methods discussed in the last section can be directly applied to the Navier-Stokes equations.

It is noted that the continuity equation requires the following compatibility condition to be satisfied by the boundary value $\vec{b}$,

$$\int_{\partial(\Omega_1 \cup \Omega_2)} \vec{b} \cdot \vec{n} = 0 \tag{4.30}$$

where $\vec{n}$ is the normal unit vector on the boundary $\partial(\Omega_1 \cup \Omega_2)$. When we solve the governing equations on both domains, the continuity equation should be satisfied on each domain, that is,

$$\nabla \cdot \vec{u} = 0 \quad \text{on} \quad \Omega_i, \ i = 1, 2 \tag{4.31}$$

or

$$\int_{\partial \Omega_i} \vec{u} \cdot \vec{n} = 0, \quad i = 1, 2. \tag{4.32}$$

This implies that across a grid interface, the mass flux should be conserved,

$$\int_{\text{interface}} \vec{u}_1 \cdot \vec{n} = \int_{\text{interface}} \vec{u}_2 \cdot \vec{n}. \tag{4.33}$$

Otherwise, a mass source or sink will be created at the interface. It is noted that Eq. (4.33) is an integral condition over the entire interface, but it should also be satisfied across any portion of the interface, that is, on a local cell-by-cell basis.

However, interpolations of any order will introduce truncation errors, which can violate Eq. (4.33). The effect of interpolations of the mass flux between the two blocks at the grid interface on the solution accuracy depends on the order of the interpolation. If necessary, based on interpolations, a conservative correction of mass flux can be applied. In the following, details of the interface treatments for the Navier-Stokes equations will be presented.

4.4 Multiblock Interface Treatments for the Navier-Stokes Equations

4.4.1 Multiblock Grid Arrangement

In the current multiblock computational method, the entire flow domain can be decomposed into several subdomains (or blocks). Within each block, a structured grid is generated. Two different blocks can either patch together or overlap over several grid layers. In the present approach, between the adjacent blocks, the grid lines can be discontinuous, and a multiple number of blocks can simultaneously overlap at the same location if necessary (Shyy 1994, Wright and Shyy 1993). In order to facilitate the exchange of information among the neighboring blocks, two layers of extended control volumes are constructed at the common interface of each of the two blocks. For example, the extended lines are assigned the indices 0 and -1 on the bottom boundary of a block and $nj + 1$ and $nj + 2$ on the top boundary of a block. These two layers are constructed from appropriate interpolations of the coordinates of the two grid lines next to the interface in the neighbor block. To illustrate this for a particular example, refer to the horizontal interface shown in Fig. 4.6. In this figure, the bottom side of a fine grid and the top side of a coarse grid are shown abutting (horizontal boundary). Cartesian grids are used for ease of illustration. For each of the two blocks, the extended lines are shown as dotted lines. For the fine (top) block shown in Fig. 4.6a, the extended lines $j = 0$ and -1 correspond to the lines $j = nj - 1$ and $nj - 2$, respectively, of the coarse (bottom) block. Likewise, the extended lines $j = nj + 1$ and $j = nj + 2$ of the coarse block correspond to the lines $j = 2$ and

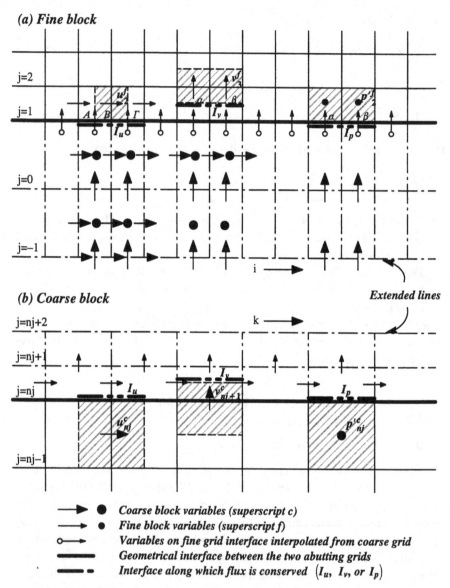

Figure 4.6 Illustration of the multiblock method using abutting grids and extended lines. (The $j = 1$ line on fine block and $j = nj$ line on coarse block are identical lines in physical domain).

$j = 3$, respectively, of the fine block. The extended control volumes are constructed by extending the grid lines for both blocks. Thus, for the coarse block, in the present case, the k line is extended (from $j = nj$ to $j = nj + 2$) to the fine block. For the fine block, the i line is extended (from $j = 1$ to $j = -1$) to the coarse block. The height of any extended control volumes is the same as the corresponding control volume of the neighboring block. The variable values from the neighboring block are stored in these extended control volumes and are used in the computation of the fluxes at or near the interface. The geometric quantities, such as the Jacobian and the metrics, are computed from the coordinates of the extended control volumes themselves.

4.4.2   Interface Treatment for the Momentum Equations

4.4.2.1   DIRECT INTERPOLATION OF DEPENDENT VARIABLES

For the momentum equations, at an internal boundary both velocity and pressure boundary conditions are needed. Let the south boundary be an internal boundary and let us consider the u-momentum equation as an example. The total momentum flux at the south boundary (see section 2.1)

$$F = \varrho V u - y_\xi p - \frac{\mu}{J}(-q_2 u_\xi + q_3 u_\eta) \tag{4.34}$$

needs to be evaluated at the grid interface. For smooth flow fields, the direct interpolation of dependent variables from the adjacent block should be satisfactory. With the extended grid points, the dependent variables u, v, and p can be interpolated from the adjacent block. A linear (or quadratic) interpolation can be used. The same approach is applicable to the boundaries on the other sides.

4.4.2.2   INTERPOLATION OF MOMENTUM FLUXES

For problems involving large flow gradients, the conservative treatment of momentum fluxes is often desirable. Here, to analyze the interpolation of momentum fluxes as the internal boundary condition at the grid interface, consider the flux at the south boundary again. As indicated by Eq. (4.34), the momentum flux F includes convection, diffusion, and pressure terms. One straightforward way is to directly interpolate flux F as a whole, from the adjacent block, and then put the flux into the source term of the momentum equation. However, the flux F includes gradient terms such as u_ξ and u_η and a direct interpolation of these gradient terms admits solutions where an arbitrary constant may be added to the solution on each grid, rendering the solution nonuniform (Wright and Shyy 1996). In the current treatment, the flux F is split into two terms, as indicated by Eq. (4.35), and the two terms are treated separately:

$$F = \underset{(i)}{[\varrho V u - y_\xi p]} - \underset{(ii)}{\left[\frac{\mu}{J}(-q_2 u_\xi + q_3 u_\eta) \right]}. \tag{4.35}$$

The first term in F is interpolated directly from the adjacent block and is put into the source term in the momentum equation. For the second term, u itself is interpolated from the adjacent block, and gradient terms u_ξ and u_η at the interface are evaluated with the u values in both blocks. In this way, the problem of nonuniqueness is eliminated, and at the same time, the convection and pressure fluxes can be interpolated directly from the adjacent block.

Now, consider the interpolation of the first term in flux F,

$$F_1 = \varrho V u - y_\xi p = \varrho(v x_\xi - u y_\xi)u - y_\xi p, \tag{4.36}$$

which includes geometric terms x_ξ and y_ξ. Since there is no relation between x_ξ and y_ξ in different blocks, it is fundamentally wrong to conduct a straightforward interpolation of F_1 between different blocks according to the form of Eq. (4.36). On the other hand, F_1 can be written in another form, namely,

$$F_1 = \frac{\varrho(vx_\xi - uy_\xi)u - y_\xi p}{\left(x_\xi^2 + y_\xi^2\right)^{1/2}} \left(x_\xi^2 + y_\xi^2\right)^{1/2} \tag{4.37}$$

$$= [\varrho(v\cos\theta - u\sin\theta)u - p\sin\theta]\left(x_\xi^2 + y_\xi^2\right)^{1/2}$$

$$= \bar{F}_1 S$$

where $\bar{F}_1$ is the normalized flux, and

$$\sin\theta = \frac{y_\xi}{\left(x_\xi^2 + y_\xi^2\right)^{1/2}} \tag{4.38}$$

$$\cos\theta = \frac{x_\xi}{\left(x_\xi^2 + y_\xi^2\right)^{1/2}} \tag{4.39}$$

$$S = \left(x_\xi^2 + y_\xi^2\right)^{1/2}. \tag{4.40}$$

Here, θ is the angle between the grid line and the x axis, and S is the length of grid face, as shown in Fig. 4.7. With the current interface arrangement, the angle θ in different blocks at the interface can be estimated directly and the normalized flux $\bar{F}_1$ can be interpolated directly. If we let the superscript 1 denote the current block and 2 denote the adjacent block, then the following expressions are obtained:

$$\bar{F}_1^{(1)} = \sum_{i=1}^{i=k} \gamma_i \bar{F}_{1i}^{(2)} = \sum_{i=1}^{i=k} \gamma_i \frac{F_{1i}^{(2)}}{S_i^{(2)}} \tag{4.41}$$

$$F_1^{(1)} = \bar{F}_1^{(1)} S^{(1)} = S^{(1)} \sum_{i=1}^{i=k} \gamma_i \frac{F_{1i}^{(2)}}{S_i^{(2)}} \tag{4.42}$$

where γ_i are the interpolation coefficients which depend on the interpolation formula adopted. With the above interpolation treatment, the conservation of flux F_1 is not strictly enforced. Later, an interpolation scheme with a locally conservative correction

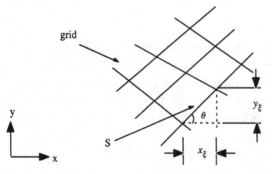

Figure 4.7   Configuration of momentum flux interpolation.

method will be introduced to strictly conserve the flux F_1. In summary, in the present interface treatment, the nonlinear part and the diffusion part of total momentum flux are treated separately. While the nonlinear part is interpolated from the adjacent block, the evaluation of the diffusion flux employs the velocity components from both blocks to avoid creating nonunique solutions.

Another method to avoid the nonuniqueness is to interpolate the variables at one side of an interface and interpolate the fluxes at the other side. For each common interface between the two neighboring blocks, we first compute fluxes in the fine block using the variables stored in the interior of the block as well as the variables from the neighbor (coarse) block stored in the extended control volumes. Thus, for example, the values $u_{i,2}^F$, corresponding to the row $j = 2$ for the fine block, are computed using fluxes calculated from the variables $(u, v, p,$ etc.) located at $j = 2, 3, \ldots$, as well as the coarse grid variables located on the extended lines $(j = 0$ and $-1)$. The variable $u_{i,2}^F$ is treated just like an interior unknown. While computing the corresponding variable near the interface $(j = nj)$ on the coarse block, namely $u_{k,nj}^C$, the flux on the common interface, I_u, is obtained by interpolating the fluxes on the common interface (I_u) in the fine block. In this way both the continuity of variables and the conservation of fluxes are maintained.

4.4.3 Interface Treatment for the Pressure-Correction Equation

Three interface boundary conditions are proposed for the pressure-correction equation. With the first treatment, the mass flux at the interface is taken as the boundary condition for the pressure-correction equation. Suppose the south face of a control volume of the pressure-correction equation is at the interface, as illustrated in Fig. 4.8. The p' equation for that control volume is as follows:

$$a_P p_P' = a_E p_E' + a_W p_W' + a_N p_N' + S_P \qquad (4.43)$$

$$a_P = a_E + a_W + a_N \qquad (4.44)$$

$$S_P = (\varrho U^*)_w - (\varrho U^*)_e + (\varrho V)_s - (\varrho V^*)_n \qquad (4.45)$$

where $(\varrho V)_s$ is the mass flux from the south face at the interface, presumably known, which is interpolated from the values on the adjacent block and is put into the source term. The term $a_S p_S'$ is no longer present in the p' equation for that control volume. Conversely, for the corresponding control volume at the interface of the adjacent block, the north face of the control volume of the pressure-correction equation is at the interface. The corresponding p' equation is as follows:

$$a_P p_P' = a_E p_E' + a_W p_W' + a_S p_S' + S_P \qquad (4.46)$$

$$a_P = a_E + a_W + a_S \qquad (4.47)$$

$$S_P = (\varrho U^*)_w - (\varrho U^*)_e + (\varrho V^*)_s - (\varrho V)_n \qquad (4.48)$$

where $(\varrho V)_n$ is interpolated and $a_N p_N'$ is no longer present. Thus, on both sides of the grid interface, the boundary conditions for the p' equation are implemented via the mass flux (ϱV) and are of the Neumann type. In other words, the pressure fields in different blocks are independent of one another. To make the pressure field compatible

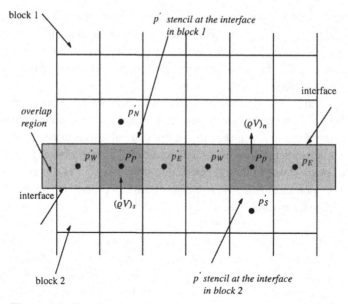

Figure 4.8   Interface layout with one grid layer overlap.

in the entire flow domain, the reference pressure in each block resulting from the Neumann boundary condition is adjusted according to the total momentum flux balance at a common line in the overlapping region between the blocks (Shyy 1994, Wright and Shyy 1993). This treatment is called method NN (meaning Neumann-Neumann condition). It is noted that when evaluating the coefficients a_E and a_W in Eq. (4.43), whose expressions are shown in Eqs. (2.35) and (2.36), the A_P^v term needs to be evaluated, which is taken as the average of the four surrounding values from the corresponding v-momentum equation, two of which are at the interface. The associated (ϱU) and (ϱV) are evaluated from the values within the block and the interpolated values in the adjacent block; the terms $\mu q_1/J$ and $\mu q_3/J$ are evaluated from the grid points within the current block and the extended fictitious points in the adjacent block.

With the second method, for the p' equation, on one side of the interface, the interpolated mass flux normal to the interface, say, $(\varrho V)_s$, is still used as the boundary condition in grid block 1, as shown in Fig. 4.9. On the other side of the interface, in the adjacent block, instead of $(\varrho V)_n$, p_N' is interpolated from grid block 1, as the boundary condition, as illustrated in Fig. 4.9. Besides the coefficients a_E and a_W discussed in the first method, another coefficient a_N needs to be evaluated at the interface, which is expressed as

$$a_N = \varrho \left(\frac{x_\xi^2}{A_P^v} + \frac{y_\xi^2}{A_P^u} \right)_n . \tag{4.49}$$

The term $\left(A_P^u \right)_n$ is the average of the values at the four surrounding u-locations. At the interface, the two locations are outside the current block, as illustrated in Fig. 4.10. To compute the A_P^u at those two locations, the terms (ϱU) and (ϱV) are interpolated at the extended grid points, and the terms $\mu q_1/J$ and $\mu q_3/J$ are evaluated using

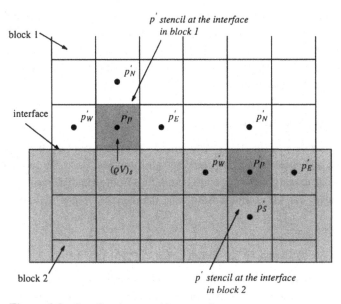

Figure 4.9   Interface layout without overlap.

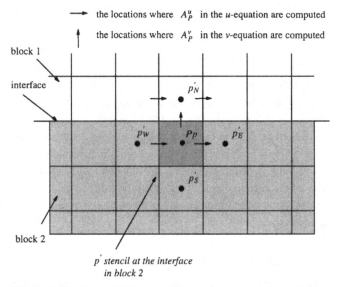

Figure 4.10   Interpolation stencil for pressure-correction equation at grid interface.

both the grid points within the current block and the extended fictitious points in the adjacent block. The same method is applied when evaluating the $(A_P^v)_n$ term at the interface. The source term S_P still has its original form given by Eq. (2.34), in which the term $(\varrho V^*)_n$ is actually computed using the v-value at the interface and the u-values within the current block and the interpolated values outside the block. In summary, on one side of the interface, the (ϱV) term is computed, while on the other side, it is interpolated. It is also noted that with the second treatment, the boundary condition is of Neumann type on one side and Dirichlet type on the other side. Since the pressure fields across the interface are tightly coupled together, the adjustment for the pressure field between blocks is no longer needed. This treatment is called method DN, standing for Dirichlet-Neumann condition. The third method is to interpolate the pressure correction p' at both sides of the interface, which is called method DD, standing for Dirichlet-Dirichlet condition.

4.4.4 Interface Treatment for Mass Flux

In the previous section, the interface treatments for the pressure-correction equation are discussed, and three interface conditions are proposed. In those interface conditions, two involve using the interpolated mass flux from the adjacent block as the interface boundary condition. In the following, the interpolation of mass flux is discussed.

4.4.4.1 LINEAR (QUADRATIC) INTERPOLATION OF MASS FLUX

With this treatment, the mass flux (ϱU) or (ϱV) is directly interpolated from the adjacent block, as discussed in the last section. However, this direct interpolation is not conservative, that is, across the interface, the total interpolated flux is not equal to the original flux. The difference between the two fluxes depends on the order of the interpolation formula.

4.4.4.2 INTERPOLATION WITH GLOBAL CORRECTION

A natural way to get the conservative flux is to globally correct the flux after the interpolation. First, a nonconservative interpolation is conducted to obtain the flux at the interface from the neighboring block. The total flux evaluated from the neighboring block is denoted as F_a, and the total interpolated flux is denoted as F_c. Second, the flux deficit $\Delta F = F_a - F_c$ is computed. Then, the flux at each current interface control volume face is added with $\Delta F/N$ (suppose there are N control volumes at the interface of the current block) so that the total flux across the interface is conserved globally. But this conservation is not locally enforced, and its effect on the solution needs to be assessed.

4.4.4.3 PIECEWISE-CONSTANT INTERPOLATION

Piecewise-constant interpolation, which is used with inviscid flows by Rai (1984), is by nature locally conservative, but with only first-order accuracy. It may cause accuracy problems for viscous flow computations.

4.4.4.4 LINEAR (QUADRATIC) INTERPOLATION WITH LOCAL CONSERVATIVE CORRECTION

With this treatment, the flux is interpolated first with a linear or quadratic formula, then, the interpolated flux is corrected locally on a cell-by-cell basis. This treatment will be discussed in detail. Suppose a portion of an interface corresponds to the width of a single control volume of the coarse grid 1, and to the width of several

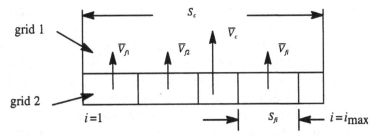

Figure 4.11 Notations used in two-block interface (subscripts c and f designate coarse and fine grid quantities, respectively).

control volumes of the fine grid 2, indexed from $i = 1$ to i_{max}. This scenario is shown schematically in Fig. 4.11, where $\bar{V}_c$ is the contravariant velocity component normal to the grid face, normalized by the control volume face length S_c, at the coarse grid control volume face; and $\bar{V}_{fi}$ represents the contravariant velocity component, normalized by the control volume face length S_{fi}, at the fine grid control volume face. From the fine grid to the coarse grid, $\bar{V}_c$ can be obtained as

$$\bar{V}_c S_c = \sum_{i=1}^{i_{max}} \bar{V}_{fi} S_{fi}. \tag{4.50}$$

Equation (4.50) is for a constant density flow. It can be easily extended to account for density variation. For the sake of simplicity, this aspect is neglected in the current discussion. In this way, the flux into the coarse grid control volume is uniquely determined from the corresponding fine grid flux conservatively. Conversely, given the coarse grid flux $\bar{V}_c S_c$, the conservation constraint yields

$$\sum_{i=1}^{i_{max}} \bar{V}_{fi} S_{fi} = \bar{V}_c S_c. \tag{4.51}$$

In this situation, conservation does not provide unique values for $\bar{V}_{fi}$. A certain distribution has to be chosen to determine each $\bar{V}_{fi}$. As a first approximation, $\bar{V}_{fi}$ is obtained using the linear (or quadratic) interpolation of the normalized contravariant velocity in the coarse grid. The interpolated value is denoted as $\bar{V}'_{fi}$. With the first approximation, Eq. (4.51) is not satisfied. Then the fine grid fluxes are scaled so that the total flux obtained is $\bar{V}_c S_c$. Accordingly, the values $\bar{V}_{fi}$ at the fine grid boundary are computed:

$$\bar{V}_{fi} S_{fi} = \frac{|\bar{V}'| S_{fi}}{\sum\limits_{i=1}^{i_{max}} |\bar{V}'| S_{fi}}. \tag{4.52}$$

From Eq. (4.52), it can be seen that Eq. (4.51), expressing flux conservation from coarse grid to fine grid, is satisfied and that the flux distribution is close to that determined by the interpolation employed.

If the grid interface is not exactly matched in the sense of a coarse control volume face adjoining an integer number of fine grid control volume faces, a split-merge procedure is devised. The control volume in grid 2 can be split into smaller subcontrol volumes. In this manner, from control volumes in grid 1 and subcontrol volumes in grid 2, the locally conservative treatment can still be applied on a cell-by-cell basis, after the interpolation and correction have been conducted. After that, the fluxes at the split subcontrol volume faces will be merged back to get the flux at the original control volume face. From grid 2 to grid 1, this same split-merge treatment is also applicable. The only extra work in this procedure is to create some arrays to store the intermediate information. Thus, the interface can be treated no matter what kind of interface arrangement is encountered. This linear (quadratic) interpolation with local correction treatment is not limited to the mass flux; it can also be applied to the conservative treatment of the momentum flux.

4.5   Solution Methods

4.5.1   General Procedure

After the governing equations are discretized on a multiblock domain, they can be solved by different solvers and solution strategies. Suppose there is a two-block domain $D = D_1 \cup D_2$. After the governing equations are discretized, we get two linear systems,

$$A_1 \Phi_1 = f_1 \tag{4.53}$$
$$A_2 \Phi_2 = f_2. \tag{4.54}$$

At the block interface, there is an interpolation relation between Φ_1 and Φ_2, say,

$$L_1 \Phi_1 = L_2 \Phi_2 \tag{4.55}$$

where L_1 and L_2 are interpolation operators. Equations (4.53)–(4.55) can be assembled into one linear system on domain $D = D_1 \cup D_2$,

$$A \Phi = f. \tag{4.56}$$

This whole linear system can be solved by different direct or iterative solvers. With this kind of strategy, the flexibility of grid generation is still retained, but the flexibility and convenience of implementation is lost. Furthermore, it is more difficult to achieve memory savings for large system of equations. A more attractive method is to solve each linear system of each block individually, with direct or iterative solvers. Between solution sweeps in each block, an updating exchange between blocks is conducted by interpolation. Suppose S_i denotes a solution sweep on block i and that I_i denotes an interpolation from block i. In a two-block case, the solution sequence could be

$$S_1 \rightarrow S_2 \rightarrow I_1 \rightarrow I_2 \tag{4.57}$$

or

$$S_1 \rightarrow I_2 \rightarrow S_2 \rightarrow I_1. \tag{4.58}$$

This way, not only the flexibility of grid generation is retained, but the implementation could be easier and more flexible also. More information about the efficiency of solution strategies on multiblock grids can be found in Meakin and Street (1988).

4.5.2 Solution Strategy for the Navier-Stokes Equations on Multiblock Grids

4.5.2.1 SOLUTION PROCEDURE FOR SINGLE-BLOCK COMPUTATIONS

For single-block computations the procedure is the following:

(1) Guess the pressure field p^*.
(2) Solve the momentum equations to obtain u^* and v^*.
(3) Solve the p' equation.
(4) Calculate p by adding p' to p^*.
(5) Calculate u and v from their started values using velocity-correction formulas and relations between velocities and contravariant velocities.
(6) Solve the discretization equations for other scalar quantities (such as temperature, concentration, and turbulence quantities) if they influence the flow field through fluid properties, source terms, etc.
(7) Treat the corrected pressure p as a new guess pressure p^*, return to step (2), and repeat the whole procedure until a converged solution is obtained.

It is noted that during this procedure, the solutions of u^*, v^*, and p' only proceed a few sweep steps, before going to solve other variables. This is because between the outer iterations, if all other variables are in the intermediate stages, solving only one variable close to convergence does not help the convergence of all the variables very much. The procedure adopted here can maintain tighter coupling between the variables in the process of iteration. Furthermore, to prevent the procedure from diverging or to accelerate convergence, we usually under-relax u^* and v^* (with respect to the previous values of u and v) while solving the momentum equations. Also, we add only a fraction of p' to p^*. We employ

$$p = p^* + \alpha_p p' \tag{4.59}$$

with $\alpha_p < 1$. The optimum value of α_p is usually problem dependent.

4.5.2.2 SOLUTION PROCEDURE FOR MULTIBLOCK COMPUTATIONS

In multiblock computations, another level of iteration is introduced, the iteration between different grid blocks. With the iterations between equations in each block and iterations between blocks, two different iteration strategies can be adopted:

(a) An outer loop is formed by iterating between blocks; an inner loop is formed by iterating between equations. Between the outer iterations, information transfer of all relevant variables is conducted.
(b) An outer loop is formed between different equations; an inner loop is formed between blocks. Between the inner loops, for each equation, information transfer is conducted.

These two methods can be illustrated as follows. For method (a),
For $bn = 1, nb$

(1) Solve the momentum equations to obtain u^* and v^*.
(2) Solve the p' equation.
(3) Calculate p by adding p' to p^*.
(4) Calculate u and v using the velocity-correction formulas.
(5) Solve the discretization equation for other scalar quantities.

end for
Repeat the whole procedure until a converged solution is obtained. For method (b),

(1) Solve the momentum equations to obtain u^* and v^*, for $bn = 1, nb$.
(2) Solve the p' equation, for $bn = 1, nb$.
(3) Calculate p by adding p' to p^*, for $bn = 1, nb$.
(4) Calculate u and v using the velocity-correction formulation, for $bn = 1, nb$.
(5) Solve the discretization equation for other scalar quantities, for $bn = 1, nb$.

Return to step (1) and repeat the whole procedure until a converged solution is obtained. Here, nb denotes the total number of blocks.

In principle, both strategies can be used. Which one gives better performance may be problem dependent. In the present method, the first one is adopted. The main consideration is that the evaluation of coefficients for the p' equation must use the stored coefficients from the momentum equations. With method (a), only the coefficients of the momentum equations from one block need to be stored. On the other hand, with method (b), the coefficients of the momentum equations from all the blocks have to be stored. Method (a) saves some memory compared to method (b).

4.5.3   Convergence Criterion

Given appropriate boundary conditions and a well-constructed computational mesh, the flow solver described above can be applied to a domain of any shape. This is true whether a global subdomain or one of a number of subdomains is being considered. When the solver is applied to a single domain, convergence is measured by how well the computed flow fields satisfy the discrete equation of motion. Specifically, residual sources can be computed for each of the momentum equations, mass continuity equation, and governing equations for other scalar variables.

If the discrete equations of motion have been satisfied exactly, then the residual sources will be zero. In practice, however, the residual sources are driven to some specified level of tolerance:

$$Res_i \leq Tol_i \qquad\qquad\qquad\qquad (4.60)$$

where Res_i and Tol_i correspond to the ith residual source and tolerance, respectively, and i corresponds to u, v, mass, and other scalar variables. When the solver is applied to a multiblock domain, one question is how a global convergence criterion should be defined and what iteration strategy should be used to achieve the global convergence.

The necessary condition for global convergence must be that in each block the solutions meet the requirements given by Eq. (4.60). However, this condition may not be a sufficient condition to guarantee global convergence; a highly converged solution can be obtained in one block before moving to a neighboring block. After applying the solver only once in each block, condition (4.60) can be satisfied. However, it is not likely that the correct global solution has been achieved. Rather, a highly converged solution would have been generated for each block based on only intermediate interface boundary conditions. Condition (4.60) is necessary for global convergence, but it is not sufficient to guarantee it. The solutions in the overlap regions must also converge to the same solution. In the present method, the interface information transfer between variables in overlapping regions is based on interpolation. At global convergence, the differences between solutions in respective overlapping regions should be on the order of interpolation truncation errors. This can be achieved by conducting only a few sweeps between equations within each block, instead of iterating all the equations to complete convergence in a particular block. Then the iteration sweeps move to other blocks, until condition (4.60) is satisfied in all the blocks.

4.6 Data Structures

4.6.1 General Interface Organization

With multiblock computational methods, because of the introduction of grid interface, certain data structures have to be devised to facilitate information transfer between different grid blocks. In the present method for a two-dimensional computation, each grid block has four sides, which are numbered in counterclockwise sequence, as shown in Fig. 4.12. Each side may be divided into several segments corresponding to different boundary condition types (e.g., physical boundary conditions or grid interface conditions). For each segment, the indices of the starting and ending grid points, the boundary condition type, and the corresponding values of dependent variables on the physical boundaries are assigned. If the segment is a grid interface, the block number of the adjacent block, the identification of overlapping direction (in ξ and η direction), and the number of the overlapping layer are assigned as well. Based on these inputs, a series of intersection tests are conducted to provide the information for interpolation coefficients and conservative treatment. All these coefficients are stored in tables for later use.

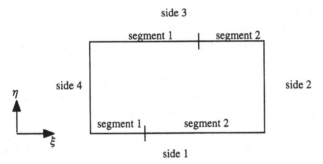

Figure 4.12 Grid interface organization for multiblock configuration.

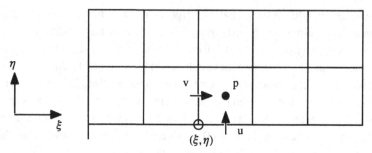

Figure 4.13   Interpolation stencil for grid points and dependent variables.

4.6.2 Interpolation

For a given grid point or a location of a dependent variable at an interface of a particular grid block, the basic information needed to facilitate the information transfer between the blocks is: (1) the adjacent block number, (2) the indices (ξ, η) of corresponding interpolating control volume, and (3) the interpolation coefficients. Among them, (1) is given by the input, and (2) and (3) will be determined by interpolation.

With the staggered grid arrangement, there are four pairs of interpolation needed for grid points and all the dependent variables: interpolations between (a) grid vertices (for geometric quantities), (b) grid face centers for velocity u and contravariant velocity U, (c) grid face centers for velocity v and contravariant velocity V, and (d) grid cell centers (for pressure and other scalar variables). These interpolation stencils are indicated in Fig. 4.13. In the present method, only two basic interpolations are conducted, for example, interpolations for cases (a) and (b). If we take the south boundary as an example and use the Cartesian grid for simplicity, we can see that the result of case (a), followed by an average procedure in η direction, can be used by case (c), the interpolation for v; similarly, the result of case (b) can be used by case (d), the interpolation for scalar variables. The same approach can be applied to the other three boundary sides.

4.6.2.1 GRID VERTEX–BASED INTERPOLATION

In the following, the grid vertex–based interpolation is described in detail, and the linear interpolation is taken as an example. Suppose there are two grid blocks. Side 3 of block 1 is patched with side 1 of block 2, as indicated in Fig. 4.14. Here, for illustration, the two grids are taken apart a small distance. Block 1 is called the current block, and block 2 is called the adjacent block. To conduct interpolation, the relative positions between grid points of block 1 and block 2 at the interface need to be identified. Therefore, the grid lines in the η direction in block 1 are extended to intersect with side 1 of block 2. In block 1, line segments are formed by connecting grid points at the interface, that is, side 3, with grid points next to the interface, as shown in Fig. 4.14. These line segments, for instance segment $\overline{ab}$, are extended to intersect side 1 of block 2. The side 1 of block 2 is formed by a series of straight line segments between two adjacent grid points. For segment $\overline{ab}$, a series of intersection

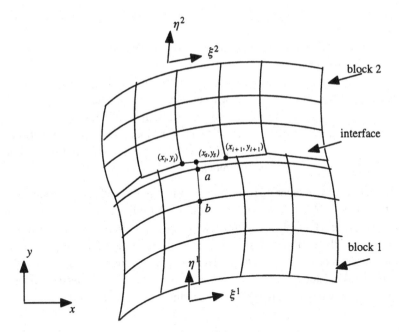

Figure 4.14   Grid vertex–based interpolation.

tests are conducted beginning from the starting point toward the ending point of side 1 of block 2. The straight line passing through point $a(x_a, y_a)$ and point $b(x_b, y_b)$ can be expressed as

$$a_1 y + b_1 x = c_1 \tag{4.61}$$

where

$$a_1 = x_b - x_a \tag{4.62}$$
$$b_1 = y_b - y_a \tag{4.63}$$
$$c_1 = y_a(x_b - x_a) - x_a(y_b - y_a). \tag{4.64}$$

Similarly, for a straight line passing through (x_i, y_i) and (x_{i+1}, y_{i+1}), at side 1 of block 2, the equation is:

$$a_2 y + b_2 x = c_2 \tag{4.65}$$

where

$$a_2 = x_{i+1} - x_i \tag{4.66}$$
$$b_2 = y_{i+1} - y_i \tag{4.67}$$
$$c_2 = y_i(x_{i+1} - x_i) - x_i(y_{i+1} - y_i). \tag{4.68}$$

The intersection point (x_s, y_s) of these two lines can be found by solving Eq. (4.61) and Eq. (4.65). If the intersection point falls within the range of (x_i, y_i) and (x_{i+1}, y_{i+1}), the corresponding interpolation coefficients are computed. The returning information is the following:

vertex_blk(IC,SN,BN)	the adjacent block number
vertex_ik(IC,SN,BN)	index number of interpolation stencil in ξ direction of the adjacent block
vertex_jk(IC,SN,BN)	index number of interpolation stencil in η direction of the adjacent block
vertex_coef(1:2, IC,SN,BN)	coefficients for linear interpolation between grid vertices of the adjacent and the current block

In these arrays, the integer IC is the index number of the grid along the side 3 of block 1; the integer SN is the side number of block 1 (in this example, SN is equal to 3); and the integer BN is the current block number. Therefore, for any grid point at the interface of the current block, the necessary information for interpolation is obtained by a series of intersection tests. The same procedure can be applied to the other three boundary sides. For stationary grid computation, these intersection tests need to be done only once; the result is stored in the arrays for later use.

4.6.2.2   GRID FACE CENTER–BASED INTERPOLATION

Grid face center (or edge)–based interpolation can be implemented in a manner similar to the grid vertex–based interpolation. However, one more step is needed. The line segments are formed by connecting the grid face center points instead of by connecting the grid vertices, and then intersect tests are conducted. The output information is stored in four arrays:

edge_blk(IC,SN,BN)	the adjacent block number
edge_ik(IC,SN,BN)	index number of interpolation stencil in ξ direction of the adjacent block
edge_jk(IC,SN,BN)	index number of interpolation stencil in the η direction of the adjacent block
edge_coef(1:2,IC,SN,BN)	coefficients for linear interpolation between grid face center of the adjacent and the current block

The definitions of the integers IC, SN, and BN for the edge-based interpolation are similar to those for the vertex-based interpolation. The interpolation procedures above can be described by the following pseudocode:

```
for BN=1, ..., Nblock
   for SN=1, ..., 4 do
      for sgn=1, ..., Nsegment
         if the segment is a interface do
               intersection_test
         endif
      endfor
   endfor
endfor
```

The subroutine intersection_test can be described as

```
for i=Indexc1, ... , Indexc2
  for j=1, ... , Jmax do
    solve linear equations of two line segments
      if intersection point found do
          get the index numbers of the grid cell in the adjacent block
          get the interpolation coefficients
      endif
  endfor
endfor
```

In the above procedure, Nblock is the total block number, Nsegment is the total segment number at a boundary side, and Jmax is the total grid number of the adjacent block side. Indexc1 and Indexc2 are the starting and ending indices of interface segments, respectively.

4.6.3   Interpolation with Local Conservative Correction

In this section, the implementation and data structure for the interpolation with local conservative correction will be described. We combine an explanation in words with a brief section of pseudocode. Again, suppose there are two grid blocks. Side 3 of block 1 is patched with side 1 of block 2, as indicated in Fig. 4.14. For simplicity, the grid layers adjacent to the interface are isolated for illustration, using a Cartesian grid, as shown in Fig. 4.15. In the current configuration, the relative position between the two blocks is very general, that is, there is no distinction between coarse and fine grid blocks. Basically, the whole procedure can be broken into three steps: grid cell splitting, flux interpolation, and flux correction.

4.6.3.1   GRID CELL SPLITTING

The first step is to split the grid cells of block 1 into a number of subcells by grid lines of block 2, so that each grid cell in block 2 matches an integer number of subcells in block 1, as illustrated in Fig. 4.16. The method of interpolating the flux with a local conservative correction can be conducted based on every grid cell flux at the interface of block 2. The output of the first step is to give the relation

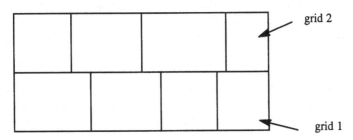

Figure 4.15   The general interface configuration between two blocks.

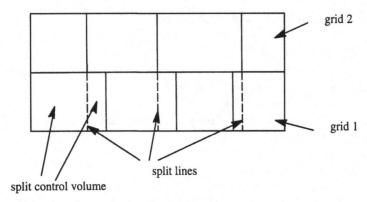

split lines

split control volume

Figure 4.16   The configuration of split cells at a grid interface.

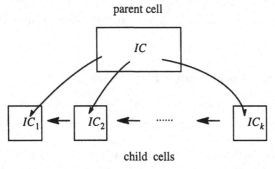

Figure 4.17   The mapping between a parent cell and the child cells.

between every grid cell (named as parent cell) at the interface of block 1 and the corresponding split subcells (named as child cells). A parent cell is identified by the index number along the interface, the side number, and the current block number. A child cell is identified by the corresponding parent cell identification, an index number indicating the sequence number of the child cell in the parent cell, and the x, y coordinates of the starting and ending points of the child cell. This relation is shown in Fig. 4.17. The output information is stored in the following three arrays:

number_child(IC,SN,BN)	total number of child cell plus 1, in a parent cell
x_child(ICELL,IC,SN,BN)	x-coordinate of the starting point of a child cell
y_child(ICELL,IC,SN,BN)	y-coordinate of the starting point of a child cell

Here, the integer IC is the index number of the parent cell along the interface of the current block, the integer SN is the side number of the interface, and the integer BN is the current block number. The integer ICELL is the index number of the sequence of the child cell in the parent cell.

The split procedure above is completed by conducting intersection tests. This procedure can be described by the following pseudocode:

```
for BN=1, ..., Nblock
   for SN=1, ..., 4 do
      for sgn=1, ..., Nsegment
         if the segment is a interface do
               subroutine split-cell
         endif
      endfor
   endfor
endfor
```

The subroutine split-cell can be described as

```
for i=Indexc1, ..., Indexc2
   initialize number_child(i,SN,BN)=0
   for j=1, ..., Jmax do
      intersection test
         if interception point found do
               increase number_child(i, SN, BN) by 1,
               ICELL=number_child(i, SN, BN)
               get x_child(ICELL,i,SN,BN), y_child(ICELL,i,SN, BN)
         endif
   endfor
endfor
```

In the above procedure, Nblock is the number of total blocks, Nsegment is the total segment number at a boundary side, and Jmax is the total grid number of the adjacent block side. Indexc1 and Indexc2 are the starting and ending indices of interface segments, respectively.

4.6.3.2   FLUX INTERPOLATION

The second step is to interpolate fluxes from the grid cells at the interface of the adjacent block (block 2) to the child cells at the interface of the current block (block 1). For mass flux transfer, this is basically a grid face center–to–grid face center interpolation, as described in the previous section on interpolation. For each child cell, the relevant information about interpolation is stored in the following arrays:

flux_bk(ICELL,IC,SN,BN)	the adjacent block number
flux_ik(ICELL,IC,SN,BN)	index number of interpolation stencil in ξ direction of the adjacent block
flux_jk(ICELL,IC,SN,BN)	index number of interpolation stencil in η direction of the adjacent block
flux_coef(1:2,ICELL,IC,SN,BN)	flux interpolation coefficients for interpolation between the grid face center of the adjacent block and the child cell center of the current block

Here, the integers IC, SN, and BN are defined the same as in the vertex-based interpolation information arrays. The integer ICELL is the child cell index in the corresponding parent cell. With the above information, the flux for each child cell can be obtained by interpolating fluxes of grid cells at the interface of the adjacent block. But the interpolated flux is not strictly conserved either on the local cell-by-cell basis or global interface basis; it is used only as a first approximation. The extent of flux conservation relies on the interpolation formula employed. Next, a correction step is introduced to conserve the flux strictly.

4.6.3.3 LOCAL FLUX CORRECTION

The third step is to make a conservative correction of the flux based on the first approximation in the second step, and on the scale of each individual grid cell at the interface of the adjacent block (block 2). After the cell split step, a grid cell at the interface of the adjacent block matches an integer number of child cells of the current block. The interpolated flux at each corresponding child cell can be scaled according to the magnitude of the first approximation, so that the total flux from the child cells is equal to the flux of the grid cell of the adjacent block exactly. The remaining question is how to identify the correspondence between a grid cell at the interface of the adjacent block and the matching child cells in the current block (block 1). With the cell split step, the relation between a parent cell in the current block and the corresponding child cells is uniquely determined, but a grid cell in the adjacent block may include child cells from several different parent cells of the current block, as shown in Fig. 4.18. The procedure to set up this unknown correspondence will be described below.

In the current block, for each interface segment (called the current segment) at a boundary side, the corresponding interface segment in the adjacent block and the

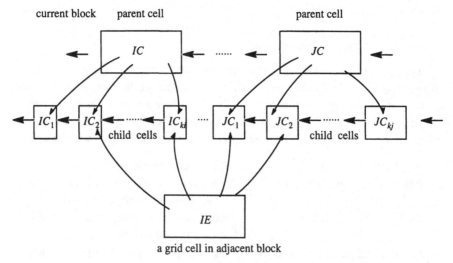

Figure 4.18 The mapping between a parent cell and the child cells in the current block, and the mapping between the child cells and a grid cell in the adjacent block.

starting and ending indices along that interface segment (called the adjacent segment) are identified first. This identification process can be accomplished by the intersection test, and the following information is provided:

seg_bk(ISG,SN,BN)	the adjacent block number
seg_start(ISG,SN,BN)	starting index number along the interface segment in the adjacent block
seg_end(ISG,SN,BN)	ending index number along the interface segment in the adjacent block
seg_side(ISG,SN,BN)	side index number of the interface segment in the adjacent block

Here, the integer BN is the current block number, SN is the side number of the current block, and ISG is the segment number of the interface segment on that side.

Then, beginning at the starting grid cell in the adjacent segment, for each grid cell, a counting procedure is conducted to calculate the number of corresponding child cells in the current segment. This counting procedure is based on another intersection test, that is, the intersection test between the line segment of the grid cell in the adjacent block along the interface direction and the line segment connecting the child cell face centers in the current block other than the interface direction. If the intersection point between the two line segments is identified, the child cell is found to belong to the grid cell in the adjacent segment and is added to that grid cell list. A particular array is created to store this information:

number_cell(IE,SE,BE)	the number of child cells included in a grid cell in adjacent segment

Here, the integer BE is the adjacent block number, SE is the side number of the adjacent block, and the integer IE is the grid cell index number at the interface of the adjacent block. The integer array number_cell determines the number of child cells each grid cell in the adjacent segment matches.

After the counting procedure, the child cells in the current segment can be put into a one-dimensional array based on the relative sequence of each child cell in the parent cell and the index sequence of the parent cells in the current segment. With this child cell sequence in the current segment, the grid cell sequence in the adjacent segment, and the correspondence between the two segments, the correspondence between the each grid cell in the adjacent segment and the child cells in the current segment can be uniquely determined, that is, each grid cell in the adjacent segment matches an integer number of child cells in the current segment. Therefore, the conservative correction of flux for each child cell can be made accordingly. Based on this correction, the flux is strictly conserved between each grid cell in the adjacent segment and the corresponding child cells, and the flux distribution is close to that determined by the interpolation method chosen. Finally, after the correction, the fluxes in the child cells are merged back to the parent cell in the current segment, and the flux distribution along the interface is uniquely determined.

4.7   Assessment of the Interface Treatments

4.7.1   Assessments of Conservative Treatments Using the N-N Boundary Condition for the Pressure-Correction Equation

In this section, the conservative and nonconservative treatments of mass flux with varying accuracies are tested and compared. The test problem is the lid-driven cavity flow with $Re = 1000$. First, a three-block discontinuous Cartesian grid configuration with different grid resolutions is used. A grid system of 41×21, 81×13, and 41×11 grid points for blocks 1, 2, and 3 respectively, as shown in Fig. 4.19, is first employed. Here, the interface between blocks 1 and 2 coincides with the cavity horizontal center line. The second grid system doubles the grid resolutions of the first grid system, and has 81×41, 161×23, and 81×21 grid points in each block, respectively. The third grid system doubles the grid resolution in the x direction of the second grid system and has 161×41, 321×23, and 161×21 grid points in the three blocks. To clarify the terminology, the multiblock grid for the whole flow domain is called "grid system" here. Each grid system consists of several blocks. The first grid system described above is denoted as the coarse grid system, the second as the median grid system, and the third as the fine grid system. The three grid systems share the same topological characteristics in each block. We have created these three grid systems to investigate the interplay of the interface treatment and overall grid resolutions. The grid layouts are not ideally suitable for the present recirculating flow; they are purposely set up to test the relative merits of different interface treatments. For all the test cases presented in this section, the second-order central difference is used for convection, diffusion, and pressure terms. Furthermore, the Neumann-Neumann interface condition is employed in all the cases studied in this subsection. The Dirichlet-Neumann and Dirichlet-Dirichlet conditions will be investigated in the next section.

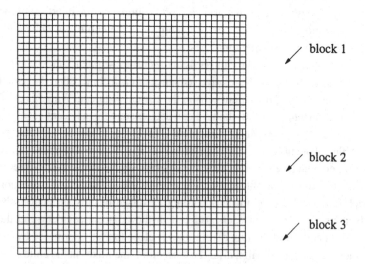

Figure 4.19   Three-block grids with 41×21, 81×13, and 41×11 nodes for blocks 1, 2, and 3, respectively.

Case 1. The Cartesian velocity components u and v and the contravariant velocity components U and V are linearly interpolated at the grid interfaces. Total momentum fluxes across the interfaces are conserved via pressure adjustment. No mass conservation is enforced across the grid interfaces. The computations are conducted over the coarse, median, and fine grid systems. For all three grid systems, the normalized residuals for u and v momenta are below 10^{-4}. The normalized mass residuals reach down to 1.5×10^{-3}, 3.3×10^{-4}, and 1.3×10^{-4} for the coarse, median, and fine grid systems, respectively, and stabilize at those levels. Figures 4.20a and 4.20b show the u-component distributions along the vertical center line and v-component distributions along the horizontal center line. They are compared with the corresponding benchmark solution reported by Ghia et al. (1982). It can be seen that the solutions for all the three grid systems have substantial discrepancies with respect to the benchmark values. Although in general the solutions improve as the grids are refined, the overall performance of all three systems are unsatisfactory. It is noted that for this problem computed with a single grid distributed uniformly, an 81×81 grid system can yield a very accurate solution (Shyy et al. 1992a). Accordingly, it is

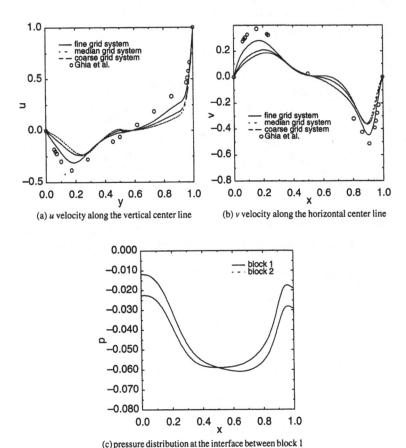

(a) u velocity along the vertical center line

(b) v velocity along the horizontal center line

(c) pressure distribution at the interface between block 1 and 2 on the fine grid system.

Figure 4.20 Solution profiles based on linear interpolation for U and V without mass conservation correction.

unsatisfactory to observe that both median and fine grid systems, even with resolutions better than the 81×81 single uniform grid, still do not yield accurate solutions. Figure 4.20c shows the interface pressure distributions computed on blocks 1 and 2. The pressure in each block has been adjusted according to the total momentum balance at the interface. The absolute value of pressure has no practical meaning. For a well-converged solution, the pressures from different blocks are expected to have the same distributions at the interface; in the current case, even with the fine grid system, the discrepancy between the two pressure distributions is obvious.

Case 2.   Both the Cartesian velocity components u and v and the contravariant velocity components U and V are quadratically interpolated, but still without enforcing mass conservation. This procedure is of a higher interpolation accuracy than that in case 1; however, neither has conservative treatment. The motivation here is to test the role played by the interpolation accuracy and to investigate for viscous flow whether conservative treatment is necessary to obtain satisfactory solutions. For all three grid systems, the residuals for momentum equations are below 10^{-4}. The mass residuals stabilize at the level of 2.3×10^{-3}, 4.3×10^{-4}, and 7.5×10^{-5} for the coarse, median, and fine grid systems, respectively. Figures 4.21a and 4.21b

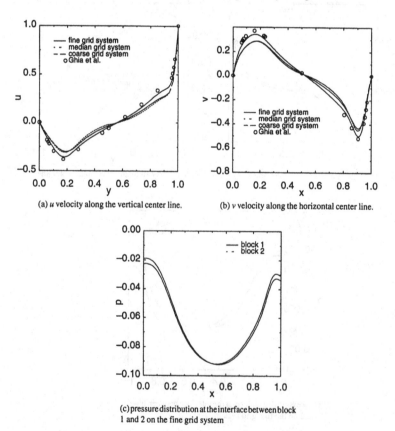

(a) u velocity along the vertical center line.

(b) v velocity along the horizontal center line.

(c) pressure distribution at the interface between block 1 and 2 on the fine grid system

Figure 4.21   Solution profiles based on quadratic interpolation for U and without mass conservation correction.

show the u-component and v-component distributions respectively for the three grid systems. Overall, these solutions are improved compared to the solutions shown in Fig. 4.20, and the solution on the fine grid system shows an obvious improvement over the coarse and median grid systems solutions. However, discrepancies still exist between the present and the benchmark solutions. Figure 4.21c shows the pressure distributions at the interface of blocks 1 and 2. The discrepancies between the two distributions still exist. From cases 1 and 2, it can be seen that the solution can be improved with grid refinement. The overall solution accuracy is better with the quadratic interface interpolation than with the linear interpolation, which is expected because the quadratic interpolation increases the order of interpolation accuracy and should reduce the nonconservative error for mass flux across the grid interfaces. However, even with the fine grid system, the solution with quadratic interpolation is still not satisfactory. Since with both interpolations the solutions are not as accurate as the single grid result, it appears that a nonconservative interface treatment for mass flux is not satisfactory when complex flow structures, such as recirculation, are present.

Case 3. Since the nonconservative linear and quadratic interpolations for the contravariant velocities cannot lead to a satisfactory solutions even for a very fine grid, a conservative interface is then tested. In this case, the Cartesian velocity components u and v are linearly interpolated. The contravariant velocities U and V are linearly interpolated first, followed by a global correction procedure for the mass flux as discussed previously. The computation is conducted for the coarse grid system only. The momentum and mass flux residuals reach the level of 10^{-5}. The results are shown in Fig. 4.22. Obviously, the solutions are not satisfactory. It appears that the conservative interface treatment conducted at the global level can not yield the correct solution either.

Case 4. The Cartesian velocity components u and v are still linearly interpolated. The contravariant velocities U and V are interpolated based on the piecewise-constant formula, which by nature is locally conservative with first-order accuracy. This treatment is implemented to investigate the effect of local conservation on the solution. The computations are computed for the three grid systems. For the coarse and median grid systems, the residuals for the momentum and mass fluxes reach down to 10^{-5}. But for the fine grid, a converged solution can not be obtained. Figures 4.23a and 4.23b present the u- and v-component distributions for the coarse and median grids. Both u and v profiles agree well with the benchmark solutions. Figures 4.23c and 4.23d exhibit the pressure distributions at the interfaces of blocks 1 and 2 for two grid systems. The pressures in the interface region obtained on different blocks conform to each other generally well, except that some oscillations appear in block 2. The cause of these nonphysical oscillations is that the mass flux distributions at the fine grid interface are assigned according to the piecewise-constant formula. This distribution results in a series of stair-step mass profiles at the interface of the fine block, forcing the pressure field to oscillate in response to the nonsmooth mass flux distribution. This same reason is also probably responsible for the nonconvergence of the fine grid solution.

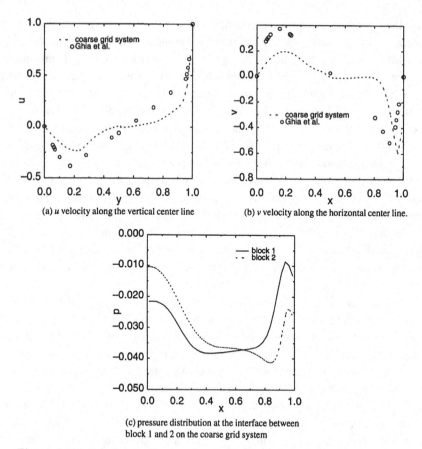

(a) u velocity along the vertical center line

(b) v velocity along the horizontal center line.

(c) pressure distribution at the interface between
block 1 and 2 on the coarse grid system

Figure 4.22   Solution profiles based on linear interpolation with global correction of mass conservation for U and V on the coarse grid system.

Case 5.   The Cartesian velocity components u and v are linearly interpolated. The contravariant velocities U and V are linearly interpolated first, followed by a local correction to maintain the cell-by-cell mass conservation across the interfaces. The tests are conducted for all three grid systems. The residuals for mass and momentum fluxes reach below the level of 10^{-5}. Figures 4.24a and 4.24b show the u- and v-component distributions. The solution on the coarse grid system shows a very small discrepancy compared to the benchmark solution; the solution on the median and fine grid systems agree well with the benchmark solution. Figure 4.24c displays the pressure distributions at the interface of blocks 1 and 2 on the fine grid system. The two pressure profiles conform to each other very well. Clearly, a treatment of interface mass flux with local conservation and certain accuracy holds a key to produce an accurate solution.

Case 6.   The Cartesian velocity components u and v are quadratically interpolated. The contravariant velocities U and V are quadratically interpolated first, followed by a local correction to maintain the cell-by-cell mass conservation across the interfaces. Again, the computations are conducted for the three grid systems and

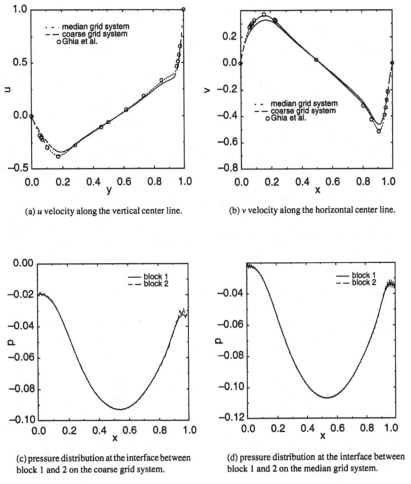

(a) u velocity along the vertical center line.

(b) v velocity along the horizontal center line.

(c) pressure distribution at the interface between block 1 and 2 on the coarse grid system.

(d) pressure distribution at the interface between block 1 and 2 on the median grid system.

Figure 4.23 Solution profiles based on piecewise-constant interpolation for U and V.

residuals go down to 10^{-5}. Figures 4.25a and 4.25b show the u- and v-component distributions. Good agreements, comparable to Case 5 (Fig. 4.24), between the current solutions and benchmark solutions are observed. Figure 4.25c presents the pressure distributions at the interface of blocks 1 and 2 on the fine grid system. The pressure distributions also conform to each other very well. For the present flow problem, both the linear and quadratic interpolations with local correction give accurate solutions. Since the discretization scheme for the interior nodes is second-order accurate, it appears that a linear interpolation (aided by the local conservative correction) treatment is sufficient.

Case 7. Finally, the same flow problem is investigated with a three-block curvilinear grid, which has 81×41, 161×23, and 81×21 grid points for blocks 1, 2, and 3, respectively, as shown in Fig. 4.26. The interface treatment is the same as that in Case 5, and the momentum and mass residuals reach down to 10^{-5}. Figure 4.27 demonstrates the u-component distribution at the vertical center line and compares

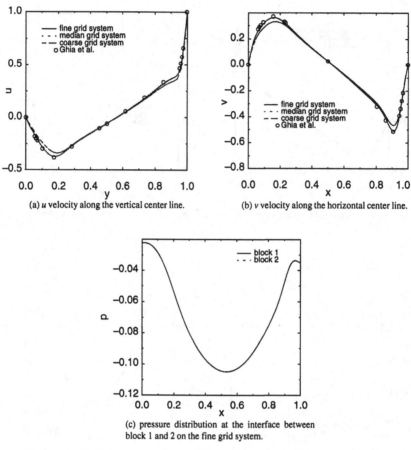

(a) *u* velocity along the vertical center line.

(b) *v* velocity along the horizontal center line.

(c) pressure distribution at the interface between block 1 and 2 on the fine grid system.

Figure 4.24 Solution profiles based on linear interpolation with local correction of mass conservation for U and V.

it with the corresponding benchmark solution. Good agreement is observed, which illustrates the capability of the current interface treatment on curvilinear grid systems.

4.7.2 Comparison of Two Different Interface Conditions for the Pressure-Correction Equation

In this section, the N-N- and D-N-type interfaces for the pressure correction are compared. The lid-driven cavity flow with $Re = 1000$ is still used as the test case, with the flow domain partitioned into two discontinuous grid blocks at the cavity horizontal center line. A grid system of 65×33 and 61×34 nodes is used for block 1 and block 2, respectively, as shown in Fig. 4.28. There is one grid layer overlap between the two blocks. The mass flux at the interface is estimated based on piecewise-linear interpolation with local conservative correction.

Figures 4.29a and 4.29b show the *u*-profile along the vertical center line and the *v*-profile along the horizontal center line for both interface methods on the overlapping grid; they are compared with the corresponding benchmark solution reported by

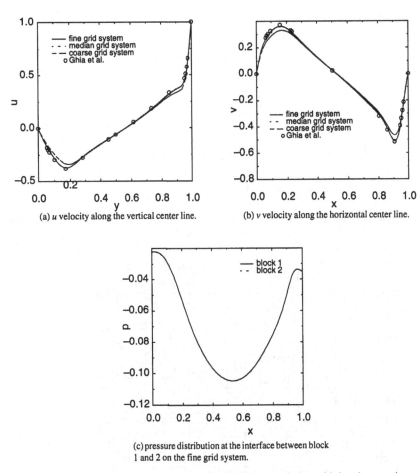

(a) u velocity along the vertical center line.

(b) v velocity along the horizontal center line.

(c) pressure distribution at the interface between block 1 and 2 on the fine grid system.

Figure 4.25 Solution profiles based on quadratic interpolation with local correction of mass conservation for U and V.

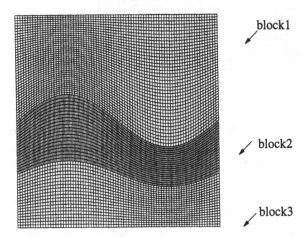

Figure 4.26 Three-block curvilinear grid with 81×41, 161×23, and 81×21 nodes for blocks 1, 2, and 3, respectively.

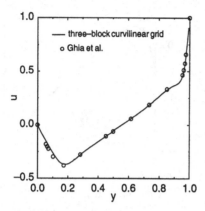

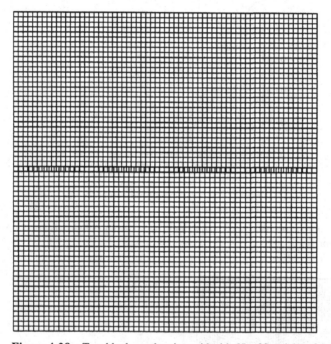

Figure 4.27 The u-velocity along the vertical center line. Linear interpolation with local correction of mass conservation for U and V.

Figure 4.28 Two-block overlapping grid with 65×33 and 61×34 nodes in each block.

Ghia et al. (1982). The pressure distributions from the two methods are shown in Fig. 4.29c. It can be seen that the solutions produced by both methods are almost identical and confirm very well with the benchmark solution. It is also noted that the discontinuous interface at the horizontal center line has no noticeable effect on the solutions.

For both methods, under-relaxation is needed to make the algorithm converge. For the method NN, the values of 0.3 and 0.5 for respectively the momentum equations and pressure correction equation are found to yield an adequate convergence rate. For the method DN, the values of 0.3 and 0.1 for the momentum equations and pressure equation are required to obtain a converged solution. Figure 4.30 compares the

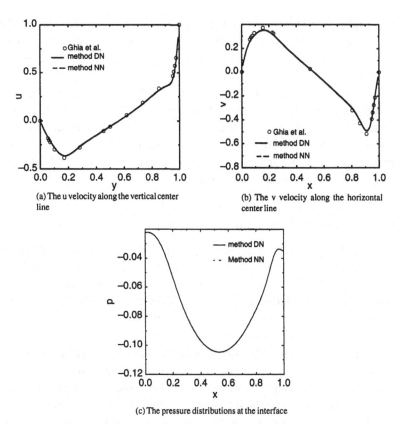

(a) The u velocity along the vertical center line

(b) The v velocity along the horizontal center line

(c) The pressure distributions at the interface

Figure 4.29 Comparison of two interface treatments on cavity flow calculations with $Re = 10^3$. (The two interface treatments result in indistinguishable solutions.)

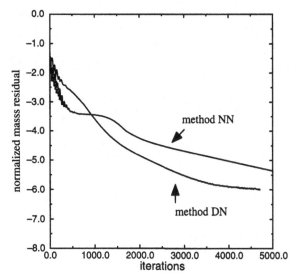

Figure 4.30 Comparison of convergence paths between the two interface methods: D-N and N-N.

convergence rates of the two methods with the relaxation factors stated above. The normalized mass residual is used as a representative quantity. The method DN appears to converge faster than the method NN, although it seems to have a more restrictive stability bound.

It is noted that for the D-N interface treatment, if the mass flux at the interface is interpolated without the local conservative correction, the solution obtained is unsatisfactory. In addition to the methods DN and NN, the D-D interface treatment for the pressure correction equation (i.e., p' is interpolated at both sides as the interface boundary condition) is also tested. For the test case above, it is found that this interface treatment can not yield a satisfactory solution.

4.8   Concluding Remarks

In this chapter, the interface conditions and the data structures proposed in the previous two chapters are investigated through numerical experiments. For the momentum equations, the dependent variables are interpolated between grid blocks by bilinear or quadratic interpolation. For the pressure correction equation, three interface treatments are investigated. It is found that with the local conservative interface mass flux treatment with certain degree of accuracy, both the Neumann-Neumann- and Neumann-Dirichlet-type boundary conditions can yield satisfactory solutions. Compared to the Neumann-Neumann-type boundary condition, the mixed Neumann-Dirichlet-type boundary condition allows better pressure coupling between grid blocks. For the cases tested, both boundary treatments give very comparable accuracy. A linear interpolation with local conservative correction method is found satisfactory for communicating mass flux across an interface.

5 Two-Equation Turbulence Models with Nonequilibrium, Rotation, and Compressibility Effects

5.1 Basic Information

In this chapter, we will discuss the issues of turbulence closures from the viewpoint of engineering calculations of complex transport problems. The two-equation models have been most popular and will be the focus in the present effort. To perform practical turbulent flow calculations at high Reynolds numbers, it is often convenient to use the Reynolds-averaged Navier-Stokes equations with constant density

$$\varrho \frac{\partial \bar{u}_i}{\partial t} + \varrho \frac{\partial}{\partial x_j}(\bar{u}_i \bar{u}_j) = -\frac{\partial p}{\partial x_i} + \frac{\partial}{\partial x_j}(2\mu S_{ij} - \varrho \overline{u'_i u'_j}) \tag{5.1}$$

$$\frac{\partial \varrho}{\partial t} + \frac{\partial}{\partial x_i}(\varrho \bar{u}_i) = 0 \tag{5.2}$$

$$S_{ij} = \frac{1}{2}\left(\frac{\partial \bar{u}_i}{\partial x_j} + \frac{\partial \bar{u}_j}{\partial x_i}\right). \tag{5.3}$$

These equations are obtained by decomposing the dependent variables into mean and fluctuating parts ($u = \bar{u} + u'$), then conducting averaging of the Navier-Stokes equations. This procedure introduces new unknowns, the so-called Reynolds-stress tensor terms, $-\varrho \overline{u'_i u'_j}$, which can be denoted by τ_{ij}, namely,

$$\tau_{ij} = -\varrho \overline{u'_i u'_j}. \tag{5.4}$$

By inspection, $\tau_{ij} = \tau_{ji}$, so that this is a symmetric tensor and thus has six independent components. For general three-dimensional constant property flows, we have four mean flow properties, viz., pressure and three velocity components. Along with the six Reynolds-stress components, we thus have ten unknowns. To close the system of the governing equations, the Reynolds-stress components have to be modeled.

One of the earliest proposals for modeling the Reynolds-stresses is Boussinesq's eddy-viscosity concept, which assumes that the turbulent stresses are proportional to the mean flow velocity gradients (Schlichting 1979). This relation can be expressed as

$$\overline{u_i' u_j'} = \frac{2}{3} K \delta_{ij} - v_t \left(\frac{\partial \bar{u}_j}{\partial x_j} + \frac{\partial \bar{u}_j}{\partial x_i} \right) \tag{5.5}$$

where K is the turbulent kinetic energy, δ_{ij} is the Kronecker delta, and v_t is the eddy viscosity. With this assumption, the task of finding the Reynolds-stresses reduces to the determination of the eddy viscosity. The eddy viscosity can be written as

$$v_t = \frac{l_0^2}{t_0} \tag{5.6}$$

where l_0 is the turbulent length scale and t_0 is the turbulent time scale quantities, which can vary dramatically in space and time for a given turbulent flow.

To solve turbulent flows in complex configurations in which the turbulent length and time scale (l_0, t_0) are not universal, the two-equation models – wherein transport equations are solved for two independent quantities that are directly related to the turbulent length and time scales – were proposed (Jones and Launder 1972). In the most common approach, the turbulent length and time scales are defined based on the turbulent kinetic energy K and the dissipation rate ε (i.e., $l_0 \propto K^{3/2}/\varepsilon$, $t_0 \propto K/\varepsilon$) with modeled transport equations for K and ε. The eddy viscosity is given by

$$v_t = C_\mu K^2 / \varepsilon \tag{5.7}$$

where C_μ is an empirical model constant. The two-equation models remove many of the problems associated with specifying the length scale in the zero- and one-equation models and have been applied to a variety of turbulent flows with various degrees of complexities. However, for some complex turbulent flows (e.g., with adverse pressure field, streamwise curvature, system rotation, etc.), the original K-ε model (Launder and Spalding 1974) cannot give satisfactory results. There are several possible factors contributing to the failures of the K-ε model for complex turbulent flows. One of them is that the coefficients are obtained based on simple flow conditions, which may not be valid for complex turbulent flows. Another weakness of the K-ε model is that it is still based on the Boussinesq hypothesis, which makes use of a scalar turbulent viscosity. The Boussinesq hypothesis inherently assumes that turbulent viscosity is isotropic. However, turbulent viscosity, unlike molecular viscosity which is a fluid property, is a flow variable. For complex turbulent flows, eddy viscosity is, in general, an anisotropic quantity strongly affected by geometry and physical mechanisms.

To solve complex turbulent flows with anisotropic turbulent stresses, the Reynolds-stress model is explored in which each component of the Reynolds-stress tensors is solved by a partial differential equation (Launder et al. 1973). But these equations are still not closed, and a significant amount of modeling is required to treat terms involving the pressure-velocity correlation and high-order moments. In reality, the success of the Reynolds-stress model is not consistent. Furthermore, compared to the eddy viscosity models, the Reynolds-stress models are more expensive to compute, because for each of the Reynolds-stress components, a partial differential equation has to be solved. This partially limits its applications in engineering flow simulations.

To reduce the expense of the Reynolds-stress models, attempts have been made to simplify the form of the Reynolds-stress transport equations. The simplification is aimed at reducing these PDEs to algebraic equations. However, the success of such models is also less than consistent. Based on the above discussion, it would seem that an improved two-equation model capable of addressing complex flow problems will be very desirable.

Different variants of the two-equation turbulence closure have been proposed in the literature (Shih 1996). Besides the K-ε model (Launder and Spalding 1974), the K-ω model has been investigated with much interest (Menter 1994, Wilcox 1993). While no model developed to date has been able to successfully predict flow characteristics of all, or even most, practically relevant fluid flows, clearly these two are among the more popular ones, in part due to the fact that they don't require exceedingly demanding computing resources. Several aspects can be briefly commented on regarding the two closure models. First, in terms of wall boundary treatments, for the K-ε model, either a wall function or a low-Reynolds-number condition can be implemented. For the K-ω model, one supplies boundary values of both turbulent kinetic energy and specific rate of dissipation of turbulent kinetic energy at the wall, thus requiring in general direct integration all the way to a solid wall. Regarding the construction of the two modeled equations in general, for the K equation the main difficulties are from pressure-velocity coupling terms. For either the ε or ω equation, the exact form is too complicated to be adopted; consequently, the entire equation is devised largely by heuristic arguments.

Between the two models, the K-ε model has been applied to more problems. A significant number of modifications have been developed for it (e.g., Shih 1996), leading to improved agreements with the experimental or direct numerical simulation data. In comparison, the K-ω model has not been tested as extensively. The works of Menter (1994) and Wilcox (1993) contain interesting summaries of the two models. For example, while there are questions about the asymptotic behavior of both models as the flow approaches the wall, the Taylor series expansion of the Navier-Stokes equations that underlies the analysis is valid only in the immediate wall proximity. Menter (1994) points out that in this region the eddy viscosity is much smaller than the molecular viscosity, and hence the asymptotic behavior of the mean flow profile is not dictated by the asymptotic form of the turbulence. Therefore, even if the turbulence model is not asymptotically consistent, the mean flow profile and the wall skin friction can still be predicted correctly. Menter also comments that using information such as turbulent kinetic energy or dissipation rate to assess model performance may not be useful since the main information a flow solver gets from the turbulence model is the eddy viscosity, which is derived by combining these variables. These observations indicate the difficulty of assessing turbulence models; however, it also seems clear that one would still prefer to see asymptotically correct trends displayed by these turbulence models, since this aspect supplies appropriate information in the boundary region that can substantially influence the predicted flow behavior in the rest of the domain, as expected from boundary value problems.

In the following, we will concentrate on the K-ε two-equation model, first discussing the basic features of the model construction, then addressing several major issues that have not been adequately handled in the original version of the model, namely the nonequilibrium circumstance under which the production and dissipation of the turbulent kinetic energy are not comparable, the effect of system rotation on turbulent flow characteristics, and the effect of compressibility resulting from high-Mach-number flows. First, the Reynolds-stress transport equations and the original K-ε transport equations are presented. The implementation of the original K-ε model for two-dimensional incompressible flows in curvilinear coordinates is detailed next. Then, issues associated with the nonequilibrium conditions, system rotation, and Mach number induced compressibility in the K-ε turbulence model for complex flows are discussed.

5.2   Turbulent Transport Equations

5.2.1   Reynolds-Stress Transport Equation

Let $\mathcal{N}(u_i)$ denote the Navier-Stokes operator for constant flow properties,viz.,

$$\mathcal{N}(u_i) = \varrho\frac{\partial u_i}{\partial t} + \varrho u_k\frac{\partial u_i}{\partial x_k} + \frac{\partial p}{\partial x_i} - \mu\frac{\partial^2 u_i}{\partial x_k \partial x_k} = 0. \tag{5.8}$$

In order to derive an equation for the Reynolds-stress tensor, we form the ensemble average

$$\overline{u_i'\mathcal{N}(u_j) + u_j'\mathcal{N}(u_i)}. \tag{5.9}$$

The full form of the Reynolds-stress equation is (e.g., see Wilcox 1993)

$$\frac{\partial \tau_{ij}}{\partial t} + \bar{u}_k\frac{\partial \tau_{ij}}{\partial x_k} = -\tau_{ik}\frac{\partial \bar{u}_j}{\partial x_k} - \tau_{jk}\frac{\partial \bar{u}_i}{\partial x_k} + \varepsilon_{ij} - \prod_{ij} + \frac{\partial}{\partial x_k}\left[\nu\frac{\partial \tau_{ij}}{\partial x_k} + C_{ijk}\right] \tag{5.10}$$

where

$$\prod_{ij} = \overline{p'\left(\frac{\partial u_i'}{\partial x_j} + \frac{\partial u_j'}{\partial x_i}\right)} \tag{5.11}$$

$$\varepsilon_{ij} = 2\mu\overline{\frac{\partial u_i'}{\partial x_k}\frac{\partial u_j'}{\partial x_k}} \tag{5.12}$$

$$C_{ijk} = \varrho\overline{u_i'u_j'u_k'} + \overline{p'u_i'}\delta_{jk} + \overline{p'u_j'}\delta_{ik} \tag{5.13}$$

are the pressure-strain correlation, the dissipation rate tensor, and the turbulent diffusion correlation, respectively.

5.2.2   K-ε Transport Equations

The transport equation for the turbulent kinetic energy $K = -\tau_{ii}/2\varrho$ is obtained by contracting Eq. (5.10)

$$\varrho\frac{\partial K}{\partial t} + \varrho\bar{u}_j\frac{\partial K}{\partial x_j} = \mathcal{P} - \varrho\varepsilon + \frac{\partial}{\partial x_j}\left[\mu\frac{\partial K}{\partial x_j} - \frac{1}{2}\varrho\overline{u_i'u_i'u_j'} - \overline{p'u_j'}\right] \tag{5.14}$$

where

$$P = \tau_{ij} \frac{\partial \bar{u}_i}{\partial x_j} \tag{5.15}$$

$$\varepsilon = \nu \overline{\frac{\partial u_i'}{\partial x_j} \frac{\partial u_i'}{\partial x_j}} \tag{5.16}$$

$$\frac{1}{2} \varrho \overline{u_i' u_i' u_j'}$$

$$\overline{p' u_j'}$$

are the turbulent production, the turbulent dissipation rate, and the turbulent diffusion and pressure diffusion terms, respectively.

A conventional approximation made to represent turbulent transport of scalar quantities is based on the notion of gradient diffusion. In analogy to the molecular transport processes, we model $-\varrho \overline{u_j' \phi'} \sim \mu_t \partial \bar{\phi} / \partial x_j$. However, there is no justification for such a model regarding the pressure diffusion term; it is simply grouped with the turbulent transport term, and together they are also treated by the gradient-transport model:

$$\frac{1}{2} \varrho \overline{u_i' u_i' u_j'} + \overline{p' u_j'} = -\frac{\mu_t}{\sigma_k} \frac{\partial K}{\partial x_j}. \tag{5.17}$$

Therefore, the modeled equation for the turbulent kinetic energy becomes

$$\varrho \frac{\partial K}{\partial t} + \varrho \bar{u}_j \frac{\partial K}{\partial x_j} = P - \varrho \varepsilon + \frac{\partial}{\partial x_j} \left[\left(\mu + \frac{\mu_t}{\sigma_k} \right) \frac{\partial K}{\partial x_j} \right] \tag{5.18}$$

where σ_k is a closure coefficient. At this point, no approximation has entered, although, of course, we hope the model is realistic enough that σ_k can be chosen to be a constant.

The exact equation for the turbulent dissipation rate ε_k can be obtained by taking the following moment of the Navier-Stokes equation (Wilcox 1993):

$$2\nu \overline{\frac{\partial u_i}{\partial x_j} \frac{\partial}{\partial x_j} (\mathcal{N}(u_i))} = 0. \tag{5.19}$$

This equation takes the form

$$\varrho \frac{\partial \varepsilon}{\partial t} + \varrho \bar{u}_j \frac{\partial \varepsilon}{\partial x_j} = P_\varepsilon - \Phi_\varepsilon + D_\varepsilon + \mu \nabla^2 \varepsilon \tag{5.20}$$

where

$$P_\varepsilon = -2\mu \overline{\frac{\partial u_k'}{\partial x_i} \frac{\partial u_k'}{\partial x_j} \frac{\partial \bar{u}_i}{\partial x_j}} - 2\mu \overline{\frac{\partial u_i'}{\partial x_k} \frac{\partial u_j'}{\partial x_k} \frac{\partial \bar{u}_i}{\partial x_j}} - 2\mu \overline{u_k' \frac{\partial u_i'}{\partial x_j} \frac{\partial^2 \bar{u}_i}{\partial x_j x_k}} \tag{5.21}$$

$$\Phi_\varepsilon = 2\mu\nu \overline{\frac{\partial^2 u_i'}{\partial x_j \partial x_k} \frac{\partial^2 u_i'}{\partial x_j \partial x_k}} + 2\mu \overline{\frac{\partial u_k'}{\partial x_i} \frac{\partial u_k'}{\partial x_j} \frac{\partial u_i'}{\partial x_j}} \tag{5.22}$$

$$D_\varepsilon = -\mu \frac{\partial}{\partial x_j} \left(\overline{u_j' \frac{\partial u_i'}{\partial x_k} \frac{\partial u_i'}{\partial x_k}} \right) - 2\nu \frac{\partial}{\partial x_j} \left(\overline{\frac{\partial p'}{\partial x_i} \frac{\partial u_j'}{\partial x_i}} \right) \tag{5.23}$$

are the production, destruction, and turbulent diffusion of dissipation, respectively. These three terms involve information not available from the Reynolds-stress transport equation, and so they need to be modeled.

By assuming that the anisotropy of dissipation tensors is proportional to the anisotropy of the Reynolds-stress tensors, due to the fact that the former follows the latter as a result of the energy cascade from large to small scales, the production of dissipations can be modeled as (Speziale 1995)

$$\mathcal{P}_\varepsilon = -C_{\varepsilon 1} \frac{\varepsilon}{K} \tau_{ij} \frac{\partial \bar{u}_i}{\partial x_j}. \tag{5.24}$$

For equilibrium flows at high Reynolds numbers, the destruction of dissipations can be modeled as

$$\Phi_\varepsilon = C_{\varepsilon 2} \varrho \frac{\varepsilon^2}{K}. \tag{5.25}$$

The turbulent diffusion of dissipation is usually grouped with molecular diffusion as

$$D_\varepsilon = \frac{\partial}{\partial x_j} \left[\frac{\mu_t}{\sigma_\varepsilon} \frac{\partial \varepsilon}{\partial x_j} \right]. \tag{5.26}$$

Therefore, the modeled equation for the dissipation rate is (Mohammadi and Pironneau 1994).

$$\varrho \frac{\partial \varepsilon}{\partial t} + \varrho \bar{u}_j \frac{\partial \varepsilon}{\partial x_j} = C_{\varepsilon 1} \frac{\varepsilon}{K} \tau_{ij} \frac{\partial \bar{u}_i}{\partial x_j} - C_{\varepsilon 2} \varrho \frac{\varepsilon^2}{K} + \frac{\partial}{\partial x_j} \left[\left(\mu + \frac{\mu_t}{\sigma_\varepsilon} \right) \frac{\partial \varepsilon}{\partial x_j} \right] \tag{5.27}$$

and $C_{\varepsilon 1}$, $C_{\varepsilon 2}$, and σ_ε are constants to be determined. Typically, $C_{\varepsilon 2}$ is determined from the experiment of the decay of isotropic turbulence, and $C_{\varepsilon 1}$ is determined experimentally based on the local equilibrium shear flow. Following such a practice, Launder and Spalding (1974) suggested that the various constants used in the original K-ε models assume the following values:

$$C_\mu = 0.09 \quad C_{\varepsilon 1} = 1.44 \quad C_{\varepsilon 2} = 1.92 \quad \sigma_k = 1.0 \quad \sigma_\varepsilon = 1.3. \tag{5.28}$$

5.3   Implementation of the K-ε Model

With the above development, the two-dimensional momentum equations can now be written as

$$\frac{\partial}{\partial t}(\varrho \phi) + \frac{\partial}{\partial x}(\varrho u \phi) + \frac{\partial}{\partial y}(\varrho v \phi)$$

$$= \frac{\partial}{\partial x} \left(\mu_{\text{eff}} \frac{\partial \phi}{\partial x} \right) + \frac{\partial}{\partial y} \left(\mu_{\text{eff}} \frac{\partial \phi}{\partial y} \right) - \left(\frac{\partial p}{\partial x} \quad \text{or} \quad \frac{\partial p}{\partial y} \right) \tag{5.29}$$

where $\phi = u$ or v, with $\mu_{\text{eff}} = \mu + \mu_t$. Here μ represents the laminar viscosity, and μ_t the turbulent viscosity.

The transport equations for K and ε, after the modeling assumptions are incorporated, can be expressed in the following form in two-dimension:

$$\frac{\partial}{\partial t}(\varrho\phi) + \frac{\partial}{\partial x}(\varrho u\phi) + \frac{\partial}{\partial y}(\varrho v\phi)$$

$$= \frac{\partial}{\partial x}\left(\Gamma\frac{\partial\phi}{\partial x}\right) + \frac{\partial}{\partial y}\left(\Gamma\frac{\partial\phi}{\partial y}\right) + R_1 + R_2 \tag{5.30}$$

where

$$\Gamma = \begin{cases} \mu + \frac{\mu_t}{\sigma_k} & \text{for the } K \text{ equation} \\ \mu + \frac{\mu_t}{\sigma_\varepsilon} & \text{for the } \varepsilon \text{ equation} \end{cases} \tag{5.31}$$

$$R_1 = \begin{cases} \mu_t R & \text{for the } K \text{ equation} \\ \frac{C_{\varepsilon 1}\mu_t\varepsilon R}{K} & \text{for the } \varepsilon \text{ equation} \end{cases} \tag{5.32}$$

with

$$R = 2\left[\left(\frac{\partial u}{\partial x}\right)^2 + \left(\frac{\partial v}{\partial y}\right)^2\right] + \left[\frac{\partial u}{\partial y} + \frac{\partial v}{\partial x}\right]^2 \tag{5.33}$$

and

$$R_2 = \begin{cases} -\varrho\varepsilon \equiv -\left(\frac{C_\mu\varrho^2 K^*}{\mu_t}\right)K & \text{for the } K \text{ equation} \\ -C_{\varepsilon 2}\varrho\frac{\varepsilon^2}{K} \equiv -\left(C_{\varepsilon 2}\varrho\frac{\varepsilon^*}{K^*}\right)\varepsilon & \text{for the } \varepsilon \text{ equation.} \end{cases} \tag{5.34}$$

where K^* and ε^* are values based on the previous iteration.

The above equations are integrated over the control volume and then discretized in a manner similar to the momentum equations, described earlier. They can be transformed to body-fitted (ξ, η) coordinates, similar to the u- and v-momentum equations. The transformation is straightforward, and will not be presented here.

5.3.1   Boundary Conditions and Wall Treatment

The treatment of inflow and outflow boundaries of the computational domain is done in the usual manner. All the variables at the inlet are specified or estimated. For example, if the inflow velocities are specified, K is estimated as a fraction of the square of the inlet velocity, whereas ε is estimated from K and a suitable length scale representing the size of turbulent eddies (usually specified as a fraction of the inlet dimension). The variables at the outflow boundaries are estimated by a zero gradient condition which is based on the assumption that the flow is convection-dominated.

Near wall boundaries, the local Reynolds numbers are low and viscous effects are important; thus the K-ε model summarized above cannot be utilized since much of the modeling argument is based on the high-Reynolds-number considerations. Moreover, steep variations in flow profiles close to the wall need a large number of grid points for proper resolution. To handle these difficulties, two approaches are commonly employed, one involving low-Reynolds-number versions of the K-ε model (Jones and

Launder 1972, Patel et al. 1985) and the other involving the so-called wall functions (Launder and Spalding 1974). The former requires a relatively fine grid resolution in the wall region. The latter is easier to implement, but the assumption of local equilibrium between the production and dissipation of the turbulent kinetic energy is not always valid. In practice, however, the wall function approach has often produced results of accuracy comparable to the low-Reynolds-number approach (Viegas et al. 1985). Moreover, for complex 3-D flows, one often has very little choice but to use wall functions due to the limitation of the number of nodes that can be employed. In the present study, the wall function approach has been employed and is described next.

Close to a solid wall, a one-dimensional Couette flow analysis, parallel to the wall, is a convenient way to develop the wall function approach. Within the wall layer, the total shear stress can be considered to be a constant. The momentum equation reduces to a simple nondimensional form given by

$$\tau = (\mu + \mu_t)\frac{\partial u^t}{\partial n} \tag{5.35}$$

where u^t is the velocity tangential to the wall (at point P, at a distance l_n normal to the wall, as shown in Fig. 5.1), and n is the direction normal to the wall. The above equation can be rewritten, using the wall shear stress and friction velocity, as

$$\frac{\tau}{\tau_w} = \left(1 + \frac{\mu_t}{\mu}\right)\frac{\partial u^+}{\partial y^+}. \tag{5.36}$$

Here, τ_w is the shear stress at the wall. The local Reynolds number (or the nondimensional normal distance of point P from the wall) y^+ is determined from

$$y^+ = \frac{\varrho u_\tau l_n}{\mu} \tag{5.37}$$

and the nondimensional velocity u^+ is given by

$$u^+ = \frac{u^t}{u_\tau} \tag{5.38}$$

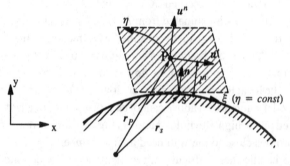

Figure 5.1   Boundary control volume adjacent to a curved wall.

where u_τ is the friction velocity at the wall:

$$u_\tau = \sqrt{\frac{\tau_w}{\varrho}}. \tag{5.39}$$

For a turbulent boundary layer over a solid flat plate, experimental evidence indicates that the wall region can be divided into two distinct zones (Hinze 1959), namely (a) the viscous sublayer ($0 < y^+ < 5$) which is dominated by molecular viscous effects and (b) the law of the wall layer ($30 < y^+ < 400$), where the shear stress is largely unchanged and is dominated by the Reynolds-stress.

A convenient engineering practice to match these two zones is to define $y^+ = 11.63$ as the point above which the flow is assumed fully turbulent and below which the flow is dominated by viscous effects. At this location the linear velocity profile of the viscous sublayer meets the logarithmic velocity profile in the law of the wall layer.

5.3.1.1 LAMINAR SUBLAYER ($y^+ < 11.63$)

In the viscous sublayer, $\mu_t/\mu \ll 1$, and the assumption $\tau = \tau_w$ leads to

$$u^+ = y^+ \Rightarrow \tau = \tau_w = \frac{\mu u'_P}{l_n}. \tag{5.40}$$

5.3.1.2 LAW OF THE WALL LAYER ($y^+ > 11.63$)

In the inertial sublayer, $\mu_t/\mu \gg 1$ and again the assumption $\tau = \tau_w$ along with

$$\mu_t = \varrho \varkappa y u_\tau \tag{5.41}$$

leads, after integrating Eq. (5.36), to

$$u^+ = \frac{1}{\varkappa} \log_e y^+ + B \equiv \frac{1}{\varkappa} \log_e(E y^+). \tag{5.42}$$

In the above, $\varkappa$ is the von Karman constant with the value 0.42. The quantity E is a function of the factors such as wall roughness, streamwise pressure gradient, etc. For a smooth impermeable wall with negligible streamwise pressure gradient, it is assigned the value 9.793. With the constant shear stress approximation in the wall layer, the transport equation for the turbulent kinetic energy (K), which is defined in Eqs. (5.30)–(5.34), reduces to the following:

$$-\overline{u'v'}\frac{du}{dy} = \varepsilon. \tag{5.43}$$

Using the definitions of Reynolds-stress given by Eq. (5.5), of ε from Eq. (5.7), and of τ from Eq. (5.35) with $\mu_t/\mu \gg 1$, we obtain the following relation:

$$\tau \equiv \tau_w = \varrho K C_\mu^{1/2}. \tag{5.44}$$

Using Eq. (5.38) and Eq. (5.39), we obtain

$$u^+ \cong \frac{u^t}{\sqrt{\tau_w/\varrho}} \Rightarrow \tau_w = \frac{u^t}{u^+}\varrho C_\mu^{1/4} K^{1/2}. \tag{5.45}$$

Then using the definitions (5.44) and (5.42), we finally obtain for the shear stress at the wall:

$$\tau_w = \frac{\varrho C_\mu^{1/4} K^{1/2} \varkappa u^t}{\ln (Ey^+)} \tag{5.46}$$

where y^+ is evaluated from

$$y^+ = \frac{l_n \sqrt{\tau_w/\varrho}}{\nu} \cong \frac{l_n \varrho C_\mu^{1/4} K^{1/2}}{\mu}. \tag{5.47}$$

Also, from Eqs. (5.44) and (5.41) and the definition of ε from Eq. (5.7), we obtain the following form for ε in the wall layer:

$$\varepsilon = \frac{C_\mu^{3/4} K^{3/2}}{\varkappa l_n}. \tag{5.48}$$

5.3.2   Implementation of the Wall Shear Stress in Momentum Equations

To illustrate the implementation of the wall shear stress for a control volume adjacent to a wall, we refer to Fig. 5.1. If the point P is in the viscous sublayer, the net viscous terms at the south control volume face are estimated in the usual manner. If the point P is in the inertial sublayer, the shear stress (parallel to the wall) is estimated from Eq. (5.46). The case of interest is where the point P is in the log layer so that the tangential velocity follows the log law normal to the wall. The resultant shear force at the south (bottom) face of the control volume shown in Fig. 5.1 is given by

$$F_{shear}^s = \tau_w A_s = \tau_s \sqrt{q_3} \tag{5.49}$$

where

$$\tau_w = \frac{\varrho_P K_P^{1/2} C_\mu^{1/4} \varkappa}{\ln E \frac{l_n K_P^{1/2}}{\nu}} u^t \tag{5.50}$$

$$u^t = u - u^n, \qquad l_n = (r_P - r_s) \cdot n. \tag{5.51}$$

In the above, l_n is the normal distance away from the wall, and u^n and u^t are the normal and tangential velocity components, respectively. The unit normal at the point s in Fig. 5.1 is $n = \nabla\eta/|\nabla\eta|$, which can be expressed as

$$(n_x, n_y) = \frac{(\eta_x, \eta_y)}{\sqrt{\eta_x^2 + \eta_y^2}}. \tag{5.52}$$

Thus, the x- and y-components of the velocity normal to the wall, $\boldsymbol{u}^n$, are given by

$$
\begin{aligned}
\boldsymbol{u}^n &= (un_x + vn_y)\boldsymbol{n} \\
&= (uy_\xi - vx_\xi)\frac{y_\xi}{q_3}\boldsymbol{i} + (-uy_\xi + vx_\xi)\frac{x_\xi}{q_3}\boldsymbol{j} \\
u_x^n &= -V\frac{y_\xi}{q_3} \qquad u_y^n = V\frac{x_\xi}{q_3}
\end{aligned}
\tag{5.53}
$$

and the x- and y-components of the velocity tangential to the wall, $\boldsymbol{u}^t$, can be written as

$$
u_x^t = \left(1 - n_x^2\right)u + n_x n_y v = u + V\frac{y_\xi}{q_3}
\tag{5.54}
$$

$$
u_y^t = \left(1 - n_y^2\right)v + n_x n_y u = v - V\frac{x_\xi}{q_3}.
\tag{5.55}
$$

Thus, the components of wall shear stress along x- and y-directions are given by

$$
(\tau_w)_x = \frac{\varrho_P K_P^{1/2} C_\mu^{1/4}\varkappa}{\ln Ey^+}u_x^t \qquad (\tau_w)_y = \frac{\varrho_P K_P^{1/2} C_\mu^{1/4}\varkappa}{\ln Ey^+}u_y^t.
\tag{5.56}
$$

The wall shear stress given by Eq. (5.50) is related to the tangential velocity as

$$
\tau_w = \mu_{\text{eff}}\frac{\partial \boldsymbol{u}^t}{\partial n}.
\tag{5.57}
$$

Since the viscous terms in the momentum equations are expressed in terms of $\partial\phi/\partial\xi$ and $\partial\phi/\partial\eta$ (where $\phi \equiv u$ or v), we need to establish a relation between the normal derivative of the tangential velocity and $\partial\phi/\partial\xi$ and $\partial\phi/\partial\eta$. Towards this end, we utilize the derivative of a quantity f, normal to a ξ (i.e., $\eta = $ const) line:

$$
\frac{\partial f}{\partial n}(\eta) = \frac{1}{J\sqrt{q_3}}\left[-q_2\frac{\partial f}{\partial\xi} + q_3\frac{\partial f}{\partial\eta}\right].
\tag{5.58}
$$

Now, taking the normal derivatives of u_x^t:

$$
\begin{aligned}
\frac{\partial u_x^t}{\partial n} &= \frac{\partial u}{\partial n} - \frac{\partial u_x^n}{\partial n} = \frac{1}{J\sqrt{q_3}}\left[-q_2\frac{\partial u}{\partial\xi} + q_3\frac{\partial u}{\partial\eta}\right] \\
&\quad - \frac{1}{J\sqrt{q_3}}\left[-q_2\frac{\partial}{\partial\xi}\left(-V\frac{y_\xi}{q_3}\right) + q_3\frac{\partial}{\partial\eta}\left(-V\frac{y_\xi}{q_3}\right)\right].
\end{aligned}
\tag{5.59}
$$

Thus,

$$
-q_2\frac{\partial u}{\partial\xi} + q_3\frac{\partial u}{\partial\eta} = J\sqrt{q_3}\frac{\partial u_x^t}{\partial n} + \left[q_2\frac{\partial}{\partial\xi}\left(V\frac{y_\xi}{q_3}\right) - q_3\frac{\partial}{\partial\eta}\left(V\frac{y_\xi}{q_3}\right)\right].
\tag{5.60}
$$

The left side of Eq. (5.60) is the net viscous term in the u-momentum equation expressed in (ξ, η) coordinates. The right side is merely a rearrangement of the net viscous stress; the first term can be interpreted as the x-component of the shear stress (parallel to the wall), and the second term as the x-component of stress normal to the wall.

Finally, the viscous terms at the south face in the u-momentum equation are computed as follows:

$$
\left[\frac{\mu_{\text{eff}}}{J}\left(-q_2\frac{\partial u}{\partial \xi}+q_3\frac{\partial u}{\partial \eta}\right)\right]_s
$$

$$
=\left[\mu_{\text{eff}}\frac{\partial u_x^t}{\partial n}\sqrt{q_3}\right]_s+\left[\frac{\mu_{\text{eff}}}{J}\left\{q_2\frac{\partial}{\partial \xi}\left(V\frac{y_\xi}{q_3}\right)-q_3\frac{\partial}{\partial \eta}\left(V\frac{y_\xi}{q_3}\right)\right\}\right]_s
$$

$$
=[(\tau_w)_x\sqrt{q_3}]_s+\left[\frac{\mu_{\text{eff}}}{J}\left\{q_2\frac{\partial}{\partial \xi}\left(V\frac{y_\xi}{q_3}\right)-q_3\frac{\partial}{\partial \eta}\left(V\frac{y_\xi}{q_3}\right)\right\}\right]_s \qquad (5.61)
$$

where $(\tau_w)_x$ is given by Eq. (5.56). Similarly, for the v-momentum equation:

$$
\left[\frac{\mu_{\text{eff}}}{J}\left(-q_2\frac{\partial v}{\partial \xi}+q_3\frac{\partial v}{\partial \eta}\right)\right]_s
$$

$$
=[(\tau_w)_y\sqrt{q_3}]_s+\left[\frac{\mu_{\text{eff}}}{J}\left\{-q_2\frac{\partial}{\partial \xi}\left(V\frac{x_\xi}{q_3}\right)+q_3\frac{\partial}{\partial \eta}\left(V\frac{y_\xi}{q_3}\right)\right\}\right]_s. \qquad (5.62)
$$

5.3.3　Implementation of the K-ε Equations Near Wall Boundaries

The turbulent kinetic energy at the point P near a wall, K_P, is obtained from the usual K-balance, but the flux expression at the wall boundary is suppressed by setting $A_S = 0$ (for the bottom boundary shown in Fig. 5.2). The evaluation of the source term for the K equation is modified by noting that the second term in R given by Eq. (5.33) can be simplified near the wall as

$$
\mu_t\left(\frac{\partial u}{\partial y}+\frac{\partial v}{\partial x}\right)^2 \cong \tau_S\left(\frac{u_P-u_S}{l_n}\right) \qquad (5.63)
$$

where $u_S = 0$ at the wall boundary, τ_S is the average of neighboring values of τ_W, and u_P is the average of the neighboring values of u. Moreover, the dissipation term in the K equation near the wall is extended to the laminar sublayer by modifying it as

$$
\varepsilon \cong \frac{c_\mu^{\frac{3}{4}}K^{\frac{3}{2}}u^+}{l_n} \qquad (5.64)
$$

where u^+ is defined by either Eq. (5.40) for the viscous sublayer or Eq. (5.42) for the inertial sublayer. Also, for lack of better knowledge, the diffusion of K to the wall is set to zero.

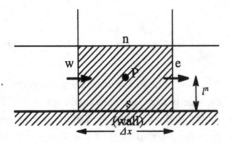

Figure 5.2　Wall treatment for K and ε.

The value of ε reaches a finite value at the walls, which is generally an unknown, unlike the value of K which falls to zero at the wall. The ε equation itself reduces to a simple form, but in the computation the value of ε near the wall boundaries is estimated from the equilibrium expression given by Eq. (5.48) and not from the usual ε balance since we have insufficient knowledge to define the coefficient A_S.

5.4 Nonequilibrium Effects

The original K-ε model detailed above is commonly employed for many different types of engineering applications. For many applications, this model yields at least competitive results with respect to other models, including both algebraic models and Reynolds-stress models. Despite this fact, the standard K-ε model has known deficiencies for certain types of flows. One notable case is for flows without equilibrium between the production and dissipation of the turbulent kinetic energy, as expressed by Eq. (5.43). In the following, we examine the performance of the standard K-ε two-equation turbulence model for certain nonequilibrium flows and assess the potential benefits of modifying the model proposed by Launder and Spalding (1974).

There are five parameters that need to be determined in the original K-ε two-equation model, namely, the turbulent Prandtl numbers for K and ε, the two coefficients regulating the magnitude of the production and dissipation in the ε equation ($C_{\varepsilon 1}$ and $C_{\varepsilon 2}$), and the coefficient regulating the magnitude of the eddy viscosity (C_μ). The coefficients $C_{\varepsilon 1}, C_{\varepsilon 2}$, and C_μ are determined largely based on the equilibrium condition, in which case the production and dissipation of K balance each other and convective/diffusive effects are negligible. However, for certain fluid flows, such as flows with rotation, adverse pressure gradient, recirculation, and large streamline curvature, flow equilibrium does not exist. When a turbulent flow is not in equilibrium, besides the length and time scales dictated by the flow configuration, dynamic characteristics create extra length and time scales according to the local flow structure. In the original model, the modeling coefficients do not respond to such structural changes because their values are fixed; variations in values of these coefficients will have to be considered when the global and local length and time scales are not comparable. An attractive approach is to extend the applicability of the original K-ε model by accounting for nonequilibrium conditions, but to make the modified model reduce back to the original model when the equilibrium condition is met.

Modifications have been proposed to the original K-ε two-equation model to account for the effect of recirculation and large mean strain rate (Lien and Leschziner 1994, Shih et al. 1995, Speziale 1995). An entirely distinct approach to modeling turbulence has been taken by Yakhot and Orszag (1986) in the context of the K-ε model, who, using Renormalization Group Theory (RNG), have been able to derive the K-ε model with much rigor. The RNG-based K-ε model is of the same form as the original K-ε model but assumes different model coefficients which are evaluated by the theory. Two of the coefficients have been modified with the corrections of Yakhot and Smith (1992). In the modified version of the RNG-based K-ε model, all the coefficients still have constant values that have been shown to be inappropriate for flows with high strain rate. A correction has been made to the ε equation (Yakhot et al. 1992). That correction, not strictly arising from RNG consideration, is designed to sensitize the ε equation to high mean strain rate so as to increase the rate of dissipation, which, in turn, can lower the level of turbulence viscosity. This correction has been found to yield improved solutions in some cases, such as the backward-facing step flow and flow in a turnaround duct (Orszag et al. 1993).

In the present section, several alternatives are investigated:

(i) Adding a time scale in the ε equation by introducing a functional form for $C_{\varepsilon 1}$, to account for nonequilibrium between production and dissipation. This modification results in what is hereafter called the nonequilibrium model. This model has been tested in a compressible recirculating flow with improved performance over the standard model (Chen and Kim 1987, Tucker and Shyy 1993). A further modification can be made by allowing a similar variation for the coefficient $C_{\varepsilon 2}$, as will be discussed later.

(ii) Employing the renormalization group (RNG) model, which can serve as a comparison to the nonequilibrium model (Orszag et al. 1993).

(iii) Using a low-Reynolds-number formulation with a modified functional form for C_μ. In the present form, a popular expression has been chosen (Launder and Sharma 1974), as shown in the equations to follow. In this modification, the key parameter is the turbulent Reynolds number, R_t, defining the characteristics of the turbulent versus laminar viscosities. It has been most frequently applied to the near wall condition; however, this modeling concept can also be extended to the nonequilibrium flow condition. This model has also been adopted and extended in turbulent flows involving phase change with substantial improvement, as reported in Shyy et al. (1992c). While other models can also be formulated in which C_μ is defined as a function of the strain rate and the streamline curvature, they are not considered in the following in order to maintain a reasonable scope.

The governing equations for the various K-ε models can be unified in the form shown below in Eqs. (5.65)–(5.72), with the constants summarized in Table 5.1.

$$\frac{\partial K}{\partial t} + \bar{u}_i \frac{\partial K}{\partial x_i} = P - \varepsilon + \frac{\partial}{\partial x_i}\left(\frac{\nu_T}{\sigma_k}\frac{\partial K}{\partial x_i}\right) \tag{5.65}$$

$$\frac{\partial \varepsilon}{\partial t} + \bar{u}_i \frac{\partial \varepsilon}{\partial x_i} = C_{\varepsilon 1}\frac{\varepsilon}{k}P - C_{\varepsilon 2}\frac{\varepsilon^2}{k} + \frac{\partial}{\partial x_i}\left(\frac{\nu_T}{\sigma_\varepsilon}\frac{\partial \varepsilon}{\partial x_i}\right) \tag{5.66}$$

$$P = 2\nu_T \bar{S}_{ij}\bar{S}_{ij} \tag{5.67}$$

$$\bar{S}_{ij} = \frac{1}{2}\left(\frac{\partial \bar{u}_i}{\partial x_j} + \frac{\partial \bar{u}_i}{\partial x_j}\right) \tag{5.68}$$

$$\nu_T = C_\mu \frac{K^2}{\varepsilon}. \tag{5.69}$$

For the low-Reynolds-number model, C_μ is no longer a constant, but instead takes

Table 5.1 *Comparison of coefficients adopted by different models*

Model	C_μ	$C_{\varepsilon 1}$	$C_{\varepsilon 2}^*$	σ_k	σ_ε
Standard	0.09	1.44	1.92	1.0	1.3
RNG	0.085	$1.42 - \frac{\eta(1-\eta/\eta_0)}{1+\beta\eta^3}$	1.68	0.7179	0.7179
Nonequilibrium	0.09	$1.15 + 0.25\frac{P}{\varepsilon}$	1.90	0.8927	1.15

$\eta = S\frac{K}{\varepsilon}$ $S = (2\bar{S}_{ij}\bar{S}_{ij})^{\frac{1}{2}}$ $\eta_0 = 4.38$ $\beta = 0.015$
*Variation in $C_{\varepsilon 2}$ will be discussed later.

the following functional form:

$$C_\mu = 0.09 f \tag{5.70}$$

where

$$f = \exp\left[\frac{-3.4}{(1 + R_t/50)^2}\right] \tag{5.71}$$

$$R_t = \frac{K^2}{\varepsilon \nu_l}. \tag{5.72}$$

In the following, we present a number of test cases to compare the relative performances of the various models described above for nonequilibrium flows.

5.5 Computational Assessment of Nonequilibrium Modifications

5.5.1 Backward-Facing Step Flow

The most frequently compared parameter in this flow problem is the length of the recirculation zone, which is around 6.5 to 7.5 times the step height for flow over a step with a height equal to one-third of the overall channel height. Detailed experimental information can be found in Kim et al. (1978). A recent work on large eddy simulation has been reported by Akselvoll and Moin (1995).

Calculations have been performed for a flow at $Re = 10^6$ with a grid consisting of 121×91 nodes. For the computations presented, a second-order upwind scheme (Shyy 1994) is used for the convection terms. Using Figs. 5.3–5.7 for reference, the following points are noted:

(i) Figure 5.3 shows the streamfunction, u- and v-velocity components, and pressure contours obtained with the standard model. A summary of the reattachment lengths predicted by the various models is given in Table 5.2. Both the RNG and nonequilibrium models improve the prediction of the reattachment point substantially; however, relatively modest differences exist between them. On the other hand, the overall flow characteristics such as

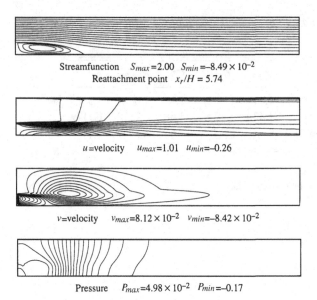

Streamfunction $S_{max}=2.00$ $S_{min}=-8.49\times10^{-2}$
Reattachment point $x_r/H = 5.74$

u=velocity $u_{max}=1.01$ $u_{min}=-0.26$

v=velocity $v_{max}=8.12\times10^{-2}$ $v_{min}=-8.42\times10^{-2}$

Pressure $P_{max}=4.98\times10^{-2}$ $P_{min}=-0.17$

Figure 5.3   Solution characteristics of standard K-ε model: $Re = 1.0 \times 10^6$, grid size $= 121 \times 91$, second-order upwind scheme.

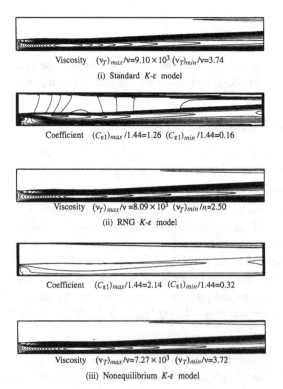

Viscosity $(\nu_T)_{max}/\nu=9.10\times10^3$ $(\nu_T)_{min}/\nu=3.74$
(i) Standard K-ε model

Coefficient $(C_{\varepsilon1})_{max}/1.44=1.26$ $(C_{\varepsilon1})_{min}/1.44=0.16$

Viscosity $(\nu_T)_{max}/\nu =8.09\times10^3$ $(\nu_T)_{min}/n=2.50$
(ii) RNG K-ε model

Coefficient $(C_{\varepsilon1})_{max}/1.44=2.14$ $(C_{\varepsilon1})_{min}/1.44=0.32$

Viscosity $(\nu_T)_{max}/\nu=7.27\times10^3$ $(\nu_T)_{min}/\nu=3.72$
(iii) Nonequilibrium K-ε model

Figure 5.4   Eddy viscosity and $C_{\varepsilon1}$ contours for the three K-ε models: $Re = 1.0 \times 10^6$, grid size $= 121 \times 91$, second-order upwind scheme.

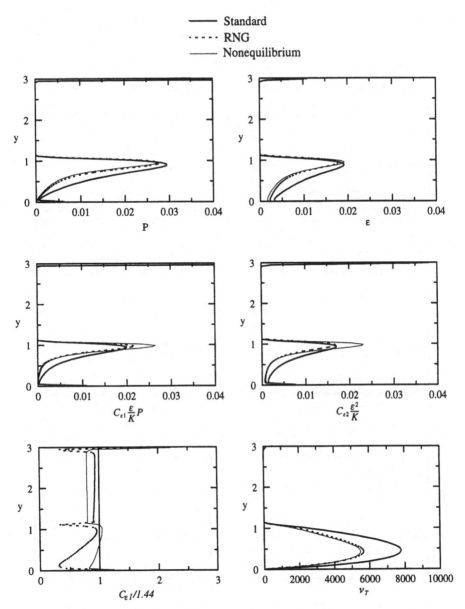

Figure 5.5 Profile comparisons of the modeling terms resulting from the different K-ε models $x/H = 3.0$.

the pressure distributions predicted by all three turbulence models are quite similar, indicating that this flow is pressure-convection dominated.

(ii) As evidenced by Fig. 5.4, substantial differences in the distribution of $C_{\varepsilon 1}$ have been predicted by the two modified models. This aspect is also clearly presented in the profile plots shown at $x_r/H = 3.0$ (Fig. 5.5) and $x_r/H = 9.0$ (Fig. 5.6). Hence, it is interesting to observe that the predicted recirculation region lengths are very close for the two models.

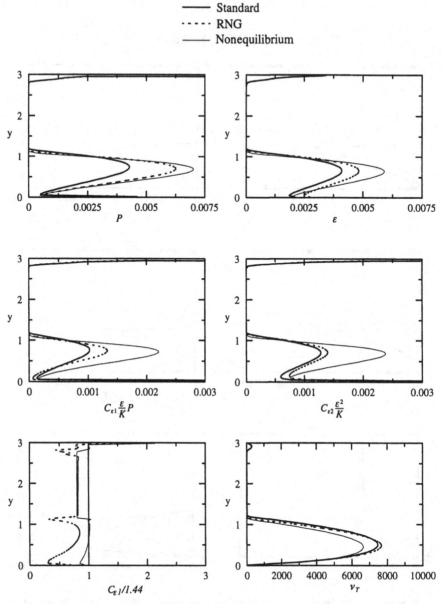

Figure 5.6   Profile comparisons of the modeling terms resulting from the different K-ε models at $x/H = 9.0$.

(iii) Figures 5.5 and 5.6 show that the production term and the dissipation term in the K equation are not equal, especially in the recirculation zone, demonstrating that the flowfield is not in equilibrium; hence, it justifies the use of modified expressions embodied in the nonequilibrium and RNG models. It is also interesting to observe that with the nonequilibrium model, and to a lesser extent with the RNG model, the source terms in both the K and ε

Table 5.2 *Reattachment lengths for backward-facing step flow*

Model	Reattachment lenght
Standard K-ε	5.74
RNG K-ε	6.78
Nonequilibrium K-ε	7.19
Low-Re nonequilibrium K-ε	7.19

equations exhibit sharper profiles, which means that the distribution of eddy viscosity is more responsive to the change in the flow structure.

(iv) Figure 5.4 shows that the standard model yields the highest eddy viscosity, while the nonequilibrium model yields the lowest. In a more detailed manner, as shown in Fig. 5.6, one observes that the RNG model does not always yield a smaller eddy viscosity compared to the standard model. However, the region where both the standard and RNG models predict comparable eddy viscosities is outside the recirculation zone, making this aspect less significant. Overall, the nonequilibrium model seems to yield consistently lower values of eddy viscosity in the whole domain.

(v) Figures 5.4–5.6 also reveal that the locations of maximum eddy viscosity show only a modest variation with respect to the x-coordinate; furthermore, they are not the same as those of K, ε, or the source/sink terms in the K-ε equations. Since μ_t is related to K^2/ε, this observation is not surprising.

(vi) Finally, the K profiles at two locations are plotted in Fig. 5.7. Combined with the ε profiles shown in Figs. 5.5 and 5.6, there is no clear trend regarding the primary quantities of the K-ε equations. The eddy viscosity trend, however, is more discernible, especially between the original and nonequilibrium models.

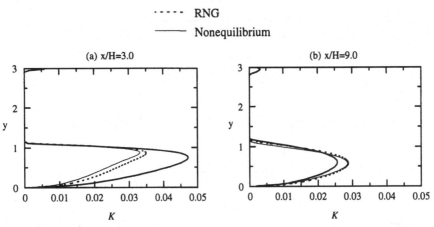

Figure 5.7 Profile comparisons of turbulent kinetic energy resulting from the different K-ε models at (a) $x/H = 3.0$ and (b) $x/H = 9.0$.

5.5.2   Hill Flow Inside a Channel

This case, as schematically shown in Fig. 5.8, is designed to investigate the characteristics of the flow over a polynomial-shaped obstacle mounted on a flat channel. Similar to the backward facing step, a recirculating region forms on the leeward side of the hill. The characteristics of this flow are well documented, and the problem was used as one of the benchmark cases for the 4th ERCOFTAC/IAHR Workshop on Refined Flow Modeling (April 3–7, 1995, Karlsruhe, Germany). The database address is *http://fluindigo.mech.surrey.ac.uk.* The inlet flow condition as well as flow profiles of mean and turbulent quantities at various locations are available in a computer data bank administered by ERCOFTAC. Detailed experimental information can be found in Almeida et al. (1993).

This flow exhibits substantial variation in the streamline curvature from both the hill and the recirculation zone behind the hill, which poses a stringent test for turbulence models. The present computations have been conducted with 151×121 grid points along the streamwise and transverse directions, respectively. The convection scheme used is based on second-order upwind differencing (Shyy et al. 1992a).

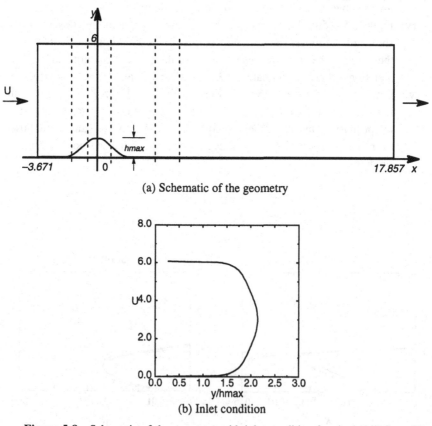

(a) Schematic of the geometry

(b) Inlet condition

Figure 5.8   Schematic of the geometry with inlet condition for single hill flow. The dashed lines indicate where comparisons are made.

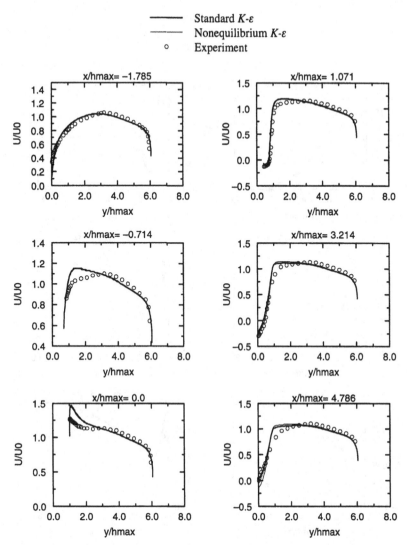

Figure 5.9 Comparisons of u-velocity among experimental, the standard K-ε model, and the nonequilibrium model results at different locations for a single hill flow; $Re = 6 \times 10^4$.

(i) Figure 5.9 compares the standard and nonequilibrium models with the experimental data for the u-velocity profiles at six locations along the hill. Substantial changes have been recorded along the streamwise location; the two models have produced very comparable agreements to measurements.

(ii) Figure 5.10 shows the v-velocity profiles at the corresponding locations. Because the v-velocity component has a smaller magnitude, the differences between the two models are more discernible. Neither model is superior in all aspects; however, the nonequilibrium model exhibits increasing maximum v-velocity magnitudes from $x/h_{\max} = 1$ to the reattachment point, located at $x/h_{\max} = 4.79$; this trend is consistent with the experimental information.

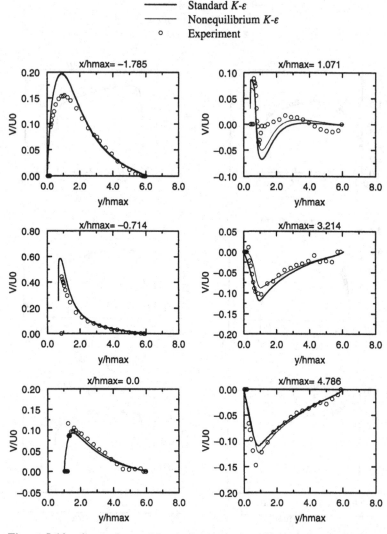

Figure 5.10   Comparisons of v-velocity among experimental, the standard K-ε model and the nonequilibrium model results at different locations for a single hill flow; $Re = 6 \times 10^4$.

(iii) Figure 5.11 shows the eddy viscosity profiles and the $C_{\varepsilon 1}$ profiles at selected locations. Consistent with the previous case, the nonequilibrium model generally shows smaller maximum eddy viscosity, and hence reduced turbulent transport. This characteristic is clearly observable in the recirculation region. Outside the recirculation region, however, the eddy viscosities predicted by the two models are in very close agreement.

(iv) Table 5.3 shows the reattachment lengths for the nonequilibrium model with the modification to $C_{\varepsilon 2}$ based on the ratio P/ε. The motivation for allowing $C_{\varepsilon 2}$ to vary according to the flow structure is provided by the observation that the nonequilibrium model substantially overpredicts the reattachment length

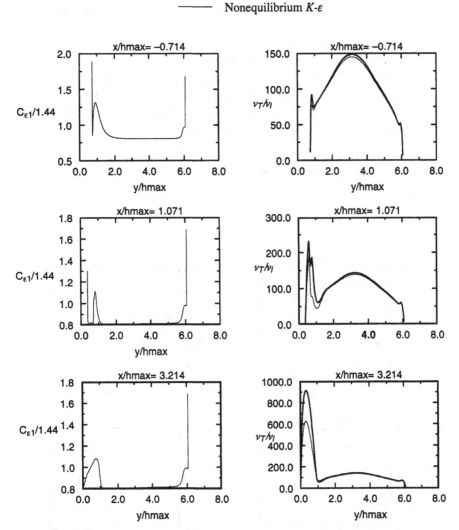

Figure 5.11 Comparisons of $C_{\varepsilon 1}$ and eddy viscosity between the standard and the nonequilibrium K-ε model results at different locations for a single hill flow; $Re = 6 \times 10^4$.

Table 5.3 *Effect of modification of $C_{\varepsilon 2}$ in nonequilibrium model on reattachment length for the hill flow*

$C_{\varepsilon 2}$	$1.35 + 0.55\left(\frac{P}{\varepsilon}\right)$	$1.45 + 0.45\left(\frac{P}{\varepsilon}\right)$	$1.55 + 0.35\left(\frac{P}{\varepsilon}\right)$	$1.65 + 0.25\left(\frac{P}{\varepsilon}\right)$	1.9	Exper.
$\frac{x_r}{h_{max}}$	4.678	4.892	5.179	5.357	5.714	4.786

for this flow. This indicates that there is too much production of ε, resulting in a very low value of ν_T. Accordingly, we have devised a modification to allow $C_{\varepsilon 2}$ to respond to the ratio P/ε in a way similar to $C_{\varepsilon 1}$. Figure 5.12 shows the u- and v-velocity profiles at three locations with two different $C_{\varepsilon 2}$. The nonequilibrium effect with these two values of $C_{\varepsilon 2}$ on the velocity profiles can be clearly observed. This exercise demonstrates that further improvement of the nonequilibrium model can be made. It appears that $C_{\varepsilon 2} = 1.45 + 0.45\left(\frac{P}{\varepsilon}\right)$ is a good choice, provided that it is larger than 1.92, the asymptotic value for isotropic turbulence.

$$C_{\varepsilon 2} = 1.90$$

$$C_{\varepsilon 2} = 1.45 + 0.45\left(\frac{P}{\varepsilon}\right)$$

○   *Experiment*

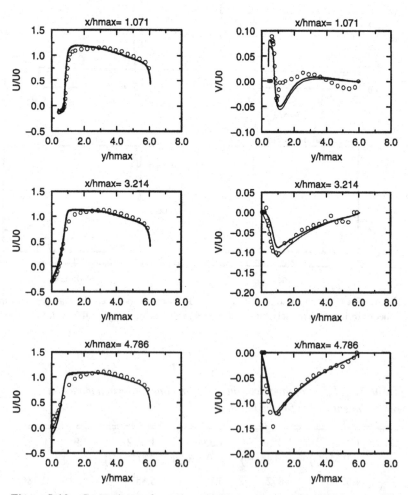

Figure 5.12   Comparisons of u- and v-velocities among experiment and nonequilibrium model with different $C_{\varepsilon 1}$ values for a single hill flow; $Re = 6 \times 10^4$.

5.5.3   3-D Diffuser Flow

This problem consists of a channel of 90-degree turn, with the cross section changing from circular at the inlet to rectangular at the outlet, and with an outlet-to-inlet area ratio of 5. The inlet flow is swirling, which causes the main flow to be prone to form recirculation regions in the core. After making the 90-degree turn, the flow may separate in the upper wall region due to the large streamline curvature associated with the turning. More details about this complicated flow problem can be found in Vu and Shyy (1990).

In the present case, we study the relative performance between the standard and the nonequilibrium models for such a diffuser, based on realistic inlet flow profiles obtained experimentally. Figure 5.13 shows the geometry and a side-view plot of the grid system which has been used. A second-order upwind convection scheme with two grid sizes, one with about 80,000 nodes and the other with about 250,000 nodes, has been used. The following points are noted about the flow solutions.

   (i) Figure 5.14 compares the distribution of the eddy viscosity in the middle side-view plane. Although both predictions seem qualitatively similar, clearly the nonequilibrium model predicts a more compact distribution than the standard model. This aspect is consistent with our observation in the previous cases. Because the nonequilibrium model is more sensitive to changes in the flow structure, it maintains lower values of eddy viscosity in the outer region. In comparison, the standard model tends to be more diffusive, in that the eddy viscosity is generally higher in much of the flow domain.

  (ii) The differences in the eddy viscosity distribution make a visible impact on the velocity field. Figure 5.15 shows that with the nonequilibrium model, convection is weaker in the central region just downstream of the inlet, which forms a blockage there and causes the flow to be stronger in the upper wall region of the elbow. The standard model, in comparison, is more uniform before the turn of the diffuser. In addition, the velocity field is less uniform in the elbow domain because it is less restricted by the core flow during the turn.

5.6   Rotational Effects

It is known from both experiment and direct numerical simulation data that for homogeneous flows, the system rotation can change the turbulence dissipation rate (Bardina et al. 1985, Shimomura 1993). It is also well known that when the axis of rotation is perpendicular to the plane of mean shear, rotation induces Coriolis force, which can have a considerable effect on the mean flow pattern as well as the turbulence structure. It was pointed out by Johnston et al. (1972) that there are two basic effects associated with boundary layers on rotating surfaces: (i) if components of the Coriolis acceleration exist parallel to the solid surface on which the layers are growing, three-dimensionality (i.e., secondary flows) will tend to develop in the mean flow field of the layers; (ii) if a component of the Coriolis acceleration is perpendicular

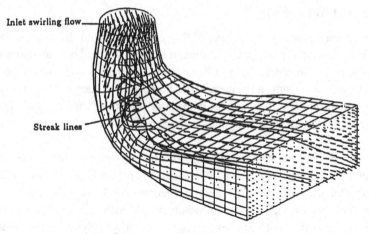

(a) a 3-D view of the flow configuration

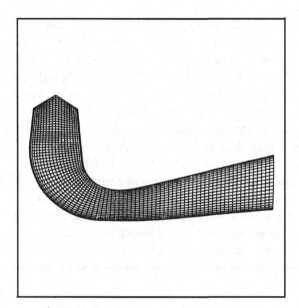

(b) Grid distribution on side–view plane

Figure 5.13   Schematic of configuration and grid layout of a draft tube.

to a solid surface, stabilizing effects are observed in the turbulence structure. Both effects are important in the flow fields of centrifugal impellers.

Three different stability-related phenomena in bounded rotating shear flows have been observed by Johnston et al. (1972): (i) rotation, by a change in the wall-layer streak burst rate, can modify the rate of turbulence production relative to dissipation and thereby modify the profiles of turbulence energy, stress, and mean velocity across the layer; (ii) rotation can increase or decrease the tendency of a hydrodynamically unstable laminar layer to undergo transition to a turbulent state, that is, rotation can promote or suppress turbulent transition; and finally (iii) for laminar boundary

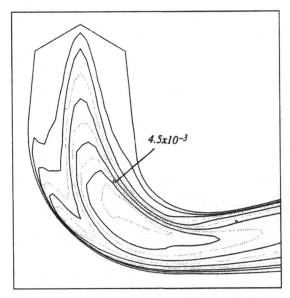

(a) Standard model

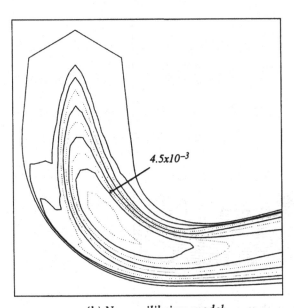

(b) Nonequilibrium model

Figure 5.14   Comparison of eddy viscosity with the standard and the nonequilibrium models.

layers and laminar channel flows, rotation may induce instability of the flow to large-scale disturbances, for example, the development of large cellular vortical models of the Taylor-Görtler type. The effect of the Coriolis force on turbulence structure of shear flows has been analyzed by Johnston et al. (1972), and the role of the gradient Richardson number has been identified. The stability-related phenomena

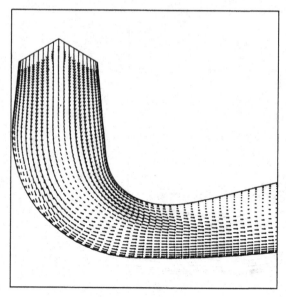

(a) Standard model

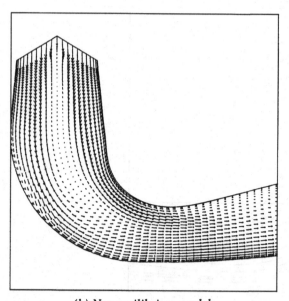

(b) Nonequilibrium model

Figure 5.15   Comparison of velocity field with the standard and the nonequilibrium models.

of rotating flows have also been analyzed by Bradshaw (1969) with the examination of other turbulent shear flows with analogous nonconservative body forces – curved-streamline flows with centrifugal forces and density stratified flows with buoyant forces normal to the plane of flow – and by Tritton (1992) with the "displaced particle analysis."

Direct and large-eddy simulations of turbulence, which require the solution of the three-dimensional time-dependent Navier-Stokes equations, may complement experimental investigation of rotating shear flows in a fruitful way, and thereby promote our understanding of the influences of rotation on the turbulence structure. The direct and large-eddy simulation of flow in a rotating channel (Kim 1983, Kristoffersen and Andersson 1993) reproduced many of the experimentally observed effects of the Coriolis forces on the mean flow and its turbulence structure. However, these flow simulations are only obtained at relatively low Reynolds numbers. For simulating high-Reynolds-number rotating turbulent flows, turbulent models with rotating effect are needed.

A number of computational studies using the Reynolds-averaged approach have been done to predict the effect of rotation on turbulence as well as mean flow quantities. The simplest turbulence model is based on the mixing length formula (Bradshaw 1969):

$$l/l_0 = 1/(1 + \beta R_i) \tag{5.73}$$

where β is a constant determined from experiments, l_0 is the mixing length at zero rotation, and R_i is the Richardson number, which is the local stability parameter; a negative value of R_i denotes an unstable region while a positive value denotes a stable region. Howard et al. (1980) and Younnis (1993) used a two-equation K-ε model with modifications based on the value of the turbulent Richardson number to incorporate the effect of rotation on turbulence structure and on the mean flow. Galmes and Lakshminarayana (1984) and Warfield and Lakshminarayana (1987) used algebraic Reynolds-stress models to account for the rotational effect and the anisotropy of the turbulence. Launder et al. (1987) used a full second moment closure to calculate fully developed rotating channel flow. In general, these methods have been successful in capturing the mean characteristics of the flow and in some cases also the turbulent quantities. However, the computational expense depends on the extent of the complexity of the turbulence models.

In the following, the rotational effect on turbulence modeling within the context of the K-ε model is investigated. First, the turbulent transport equation for Reynolds stress in the rotational frame is presented. The rotational effect analysis with the K-ε model is detailed. Second, a physical analysis of effect of rotation on turbulence structure is conducted. Based on the these analyses, a K-ε model with rotational effects in general three-dimensional coordinates is proposed. The proposed model is tested with two flow cases: two-dimensional channel flow and backward-facing step flow.

5.6.1 The Turbulent Transport Equations with Rotational Effects

If the coordinate system is attached to the rotational system, then the momentum equation is expressed as below:

$$\varrho \frac{\partial \vec{q}}{\partial t} + \varrho \vec{q} \cdot \nabla \vec{q} = -\nabla P + \mu \nabla^2 \vec{q} - 2\varrho \vec{\Omega} \times \vec{q} - \varrho \vec{\Omega} \times (\vec{\Omega} \times \vec{r}). \tag{5.74}$$

Here, $\vec{q}$ is the velocity vector, and $\vec{\Omega}$ is the angular velocity of the coordinate system. Equation (5.74) can be written in the nondimensional form:

$$\frac{\partial \vec{q}}{\partial t} + \vec{q} \cdot \nabla \vec{q} = -\nabla P + \frac{1}{Re} \nabla^2 \vec{q} - 2R_0 \vec{\Omega} \times \vec{q} - R_0^2 \vec{\Omega} \times (\vec{\Omega} \times \vec{r}). \qquad (5.75)$$

Here, for simplicity, the same symbols are used for both the dimensional and nondimensional variables. $Re = \frac{\varrho U_0 l_0}{\mu}$ is the Reynolds number, U_0 is the velocity scale, and l_0 is the length scale. $R_0 = \frac{\Omega_0 l_0}{U_0}$ is the rotational number, and Ω_0 is the angular velocity scale. Clearly, if $R_0 \sim O(1)$, then the Coriolis force and centrifugal force will have noticeable effects on flows through the momentum equation.

For turbulent flows, the system rotation affects flow fields not only through the momentum equation but also by modifying the turbulent transport process. This can be seen through the Reynolds-stress equation. If the momentum equation, Eq. (5.74), is written in the tensor form

$$\varrho \frac{\partial u_i}{\partial t} + \varrho u_k \frac{\partial u_i}{\partial x_k}$$

$$= -\frac{\partial p}{\partial x_i} + \mu \frac{\partial^2 u_i}{\partial x_k \partial x_k} - 2\varrho e_{ipk} \Omega_p u_k - \varrho e_{irs} \Omega_r (e_{spk} \Omega_p r_k), \qquad (5.76)$$

the corresponding Reynolds-stress equations can be written as

$$\frac{\partial \tau_{ij}}{\partial t} + \bar{u}_k \frac{\partial \tau_{ij}}{\partial x_k}$$

$$= -\tau_{ik} \frac{\partial \bar{u}_j}{\partial x_k} - \tau_{jk} \frac{\partial \bar{u}_i}{\partial x_k} + \varepsilon_{ij} - \prod_{ij} + \frac{\partial}{\partial x_k} \left[\nu \frac{\partial \tau_{ij}}{\partial x_k} + C_{ijk} \right] + Rot_{ij} \qquad (5.77)$$

where

$$Rot_{ij} = \underset{(i)}{-2e_{ipk} \Omega_p \overline{u'_j u'_k}} - \underset{(ii)}{2e_{jpk} \Omega_p \overline{u'_i u'_k}}. \qquad (5.78)$$

Here, Rot_{ij} is the extra term caused by Coriolis force. It can be seen that the Reynolds-stress component τ_{ij} is affected by the interaction between the angular velocity and other Reynolds-stress components, and τ_{ij} in turn affects the velocity field. Whereas nominally centrifugal force has no direct effect on τ_{ij}, as indicted by Eq. (5.77), it can affect τ_{ij} by changing the mean flow field through the momentum equations.

The turbulent kinetic energy equation can be obtained by taking the trace of Eq. (5.77). In fact, the trace of term (i) in Eq. (5.78) is

$$trace\ of\ (i) = -2\varrho[e_{123}\Omega_2\overline{u'_1 u'_3} + e_{132}\Omega_3\overline{u'_1 u'_2} + e_{231}\Omega_3\overline{u'_2 u'_1} + e_{213}\Omega_1\overline{u'_2 u'_3}$$

$$+ e_{312}\Omega_1\overline{u'_3 u'_2} + e_{321}\Omega_2\overline{u'_3 u'_1}]$$

$$= -2\varrho[\Omega_2\overline{u'_1 u'_3} - \Omega_3\overline{u'_1 u'_2} + \Omega_3\overline{u'_2 u'_1} - \Omega_1\overline{u'_2 u'_3}$$

$$+ \Omega_1\overline{u'_3 u'_2} - \Omega_2\overline{u'_3 u'_1}]$$

$$= 0. \qquad (5.79)$$

Similarly, the trace of term (ii) in Eq. (5.78) is also zero. Accordingly, the turbulent kinetic energy equation is seemingly unchanged by the system rotation:

$$\varrho \frac{\partial K}{\partial t} + \varrho \bar{u}_j \frac{\partial K}{\partial x_j} = \mathcal{P} - \varrho \varepsilon + \frac{\partial}{\partial x_j} \left[\mu \frac{\partial K}{\partial x_j} - \frac{1}{2} \varrho \overline{u'_i u'_i u'_j} + \overline{p' u'_j} \right]. \tag{5.80}$$

Although the system rotation has no direct effect on the turbulent kinetic energy equation, its effect is implicitly included in $\mathcal{P}$ and $\left(\frac{1}{2} \varrho \overline{u'_i u'_i u'_j} + \overline{p' u'_j} \right)$ terms.

With system rotation, the equation for the turbulent kinetic energy dissipation rate is

$$\varrho \frac{\partial \varepsilon}{\partial t} + \varrho \bar{u}_j \frac{\partial \varepsilon}{\partial x_j} = \mathcal{P}_\varepsilon - \Phi_\varepsilon + \mathcal{D}_\varepsilon + \mu \nabla^2 \varepsilon + Rot(\varepsilon) \tag{5.81}$$

where

$$Rot(\varepsilon) = -4\nu e_{ipk} \Omega_p \overline{\frac{\partial u'_i}{\partial x_j} \frac{\partial u'_k}{\partial x_j}} \tag{5.82}$$

is the term caused by system rotation. Expanding this term, we get

$$Rot_{123} = -4\nu \Omega_2 \overline{\frac{\partial u'_1}{\partial x_j} \frac{\partial u'_3}{\partial x_j}} \tag{5.83}$$

$$Rot_{132} = 4\nu \Omega_3 \overline{\frac{\partial u'_1}{\partial x_j} \frac{\partial u'_2}{\partial x_j}} \tag{5.84}$$

$$Rot_{231} = -4\nu \Omega_3 \overline{\frac{\partial u'_2}{\partial x_j} \frac{\partial u'_1}{\partial x_j}} \tag{5.85}$$

$$Rot_{213} = 4\nu \Omega_1 \overline{\frac{\partial u'_2}{\partial x_j} \frac{\partial u'_3}{\partial x_j}} \tag{5.86}$$

$$Rot_{312} = -4\nu \Omega_1 \overline{\frac{\partial u'_3}{\partial x_j} \frac{\partial u'_2}{\partial x_j}} \tag{5.87}$$

$$Rot_{321} = 4\nu \Omega_2 \overline{\frac{\partial u'_3}{\partial x_j} \frac{\partial u'_1}{\partial x_j}} \tag{5.88}$$

$$Rot(\varepsilon) = Rot_{123} + Rot_{132} + Rot_{231} + Rot_{213} + Rot_{312} + Rot_{321} = 0. \tag{5.89}$$

Interestingly, the system rotation has seemingly no direct effect on the turbulence dissipation rate equation, either. From the analysis above, it can be seen that the original K-ε model does not explicitly account for the rotational effect on the turbulence transport process. To simulate rotating turbulent flows with the K-ε model, modifications have to be made to account for the explicit effect of Coriolis force on the turbulent transport process. In the next section, theoretical analyses will be conducted to investigate the mechanisms of Coriolis force affecting the fluid flow stability characteristics and the turbulent transport process, and to identify the key parameters in the process.

5.6.2   Displaced Particle Analysis

The role of rotation on the stability of a shear flow is given by Tritton (1992) with the so-called "displaced particle analysis." His argument related originally to the stability of laminar flows. However, it can also be an indication of the processes occurring with turbulent flows.

The specific configuration under consideration is a unidirectional shear flow in a rotational reference frame, with the vorticity associated with the shear parallel to that of the system rotation. We choose Cartesian coordinates, so that the mean flow is in the x-direction with the speed varying in the y-direction, $u = (u(y), 0, 0)$. The system rotates with an angular velocity Ω about the z-axis. The shear vorticity is

$$\zeta = -\frac{\partial u}{\partial y} \tag{5.90}$$

and the effect of the rotation is indicated by the ratio of the background vorticity to the shear vorticity:

$$S = \frac{2\Omega}{\zeta}. \tag{5.91}$$

This can be seen in Fig. 5.16 with positive $\frac{\partial u}{\partial y}$ and positive Ω so that S is negative. Suppose that a small perturbation leads to a fluid particle being displaced a small distance δ in the y-direction. In its undisplaced position it has longitudinal velocity u_1 and when displaced it has velocity u'_1. There is a Coriolis force $2\varrho\Omega u'_1$ acting on it as shown in Fig. 5.16 (ϱ is fluid density). Similar Coriolis force $2\varrho\Omega u_2$ acts on an undisplaced particle at its new position and thus will be balanced by a pressure gradient in the y-direction. That pressure gradient also acts on the displaced particle. Hence, the net force tends to displace the particle further if $u'_1 < u_2$ and to return it towards its original position if $u'_1 > u_2$, $u'_1 \neq u_1$ because of the Coriolis force that acted while the particle was moving with velocity v in the y-direction,

$$u'_1 - u_1 = \int 2\Omega v\, dt = 2\Omega\delta. \tag{5.92}$$

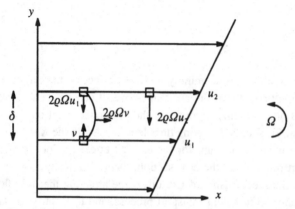

Figure 5.16   Forces on displaced and undisplaced "particles."

Hence,

$$u'_1 - u_2 = u'_1 - u_1 + u_1 - u_2$$
$$= 2\Omega\delta - \frac{\partial u}{\partial y}\delta$$
$$= -\frac{\partial u}{\partial y}(S + 1)\delta$$
$$= \zeta(S + 1)\delta. \tag{5.93}$$

The net effect of the Coriolis and pressure forces on the displaced particle is a force

$$2\varrho\Omega(u_2 - u'_1) = -\varrho\zeta S(S + 1)\delta. \tag{5.94}$$

Remembering that we are considering negative ζ (positive $\frac{\partial u}{\partial y}$) and negative S, we see that the particle tends to be displaced further if $S > -1$, and restored if $S < -1$. Corresponding considerations for positive S show that the particle always tends to be restored in this case. Thus there is a destabilizing mechanism, to which the shear and Coriolis effects are both intrinsic, occurring in the range $-1 < S < 0$.

Bradshaw (1969) defined an equivalent gradient "Richardson number" as

$$B = S(S + 1) = \frac{-2\Omega(\partial u/\partial y - \Omega)}{(\partial u/\partial y)^2}. \tag{5.95}$$

He first recognized the role of this dimensionless parameter in rotating flows by making an analogy with system rotation, streamline curvature, and buoyancy in turbulent shear flows. Although Tritton's analysis was initially related to the stability of laminar shear flows rather than to turbulence, both lines of argument led to the same main result, showing that this fluid particle stability mechanism could be an indication of the effect of Coriolis force on turbulent shear flows. In summary, the effect of rotation is destabilizing when

$$-1 < S < 0, \tag{5.96}$$

that is,

$$B < 0. \tag{5.97}$$

Maximum destabilization occurs at $S = -\frac{1}{2}$ (i.e., $B = -\frac{1}{4}$), and restabilization of the flow may be expected when $S < -1$. On the other hand, positive S is always associated with stabilized flow.

5.6.3 Simplified Reynolds-Stress Analysis

The rotational effect on turbulent flow has been analyzed with the simple boundary layer–type flow (Johnston et al. 1972). It is useful to consider the rotation effect by looking at the equations for the rates of increase of the turbulent correlations $\overline{u'^2}, \overline{v'^2}, \overline{w'^2}$, and $-\overline{u'v'}$ as we follow time-mean particle motion. For the two-dimensional boundary layer case, the Reynolds-stress equations are

$$D(-\overline{u'v'})/Dt = \overline{v'^2}\partial\bar{u}/\partial y + (\overline{u'^2} - \overline{v'^2})2\Omega + [O.T.] \tag{5.98}$$

$$D(\overline{u'^2})/Dt = -\overline{u'v'}2\partial\bar{u}/\partial y - (-\overline{u'v'})4\Omega + [O.T.] \tag{5.99}$$

$$D(\overline{v'^2})/Dt = +0 + (-\overline{u'v'})4\Omega + [O.T.] \tag{5.100}$$

$$D(\overline{w'^2})/Dt = +0 + 0 + [O.T.]. \tag{5.101}$$

The terms on the left-hand sides are the "advection" or "mean transport" terms. Just to the right of the equals signs are the production terms appearing in a nonrotating frame. Next on the right are the rotation "production" terms, which enter the Eqs. (5.98)–(5.101) for rotating flows as a result of the Coriolis accelerations. Finally, all other terms are lumped together and designated [O.T.]. These terms include rate of viscous transport and dissipation, diffusion and convection by turbulence, and the fluctuating pressure-strain interaction terms. All the [O.T.] terms are identical to their respective zero-rotation forms.

The equation for development of turbulent kinetic energy

$$K = \frac{1}{2}(\overline{u'^2} + \overline{v'^2} + \overline{w'^2}) \tag{5.102}$$

is

$$\frac{DK}{Dt} = (-\overline{u'v'})\frac{\partial\bar{u}}{\partial y} + [O.T.] = P + [O.T.]. \tag{5.103}$$

Production of turbulence energy P is not explicitly dependent on Ω but is implicitly affected by rotation through the effects Ω has on Reynolds-stress and mean shear.

For boundary layers and channel flows, peak levels of turbulence energy and stress are achieved in the regions close to the solid walls, that is, the wall layers. In the wall layers the production and dissipation terms in Eqs. (5.98)–(5.101) are very large. The excess energy or stress produced locally is diffused by the nondissipation parts of the [O.T.] terms. The advection terms are usually very small and may be neglected in the wall layers. Therefore, the production terms are the only terms that can cause a local gain in the level of energy or stress. It is reasonable to assume, as a first approximation, that effects which increase (decrease) the approximate net production rates will also lead to an increase (decrease) in levels of turbulence energy and stress.

Let us apply this idea to the rotation-boundary layer case (see Fig. 5.17) where $\partial\bar{u}/\partial y$ is positive and the usual condition exists (i.e., $-\overline{u'v'} > 0$ and $\overline{u'^2} - \overline{v'^2} > 0$). Also assume, for the moment, that the rotation rate is positive, $\Omega > 0$. It is seen from the rotation production terms in Eq. (5.98) that a positive rotation causes an increase in the net rate of production of turbulence stress. Consequently we expect that a layer with positive rotation will have a higher level of stress $-\overline{u'v'}$ than an otherwise equivalent stationary layer. If $-\overline{u'v'}$ is increased with positive rotation then the production of turbulence energy P will be higher and consequently the level of energy K will increase. The direction of rotation is arbitrary. If it is negative,

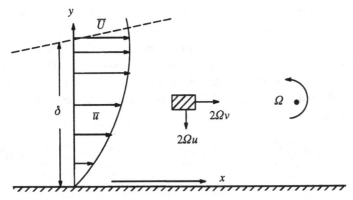

Figure 5.17 Mean velocity for a two-dimensional boundary layer on a rotating surface. Vectors on particle are Coriolis forces.

$\Omega < 0$, the same line of argument will indicate that rates of production of $-\overline{u'v'}$ and K will tend to decrease relative to conditions that would exist in an equivalent stationary flow.

In summary, the examination of the production terms in Eqs. (5.98)–(5.101) leads to the conclusion that in wall layers the sign, and the magnitude, of rotation effects might be controlled by the local dimensionless parameter

$$S = -\frac{2\Omega}{(\partial\bar{u}/\partial y)} \tag{5.104}$$

so that when $S > 0$, decrease of $-\overline{u'v'}$ and K is expected, but when $S < 0$, increase of $-\overline{u'v'}$ and K is expected.

It is noted that, in the above analysis, the effect of $[O.T.]$ terms is neglected. From the direct numerical simulation data (Kristoffersen and Andersson 1993), it can be seen that for the three-dimensional turbulent duct flow, $[O.T.]$ terms have some effects on turbulence structure, which is obvious for the correlation $\overline{w'^2}$ in the third direction. It is also clear that $-\overline{u'v'}$ and Ω affect each other. However, the result of the direct numerical simulation largely confirmed the above analysis. In the next section, the above analysis will be used to modify the K-ε equations to accommodate the system rotation effect.

5.6.4 Proposed Rotational Modifications

From the above heuristic analyses, it can be seen how rotation affects the turbulent kinetic energy and Reynolds-stress and the role of the Richardson number. A consistent approach with the K-ε model is to make the coefficients of the source terms in turbulent transport equations depend on the Richardson number. Except for the energy diffusion term, the turbulent kinetic energy equation is treated exactly as in the K-ε model. The dissipation equation seems to be the obvious place to effect adjustments to the transport of length scales. The ε equation itself is heuristically derived and therefore offers some room for refinement. On the ground of seeking the simplest possible form, Launder et al. (1977) have assumed that the effects of streamline curvature on the length scale can be accommodated by making the effective value of $C_{\varepsilon 2}$ depend on the Richardson number. Howard et al. (1980), using the Bradshaw analogy (1969) between streamline curvature and system rotation, adopted the model of Launder et al. (1977) for streamline curvature to rotating straight duct flow, making $C_{\varepsilon 2}$ depend on the rotational Richardson number. They used two forms of Richardson number:

$$(i) \quad R_i = \frac{-2\Omega\left[\left(\frac{\partial u}{\partial y} - 2\Omega\right)\right]}{\left(\frac{\partial u}{\partial y}\right)^2} \qquad (5.105)$$

or

$$(ii) \quad R_i = \frac{-2\Omega}{\frac{\partial u}{\partial y}}. \qquad (5.106)$$

Following Launder et al. (1977), the turbulent Richardson number is defined by replacing the mean flow time-scale represented by the denominator $(\partial u / \partial y)$ with a turbulent time-scale (K/ε). Thus the corresponding Richardson number is

$$(i) \quad R_{it} = -2\Omega\left(\frac{K}{\varepsilon}\right)^2\left(\frac{\partial u}{\partial y} - 2\Omega\right) \qquad (5.107)$$

or

$$(ii) \quad R_{it} = -2\Omega\left(\frac{K}{\varepsilon}\right)^2\frac{\partial u}{\partial y}. \qquad (5.108)$$

Accordingly,

$$C_{\varepsilon 2} = 1.92(1 + C_c) \qquad (5.109)$$

where

$$C_c = 0.2 R_{it}. \qquad (5.110)$$

The coefficient of 0.2 is from Launder et al. (1977). In the following discussion, Eq. (5.108) is adopted, because it appears to produce solutions in better agreement with the experimental result (Howard et al. 1980). However, the formulation above is tested only for straight channel/duct flows. It is not coordinate-invariant. In the present context, it is generalized to

$$R_{it} = -2\Omega\left(\frac{K}{\varepsilon}\right)^2\left(\frac{\partial u}{\partial y} - \frac{\partial v}{\partial x}\right). \qquad (5.111)$$

Here, $\left(\frac{\partial u}{\partial y} - \frac{\partial v}{\partial x}\right)$ is the mean vorticity. For three-dimensional flow,

$$R_{it} = -2\left(\frac{K}{\varepsilon}\right)^2\left[\Omega_x\left(\frac{\partial v}{\partial z} - \frac{\partial w}{\partial y}\right) + \Omega_y\left(\frac{\partial w}{\partial x} - \frac{\partial u}{\partial z}\right) + \Omega_z\left(\frac{\partial u}{\partial y} - \frac{\partial v}{\partial y}\right)\right] \qquad (5.112)$$

or

$$R_{it} = 2\left(\frac{K}{\varepsilon}\right)^2 e_{ipj}\Omega_p \frac{\partial u_i}{\partial x_j}. \tag{5.113}$$

The final form of modified K-ε equations are

$$\frac{\partial K}{\partial t} + \bar{u}_i \frac{\partial K}{\partial x_i} = P - \varepsilon + \frac{\partial}{\partial x_i}\left(\frac{\nu_T}{\sigma_K}\frac{\partial K}{\partial x_i}\right) \tag{5.114}$$

$$\frac{\partial \varepsilon}{\partial t} + \bar{u}_i \frac{\partial \varepsilon}{\partial x_i} = C_{\varepsilon 1}\frac{\varepsilon}{K}P - C_{\varepsilon 2}(1 + C_c)\frac{\varepsilon^2}{K} + \frac{\partial}{\partial x_i}\left(\frac{\nu_T}{\sigma_\varepsilon}\frac{\partial \varepsilon}{\partial x_i}\right). \tag{5.115}$$

5.7 Computational Assessment of Rotational Modifications

5.7.1 Rotating Channel Flow

In most practical rotating turbulent flow situations, Coriolis forces associated with rotation act directly on both the mean motion and turbulent fluctuations. Their action on the mean flow induces a secondary flow while their effect on turbulence affects the mixing process. Since secondary motions alter the turbulent stress field, and modifications to the turbulence structure affects the mean velocity profiles, the two effects become inextricably entwined. However, in one case the modification to the turbulence by Coriolis force can be looked at in isolation: that of flow between infinite rotating parallel planes, the mean Coriolis force in the direction normal to the plane being everywhere balanced by the pressure gradient. Johnston et al. (1972) conducted experiments on water flowing in a rotating channel with an aspect ratio of height to width of 7:1. This high-aspect-ratio channel reduced the effect of the secondary flow on the mainstream velocity profile in order to isolate other effects of rotation. In the test section of the channel, fully developed flow was indicated by the measurement. The velocity profiles were measured at the channel center plane, where the flow was observed to be two dimensional.

In the following, a two-dimensional rotating channel flow simulation is conducted. The configuration is schematically shown in Fig. 5.18. It is a straight, radially rotating channel with length to width ratio of 30. For this particular flow, the no-slip condition on both top and bottom walls causes the velocity gradients in the upper and lower wall regions to be of opposite signs, which, in turn, creates opposite effects in the ε equation. For the flow tested, the Reynolds number is 11,500 and the rotation number, as defined in Eq. (5.75), varies from 0 to 0.21. Uniform flow is

Figure 5.18 Schematic of the rotating duct (rotational number $R_0 = \Omega_b/U_0$). The total channel length x/b is 30.

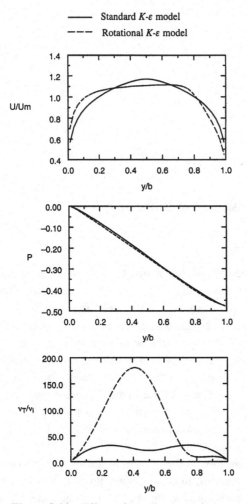

Figure 5.19   Effects of coordinate rotation on predicted profiles of u-velocity, pressure, and normalized eddy viscosity at $x/b = 26$. ($Re = 1.15 \times 10^4$, $R_0 = 0.21$.)

specified at the entrance, and zero gradient of dependent variables is applied at the outlet. A grid system of 121×41 nodes with the second-order upwind scheme was found to be adequate. The following points are noted:

(i) Figure 5.19 shows the u-velocity, pressure, and eddy viscosity profiles near the channel exit predicted with the standard K-ε model with and without the rotational modification in the ε equation. Without the rotational modification, the velocity profile is symmetric to the center line, as expected. However, with the rotational modification, the velocity profile is asymmetric due to the aforementioned sign reversal of the velocity gradient between the upper and lower wall regions. The velocity gradient is negative in the upper wall region, creating a smaller sink term in the ε equation. On the other hand, the positive velocity gradient in the lower region alters the balance among the different terms of the ε equation and the eddy viscosity in an opposite way. It should be noted that the quantitative influence of the Richardson number on ε is not straightforward to assess because the production and dissipation terms are also functions of K and ε.

(ii) The pressure profiles predicted with the two turbulence models are very close, again indicating that the effect of turbulent variables on pressure is small in this configuration.

(iii) Figure 5.20 compares the solutions obtained with the coefficients of the nonequilibrium and standard models with the rotational modification made

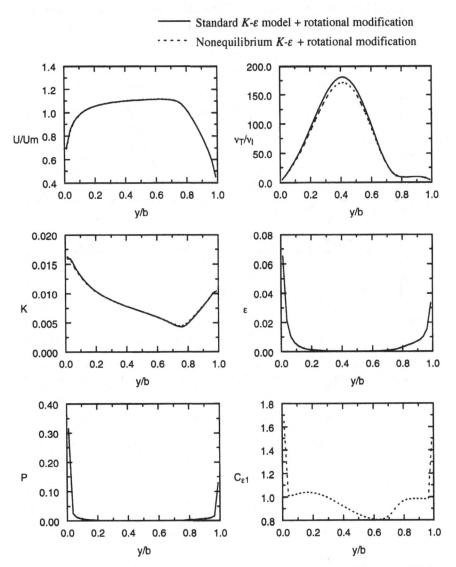

Figure 5.20 Comparisons of predictions based on the standard and the nonequilibrium model coefficients, both with rotational modification, at $x/b = 26$. ($Re = 1.15 \times 10^4$, $R_0 = 0.21$.)

to both models. The mean velocity profiles and production and dissipation terms are virtually indistinguishable between the two models, in spite of the nonuniform profile of $C_{\varepsilon 1}$ and the differences in eddy viscosity. The reason that the effect of span-wise rotation dominates over the $C_{\varepsilon 1}$ variation is that the production of K is negligible except in the wall regions. One should note that the present flow problem is a simple channel flow and hence is not expected to be sensitive to the nonequilibrium modification. Because of this, we can separate the influences of nonequilibrium between the production and consumption of K and ε and span-wise rotation.

(iv) Figure 5.21 presents the effect of different rotational numbers on the predicted flow structures with the rotational modifications made to the nonequilibrium model. Consistent with the results reported by Johnston et al. (1972),

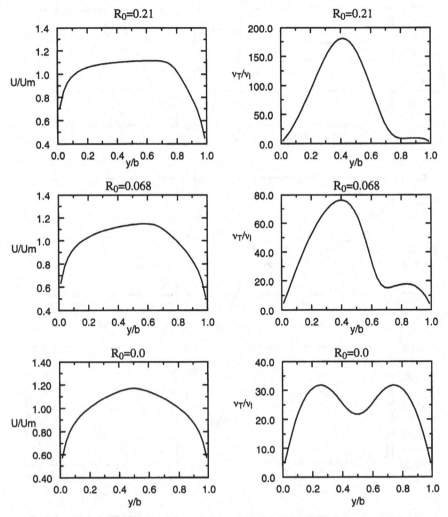

Figure 5.21 Effects of rotation number R_0 on u-velocity and normalized viscosity at $x/b = 26$. ($Re = 1.15 \times 10^4$.)

with no rotation, the velocity profile is symmetric; with the increasing rotational number, progressive departure from the symmetric profile is observed. Furthermore, the maximum eddy viscosity also increases as the rotational number becomes higher. However, the location of the peak eddy viscosity with both nonzero rotational numbers is unchanged. This location is also different from the location of the maximum u-velocity.

5.7.2 Rotating Backward-Facing Step Flow

In this section, a two-dimensional rotating backward-facing step turbulent flow is investigated. The rotation axis is perpendicular to the flow plane and is placed at the left lower corner of the flow domain. The flow field of the nonrotating backward-facing step flow consists of two distinct regions: (1) a recirculating region and (2) an attached flow region. The system rotation can modify the flow structure in both regions. In this configuration, the flow is not symmetric, and the flow direction interacts with the rotational force to have different effects on the flow field. Here, a comparison of the standard K-ε model and the rotational K-ε model with rotating recirculating flow will be conducted. The effect of rotational direction on different flow regions will be investigated.

The test case considered in the following is turbulent flow over a backward facing step with the expansion ratio E (step height to overall channel height) of 1/3 and the Reynolds number of 132,000 based on the inlet flow mean velocity and the channel height. The channel length is 24 times that of the step height. A 121×91 mesh system is used. The second-order upwind scheme is used for the convection terms.

Two rotational speeds are chosen, $\Omega = 0.21$ and $\Omega = -0.21$. Here, the counterclockwise rotation is defined as positive rotation. For each rotational speed, both the standard K-ε model and the rotational K-ε model will be tested. For each case, the streamlines, pressure distribution, velocity, vorticity, and turbulent viscosity profiles at three selected locations will be presented and compared. First of all, a nonrotating flow is computed. The reattachment length is $x/H = 5.4$. According to this result, the three locations selected for comparison are $x/H = 2.0, x/H = 6.0$, and $x/H = 20.0$. The first location is inside the recirculating zone of the nonrotating case. The second location is just behind the reattachment point, and the third location is far behind the reattachment point and is in the attached flow zone. The purpose is to assess the rotational effect on these three different regions.

5.7.2.1 COUNTERCLOCKWISE (POSITIVE) ROTATION, $\Omega = 0.21$

For comparison, the streamlines and pressure distributions for three analyses are presented, namely, the nonrotating flow, the rotating flow with standard K-ε model (SMD), and the rotating flow with rotational K-ε model (RMD). Figure 5.22 gives the streamlines for the three flows. The reattachment length for rotating flow with SMD is the same as that of the nonrotating flow, $x/H = 5.4$. As already discussed, the standard K-ε model does not respond to body force. On the other hand, for rotating flow with RMD, the reattachment length is $x/H = 3.2$, which is greatly reduced. The pressure distributions are presented in Fig. 5.23. It can be seen that, for non-

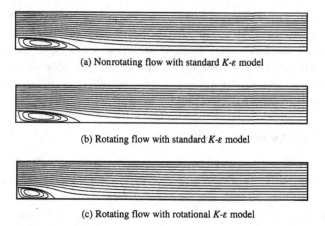

(a) Nonrotating flow with standard K-ε model

(b) Rotating flow with standard K-ε model

(c) Rotating flow with rotational K-ε model

Figure 5.22   Streamlines for nonrotating and rotating flows with different K-ε models. $\Omega = 0.21$ for rotating flow.

rotating flow and rotating flow with SMD, although there is no obvious difference in streamlines, there are substantial differences in pressure distributions. This is because in the rotating flow the Coriolis force and the centrifugal force caused by the system rotation are absorbed by the pressure field. For rotating flow with RMD, there are obvious differences in the pressure distribution compared with the rotating flow with SMD, especially in the recirculating region. This is because with RMD the rotational effect on turbulent structure changes the velocity field as well as the pressure field.

The profile comparisons for the mean u- and v-velocity components, vorticity $\omega = \partial v/\partial x - \partial u/\partial y$, and eddy viscosity at the 3 locations are shown in Figs. 5.24, 5.25, and 5.26, respectively. It is noted that, for both nonrotating flow and rotating flow with SMD, the profiles are identical. The standard K-ε model can not predict the rotational effect on turbulent structure and velocity field in this configuration, as

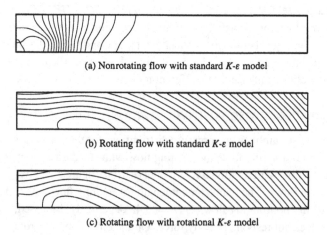

(a) Nonrotating flow with standard K-ε model

(b) Rotating flow with standard K-ε model

(c) Rotating flow with rotational K-ε model

Figure 5.23   Pressure distributions for nonrotating and rotating flows with different K-ε models. $\Omega = 0.21$ for rotating flow.

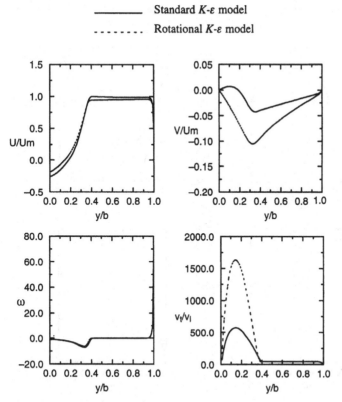

Figure 5.24   Profile comparisons of selected quantities resulting from the different K-ε models at $x/H = 2.0$. $\Omega = 0.21$.

in the case of fully attached channel flow presented in the last section. Hence, only the results of rotating flow with SMD and RMD are presented.

Figure 5.24 shows the profiles at location 1 ($x/H = 2.0$), which is in the recirculating region. From the u profile at this location, it can be seen that the flow region can be divided into two zones: (1) a recirculating zone, which is from $y/b = 0.0$ to around 0.4 and (2) an attached flow zone, which is from $y/b = 0.4$ to 1.0. In the recirculating zone, the vorticity $\omega = \frac{\partial v}{\partial x} - \frac{\partial u}{\partial y}$ is negative, and Ω is positive. $S = \frac{\Omega}{(\partial v/\partial x - \partial u/\partial y)}$ is negative. So, the turbulent stress $-\overline{u'v'}$ will be increased according to the analysis and the model discussed before. This can be observed from the profile of the eddy viscosity. In the recirculating zone, the peak value of eddy viscosity is approximately tripled with RMD compared with the result of SMD. In the attached flow zone, ω is almost zero, and there is almost no difference between the eddy viscosity with the two different models. With the rotational effect on eddy viscosity, both the u-velocity and v-velocity are changed. Near the lower wall, the u-velocity is negative. With the rotational model, the magnitude of u-velocity is reduced. The v-velocity near the lower wall is changed from positive to negative with the rotational model.

The profiles at location 2 ($x/H = 6.0$) are presented in Fig. 5.25. For non-rotating flow and rotating flow with SMD, this location is just behind the simulated

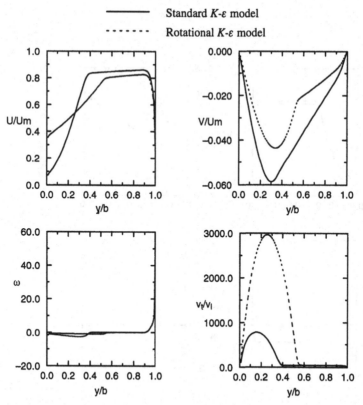

Figure 5.25   Profile comparisons of selected quantities resulting from the different K-ε models at $x/H = 6.0$. $\Omega = 0.21$.

reattachment point. At the pressure side of this location (i.e., near the lower wall), the vorticity is still negative, which is opposite to the direction of rotation. With the rotational model, the eddy viscosity is greatly enhanced compared to the result with SMD. Correspondingly, the u-velocity near the lower wall is increased with the increased eddy viscosity. The magnitude of velocity at this location is reduced.

Figure 5.26 shows the profiles at location 3 ($x/H = 20.0$). At this location, the flow is more developed and closer to the fully developed channel flow compared to the previous two locations. With SMD, the u-velocity has a large deficit near the lower wall. With RMD, this deficit is reduced because of the increased eddy viscosity. Qualitatively, at this location, there are some similarities between this flow and the rotating channel flow presented in the last section regarding the rotational effect on turbulent viscosity and the u-velocity profile.

Overall, it can be seen that, with the rotational model, the turbulent viscosity near the lower wall (pressure side) is increased, regardless of whether it is in the recirculating region or the attached flow region. The direction and the strength of the system rotation with respect to those of flow vorticity affect the turbulent flow structure. The increased eddy viscosity near the lower wall helps reduce the reattachment length.

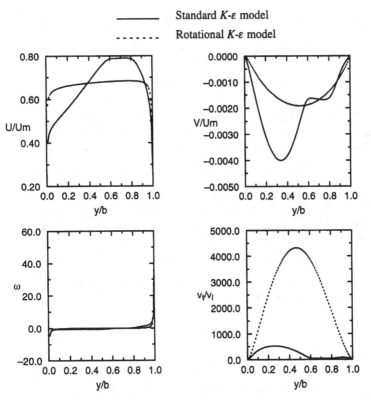

Figure 5.26 Profile comparisons of selected quantities resulting from the different K-ε models at $x/H = 20.0$. $\Omega = 0.21$.

5.7.2.2 CLOCKWISE (NEGATIVE) ROTATION, $\Omega = -0.21$

Figure 5.27 shows the streamlines of the three flows. Again, the nonrotating flow and the rotating flow with SMD have the same reattachment length. For the rotating flow with RMD, the predicted reattachment length is $x/H = 15.6$, which is greatly increased compared with that of the nonrotating flow. The pressure distributions for the three simulated flows are presented in Fig. 5.28. With SMD, the pressure distribution for the rotating flow is, again, quite different from that of the nonrotating flow, although the streamlines are very similar. The Coriolis force for the momentum equations is balanced by the pressure gradient, and the velocity field is left unchanged. For the rotating flow with RMD, the pressure field is different from that of the flow with SMD. The rotational effect on the velocity as well as the pressure field is observed.

The profiles at the three locations are presented in Figs. 5.29, 5.30, and 5.31, respectively. Again, the difference between the nonrotating flow and the rotating flow with SMD is indistinguishable. Only the latter ones are presented along with those of flow with RMD.

Figure 5.29 shows the result at location 1. At the suction side (i.e., the lower wall), the vorticity ω is negative, and the rotating speed Ω is also negative. $S = \frac{\Omega}{\omega}$ is positive. The Reynolds-stress $-\varrho\overline{u'v'}$ or eddy viscosity is expected to decrease.

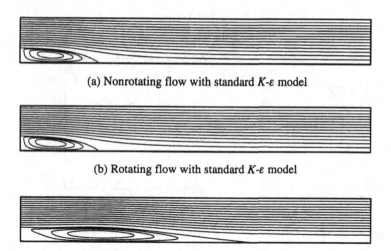

(a) Nonrotating flow with standard K-ε model

(b) Rotating flow with standard K-ε model

(c) Rotating flow with rotational K-ε model

Figure 5.27   Streamlines for nonrotating and rotating flows with the different K-ε models. $\Omega = -0.21$ for rotating flow.

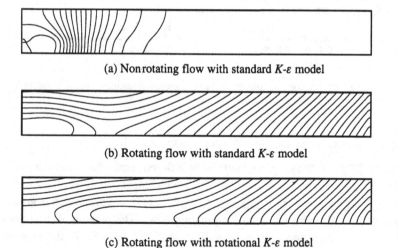

(a) Nonrotating flow with standard K-ε model

(b) Rotating flow with standard K-ε model

(c) Rotating flow with rotational K-ε model

Figure 5.28   Pressure distributions for nonrotating and rotating flows with the different K-ε models. $\Omega = -0.21$ for rotating flow.

This is observed in the eddy viscosity profile with the rotational model: The eddy viscosity is greatly reduced at the suction side. At the pressure side, the vorticity is very small and close to zero. Correspondingly, there is almost no change in eddy viscosity. The u-velocity profiles are observed to have obvious differences at the suction side between the two turbulence models. With RMD, the u-velocity near the lower wall is closer to a stagnation state. With SMD, the v-velocity is negative across the channel, except for a small region near the lower wall. With RMD, the v-velocity is positive across the channel.

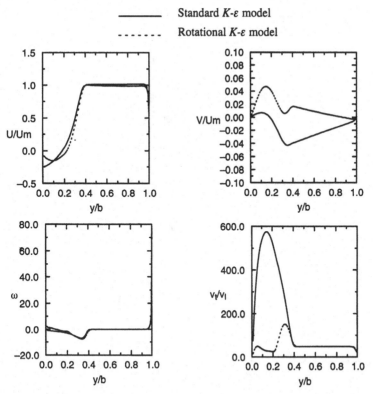

Figure 5.29 Profile comparisons of selected quantities resulting from the different K-ε models at $x/H = 2.0$. $\Omega = -0.21$.

The profiles at location 2 are displayed in Fig. 5.30. With RMD, the peak value of the eddy viscosity on the suction side is reduced by almost a factor of 4. From the u-velocity profile, it can be seen that, with RMD, the u-velocity near the lower wall is negative, which implies that this location is still in the recirculating region. With SMD, this location is outside the recirculating region, and the flow is fully attached.

Figure 5.31 presents the profiles at location 3. With RMD, the eddy viscosity is greatly reduced at the suction side, and increased at the pressure side, at this location. For the u-velocity profile, it is observed that, near the suction side, the u-velocity deficit is increased due to the reduction of the turbulent viscosity. Near the pressure side, the u-velocity deficit is reduced due to the increase of turbulent viscosity. Overall, the negative rotation decreases the turbulent viscosity on the suction side and increases it on the pressure side. On the suction side, the reduction of turbulent viscosity increases the reattachment length of the recirculating zone.

In summary, for the backward-facing step flow, the standard K-ε model can not predict the rotational effect on the velocity field. The Coriolis force is balanced by the pressure field change and does not affect the velocity field. The rotational K-ε model can predict the rotational effect on velocity fields for both attached and recirculating flows. The system rotation can increase or decrease the reattachment length depending on the rotational direction.

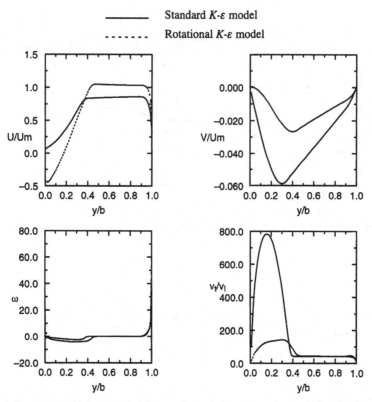

Figure 5.30   Profile comparisons of selected quantities resulting from the different K-ε models at $x/H = 6.0$. $\Omega = -0.21$.

5.8   Compressibility Effects

The effect of compressibility on the turbulence structure is an important but difficult issue in turbulence modeling. Modeling issues in both the production and dissipation of turbulent kinetic energy need to be addressed to account for Mach number effects. Several proposed treatments dealing with the dilatation dissipation and the pressure dilatation correlation are discussed in the context of the two-equation model. Additional modifications for the turbulent mass flux and the baroclinic effect, proposed by Krishnamurty and Shyy (1996), will also be discussed. This section is written in collaboration with V.S. Krishnamurty, and a more detailed account of the modifications and their impact in predicting flowfields of increased complexity can be found in Krishnamurty and Shyy (1996) and Krishnamurty (1996).

Compressibility refers to the fluctuations in the volume of the fluid cell corresponding to the fluctuations in pressure. The effect compressibility has on the turbulence structure (i.e., the various correlation coefficients and energy spectrum etc.) is of importance in the accurate modeling of compressible flows. Compressible turbulent free shear layers are characterized by a marked reduction in growth rates (in comparison to incompressible shear layers) but retain the structure of their incompressible counterparts.

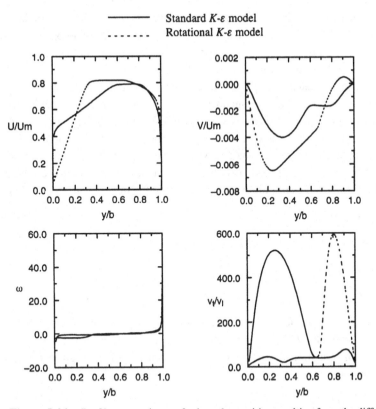

——— Standard K-ε model

- - - - - - Rotational K-ε model

Figure 5.31 Profile comparisons of selected quantities resulting from the different K-ε models at $x/H = 20.0$. $\Omega = -0.21$.

A parameter that could be used to characterize the importance of compressibility on the turbulent fluctuations is the turbulent Mach number, defined as the ratio of a characteristic speed of turbulence to the wave speed. Thus, $M_t = \frac{\sqrt{2K}}{a}$, where K is the turbulent kinetic energy and a is the speed of sound. M_t, a nondimensional parameter, is a field variable representing the propagation of information by turbulent fluctuations in comparison to the acoustic propagation. DNS results have also helped in the formulation of modifications to account for the observed effects of compressibility as a function of the turbulent Mach number. In computing flow fields involving a certain degree of complexity, the two-equation models offer a viable alternative in terms of computational ease (Speziale 1991). Here, as in the previous sections, the K-ε model is used to achieve turbulence closure.

5.8.1 Governing Equations

In the solution of turbulent flow fields, the variables are usually split into a mean and a fluctuating part. The mean can be defined in one of two ways, Reynolds average or Favre average, though both were first proposed by Reynolds (1895). In solving for compressible flows, the use of Reynolds average introduces correlations involving density fluctuations, and the modeling of these correlations is difficult. In order to overcome this, a combination of Reynolds and mass-weighted Favre averages is used.

The advantage in doing this is that the governing equations bear a close resemblance to their incompressible counterparts. Therefore incompressible methodologies can, with little modifications, be used in solving for compressible turbulent flows. Modeling of the correlations can be accomplished in a variety of ways.

The governing equations are the equations representing conservation of mass, momentum, and energy. The averaged form of the equations are given below, where density and pressure are Reynolds-averaged, while-Favre averaging is used to define the mean of the velocity components and temperature. A tilde denotes a Favre-averaged quantity, $''$ denotes fluctuations with respect to the Favre mean, an overbar denotes a Reynolds average, and $'$ denotes fluctuations with respect to the Reynolds mean. The variables are split up into their mean and fluctuating part as given below:

$$
\begin{aligned}
\varrho &= \bar{\varrho} + \varrho' \\
u_i &= \tilde{U}_i + u_i'' \\
T &= \tilde{T} + T'' \\
p &= P + p'
\end{aligned}
\tag{5.116}
$$

where ϱ, u_i, T, and p refer to the instantaneous density, velocity, temperature, and pressure, respectively. P refers to the Reynolds-averaged pressure. The exact form of the governing equations, using the mixed averaging technique, is given as follows:

Continuity

$$
\frac{\partial}{\partial t}(\bar{\varrho}) + \frac{\partial}{\partial x_j}\left(\bar{\varrho}\tilde{U}_j\right) = 0
\tag{5.117}
$$

Momentum

$$
\frac{\partial}{\partial t}(\bar{\varrho}\tilde{U}_j) + \frac{\partial}{\partial x_j}(\bar{\varrho}\tilde{U}_i\tilde{U}_j) = -\frac{\partial P}{\partial x_i} + \frac{\partial}{\partial x_j}\left[\Sigma_{ij} + \overline{\sigma_{ij}''} - \overline{\varrho u_i'' u_j''}\right]
\tag{5.118}
$$

Energy

$$
\frac{\partial}{\partial t}(\bar{\varrho}\tilde{E}) + \frac{\partial}{\partial x_j}(\bar{\varrho}\tilde{U}_j\tilde{H}) = \frac{\partial}{\partial x_j}\left[\tilde{U}_j\left(\Sigma_{ij} + \overline{\sigma_{ij}''} - \overline{\varrho u_i'' u_j''}\right) + \overline{u_i''\Sigma_{ij}} + \overline{u_i''\sigma_{ij}''}\right.
$$
$$
\left. - \tilde{q}_j - \overline{q_j''} - \overline{\varrho u_j'' h''} - \overline{\varrho u_j''\left(\frac{1}{2}u_i'' u_i''\right)}\right]
\tag{5.119}
$$

Turbulent Kinetic Energy (TKE)

$$
\frac{\partial}{\partial t}(\bar{\varrho}K) + \frac{\partial}{\partial x_j}(\bar{\varrho}\tilde{u}_i K) = -\overline{\varrho u_i'' u_j''}\frac{\partial\tilde{U}_i}{\partial x_j} - \overline{u_i''}\frac{\partial P}{\partial x_i} + \overline{u_i''}\frac{\partial\Sigma_{ij}}{\partial x_j}
$$
$$
+ \frac{\partial}{\partial x_j}\left[\overline{u_i''\sigma_{ij}''} - \overline{\varrho u_j''\left(\frac{1}{2}u_i'' u_i''\right)} - \overline{p' u_j''}\right]
$$
$$
+ \overline{p'\frac{\partial u_i''}{\partial x_i}} - \overline{\frac{\partial u_i''}{\partial x_j}\sigma_{ij}''}
\tag{5.120}
$$

where

$$\Sigma_{ij} = 2\mu S_{ij} + \lambda \tilde{U}_{k,k}\delta_{ij}; \quad S_{ij} = \frac{1}{2}(\tilde{U}_{i,j} + \tilde{U}_{j,i}) \tag{5.121}$$

$$\overline{\sigma''_{ij}} = 2\mu s_{ij} + \lambda \overline{u''_{k,k}}\delta_{ij}; \quad s_{ij} = \frac{1}{2}(\overline{u''_{i,j}} + \overline{u''_{j,i}}) \tag{5.122}$$

where μ is the coefficient of molecular viscosity, and λ is the coefficient of bulk viscosity. In the equations above, a comma is used to denote a derivative.

Following the assumption that there are regions in the flow field where molecular properties are more important, terms such as $\overline{\sigma''_{ij}}$, $\overline{q''_j}$, $\overline{u''_i \Sigma_{ij}}$, and $\overline{u''_i \frac{\partial \Sigma_{ij}}{\partial x_j}}$ in the above equations are neglected. Rewriting the modified governing equations, we have

Continuity

$$\frac{\partial}{\partial t}(\bar{\varrho}) + \frac{\partial}{\partial x_j}(\bar{\varrho}\tilde{U}_j) = 0 \tag{5.123}$$

Momentum

$$\frac{\partial}{\partial t}(\bar{\varrho}\tilde{U}_j) + \frac{\partial}{\partial x_j}(\bar{\varrho}\tilde{U}_i\tilde{U}_j) = -\frac{\partial P}{\partial x_i} + \frac{\partial}{\partial x_j}\left[\Sigma_{ij} - \underbrace{\overline{\varrho u''_i u''_j}}_{A}\right] \tag{5.124}$$

Energy

$$\frac{\partial}{\partial t}(\bar{\varrho}\tilde{E}) + \frac{\partial}{\partial x_j}(\bar{\varrho}\tilde{U}_j\tilde{H}) = \frac{\partial}{\partial x_j}\left[\tilde{U}_j(\Sigma_{ij} + \underbrace{\overline{\sigma''_{ij}}}_{B} - \underbrace{\overline{\varrho u''_i u''_j}}_{C}) + \underbrace{\overline{u''_i \sigma''_{ij}}}_{D}\right.$$

$$\left. - \tilde{q}_j - \underbrace{\overline{\varrho u''_j h''}}_{E} - \underbrace{\overline{\varrho u''_j \left(\frac{1}{2}u''_i u''_i\right)}}_{F}\right] \tag{5.125}$$

Turbulent Kinetic Energy

$$\frac{\partial}{\partial t}(\bar{\varrho}K) + \frac{\partial}{\partial x_j}(\bar{\varrho}\tilde{U}_i K) = -\underbrace{\overline{\varrho u''_i u''_j}\frac{\partial \tilde{U}_i}{\partial x_j}}_{G} - \underbrace{\overline{u''_i}\frac{\partial P}{\partial x_i}}_{H} - \underbrace{\overline{\frac{\partial u''_i}{\partial x_j}\sigma''_{ij}}}_{I}$$

$$+ \frac{\partial}{\partial x_j}\left[\underbrace{\overline{u''_i \sigma''_{ij}} - \overline{\varrho u''_j \left(\frac{1}{2}u''_i u''_i\right)} - \overline{p' u''_j}}_{J}\right]$$

$$+ \underbrace{\overline{p'\frac{\partial u''_i}{\partial x_i}}}_{K} \tag{5.126}$$

where $K = \frac{\overline{\varrho u''_i u''_i}}{\bar{\varrho}}$ represents the turbulent kinetic energy.

In the above equations, the terms that need to be modeled are indicated by an underscore and are denoted as A through K. The current modeling procedure (of these terms) based on the K-ε model is given below.

5.8.1.1   TERMS A, C, AND G

These are the Reynolds-stress terms. They are modeled as

$$-\overline{\varrho u_i'' u_j''} \simeq 2\mu_t S_{ij} + \lambda_t \frac{\partial \tilde{U}_k}{\partial x_k} \delta_{ij} - \frac{2}{3} \varrho K \delta_{ij} \tag{5.127}$$

where S_{ij} is defined in Equation (5.121). μ_t is the eddy viscosity and λ_t is the coefficient of bulk viscosity. Throughout the rest of this section we will use capitalized notation to indicate averaged quantities (except for density, where ϱ will denote the average value). Implicit in the notation used is that the pressure and density are Reynolds-averaged, while the velocity components and temperature are Favre-averaged.

5.8.1.2   TERMS D, F, AND J

These terms represent the diffusion of energy due to turbulent fluctuations and are modeled as

$$\overline{u_i'' \sigma_{ij}''} - \overline{\varrho u_j'' \left(\frac{1}{2} u_i'' u_i'' \right)} \simeq \frac{\mu_t}{\sigma_k} \frac{\partial K}{\partial x_j}. \tag{5.128}$$

The effect of the term $\overline{p' u_j''}$ on the rate of change of TKE is not explicitly accounted for and is included in the model for the diffusion terms given in Equation (5.128).

5.8.1.3   TERM E

This represents the turbulent heat flux. Using the Reynolds' analogy (Schlichting 1979), this term is modeled as

$$-\overline{\varrho u_j'' h''} = \frac{\mu_t}{\mathrm{Pr}_t} \frac{\partial T}{\partial x_j} \tag{5.129}$$

where Pr_t is the turbulent Prandtl number and is usually specified to be equal to 0.9.

5.8.1.4   TERM H

This represents a production mechanism for the turbulent kinetic energy due to the interaction of the turbulent fluctuations with the mean "energy field." Modifications for this term have been recently proposed by Krishnamurty and Shyy (1996) and will be discussed in a later subsection.

5.8.1.5   TERM I

This term represents the rate of dissipation of turbulent kinetic energy due to molecular effects and is solved for via a transport equation. That is,

$$\overline{\frac{\partial u_i''}{\partial x_j} \sigma_{ij}''} = \varrho \varepsilon. \tag{5.130}$$

We will return to this definition of ε when we discuss the dissipative effects of compressibility.

In the K-ε-based modeling, transport equations for K and ε are solved. With these, a characteristic velocity scale and a length scale can be identified resulting in the following definition for eddy viscosity as already discussed:

$$\mu_t = C_\mu \frac{\varrho K^2}{\varepsilon} \quad \text{where } C_\mu = 0.09. \tag{5.131}$$

5.8.2 Modeling of Compressibility Effects

In the equations governing the compressible turbulent flow field, terms can be identified that are of relevance and different from those of incompressible flows (see Table 5.4). From the governing equations given above, the terms that are unique to compressible, turbulent flows (and not accounted for in "incompressible" models) are: $\overline{\sigma''_{ij}}$, $\overline{p' \frac{\partial u''_i}{\partial x_i}}$, and $\overline{u''_i \frac{\partial P}{\partial x_i}}$ in addition to the dilatational effects on the rate of dissipation

Table 5.4 *Terms representing the effect of compressibility on the structure of turbulence.*

Terms	Proposed modifications	Advantages/Drawbacks
1. Dilatation Dissipation (ε_d)	(a) Sarkar et al. (1991) (b) Zeman (1990) (c) El Baz and Launder (1993)	Zeman's modification requires information regarding the Kurtosis of the fluctuations, which is not available in general. Both models do a good job in predicting the growth rates of mixing layers but fail to predict the correct variations in skin friction coefficient in the case of wall boundary layers (Wilcox 1992).
2. Pressure Dilatation ($\overline{p'd''}$)	(a) Sarkar (1992) (b) Zeman (1992) (c) El Baz and Launder (1993) (d) Krishnamurty and Shyy (1996) (e) Rubesin (1990)	Models (a) and (c) were intended for the mixing layer. When applied to boundary layers, they yield reduced levels of TKE which is an undesirable outcome (Huang et al. 1994). Model (b) improves the prediction of log-law profiles. Model (e) makes the system of equations very stiff and difficult to solve (Huang et al. 1994).
3. Turbulent Mass Flux ($\overline{u''_i}$)	(a) Rubesin (1990) (b) Ristorcelli (1993) (c) Krishnamurty and Shyy(1996)	Model (a) introduces an ad hoc assumption regarding the fluctuating enthalpy and requires prescription of a polytropic constant. Model (b) has not been tested.
4. Enthalpic Production ($\overline{u''_i} \partial P / \partial x_i$)	Krishnamurty and Shyy (1996)	Model for $\overline{u''_i}$ enables us to compute this production term, but the model still needs to be validated, in particular for a flow field where compressibility is an important issue.

of TKE. $\overline{\sigma''_{ij}}$ is purely a result of Favre averaging and at low Mach numbers does not represent compressibility effects (Lele 1994). Therefore, in order to close the system of equations, we need to suitably account for $\overline{p'\frac{\partial u''_i}{\partial x_i}}$ and $\overline{u''_i\frac{\partial P}{\partial x_i}}$. We henceforth will refer to the first term as the pressure dilatation term and the second term as the enthalpic production term. Before we discuss the modifications that have been proposed for the pressure dilatation term, let us consider the effects of compressibility on the rate of dissipation of TKE, ε.

5.8.2.1 DILATATION DISSIPATION

From a DNS analysis of compressible flows, Sarkar et al. (1991) and Zeman (1990) concluded that the effect of compressibility on the turbulence structure was a dissipative one. Compressibility introduces an extra amount of dissipation (of the turbulent fluctuations) due to the nondivergent nature of the velocity fluctuations, as can be seen by examining the definition of the rate of dissipation of TKE. Following the definition of ε given above in Equation (5.130), we get (Sarkar et al. 1991, Zeman 1990)

$$\varrho\varepsilon = \overline{\frac{\partial u''_i}{\partial x_j}\sigma''_{ij}}. \tag{5.132}$$

The fluctuating component of the viscous stresses, $\overline{\sigma''_{ij}}$, is defined as in Equation (5.122). Assuming the fluctuations in molecular viscosity are negligible, we can write

$$\varrho\varepsilon = 2\mu\overline{(s_{ij}s_{ij})} - \frac{2}{3}\mu\overline{(u''_{k,k}u''_{k,k})} \tag{5.133}$$

where the comma is used to denote a derivative. Denoting the dilatation of the velocity fluctuations as $d'' = \frac{\partial u''_k}{\partial x_k}$, we can write

$$\varrho\varepsilon = \mu\left(2\overline{s_{ij}s_{ij}} - \frac{2}{3}\overline{d''^2}\right) \tag{5.134}$$

where μ represents the coefficient of molecular viscosity. If we define the fluctuating vorticity vector as $\omega''_p = (u''_{ij} - u''_{j,i})$ we can obtain the relationship

$$\overline{s_{ij}s_{ij}} = \frac{1}{2}\overline{\omega''_p\omega''_p} + \overline{u''_{i,j}u''_{j,i}}. \tag{5.135}$$

Substituting this relationship into Eq. (5.134), we obtain

$$\varrho\varepsilon = \mu\left(\overline{\omega''_p\omega''_p} + 2\overline{u''_{i,j}u''_{j,i}} - \frac{2}{3}\overline{d''^2}\right). \tag{5.136}$$

The second term on the right side of Eq. (5.136) satisfies the relation (Sarkar et al. 1991)

$$\overline{u''_{i,j}u''_{j,i}} = \overline{(u''_i u''_j)}_{,ij} - 2\overline{(u''_{i,i}u''_j)}_{,j} + \overline{u''_{i,i}u''_{j,j}}. \tag{5.137}$$

For homogeneous turbulent flows, this relationship reduces to

$$\overline{u''_{i,j}u''_{j,i}} = \overline{u''_{i,i}u''_{j,j}} = \overline{d''^2}. \tag{5.138}$$

Combining this with Eq. (5.136), we obtain

$$\varrho\varepsilon = \mu\left(\overline{\omega''_p\omega''_p} - \frac{4}{3}\overline{d''^2}\right). \tag{5.139}$$

The dissipation rate (of TKE) in compressible turbulent flows can therefore be written as a sum of a "solenoidal" dissipation rate (the first term on the right-hand side of Eq. (5.139)) and a "dilatational dissipation" rate. Thus,

$$\varrho\varepsilon = \varrho(\varepsilon_s + \varepsilon_d) \tag{5.140}$$

where

$$\varrho\varepsilon_s = \mu(\overline{\omega''_p\omega''_p}); \qquad \varrho\varepsilon_d = \frac{4}{3}\mu\overline{d''^2}. \tag{5.141}$$

The solenoidal dissipation rate can be thought of as the dissipation due to the regular process of cascade of energy to the smaller scales, and in the absence of dilatational effects it can be considered to be equivalent to the "incompressible" dissipation rate. The dilatational dissipation (also referred to as compressible dissipation) is due to the nondivergent nature of the velocity fluctuations.

Zeman Modification   Zeman (1990), in his study of decaying compressible turbulence, observed the presence of eddy shocklets in the flow field. Assuming that these eddy shocklets directly affected the dilatational dissipation rate, ε_d, but not the solenoidal dissipation rate, he proceeded to model this dissipation rate as a function of a probability density function of the fluctuations in velocity. He assumed that the variance of these fluctuations was equal to a nondimensional parameter called the turbulent Mach number, M_t, which is defined as $M_t = \frac{\sqrt{2K}}{c}$. The turbulent Mach number is a field quantity and represents a ratio of the propagation of information by turbulence to acoustic propagation, with the turbulent kinetic energy providing a characteristic velocity scale at which turbulent fluctuations transfer information. The model proposed by Zeman (1990) for the dilatational dissipation rate is given as

$$\varepsilon_d = c_d\varepsilon_s F\{M_t, k\} \tag{5.142}$$

where c_d is an adjustable constant of order one and k is the Kurtosis of the fluctuations. The function $F\{M_t, k\}$ is given as

$$F\{M_t, k\} = \left[\frac{1}{M_t^4}\int_1^\infty \left(\frac{m_1^2 - 1}{m_1}\right)^3 p(m_1)dm_1\right] \tag{5.143}$$

where $p(m_1)$ is an assumed non-Gaussian pdf. Also, $m_1 = \frac{u}{a}$, where u is the velocity fluctuation ahead of the shocklet, and a is the sonic speed. This expression was further approximated (for computational ease) as

$$F(M_t) = 1 - \exp\{-[(M_t - 0.1)/0.6]^2\}$$

and

$$F(M_t) = 0, \quad \text{if } M_t < 0.1. \tag{5.144}$$

Sarkar et al. Modification   Sarkar et al. (1991) considered the evolution of the turbulence fluctuations on an acoustic time scale. They considered the effect of varying compressibility, based on the turbulent Mach number, M_t, on the rate of dissipation of TKE, ε. Analysis of decaying compressible turbulence indicated that the impact of varying compressibility on the solenoidal component of the dissipation rate, ε_s, was negligible in comparison with the effect on the dilatational component, ε_d. They also observed that the ratio of dilatational dissipation rate to the solenoidal dissipation rate $\left(\chi = \frac{\varepsilon_d}{\varepsilon_s} \right)$ varied directly as the square of the turbulent Mach number, $M_t = \frac{\sqrt{2K}}{a}$. Based on an analysis of the evolution of the fluctuations on an acoustic time scale, they proposed a model for the dilatational dissipation rate, which is given as

$$\varepsilon_d = \alpha_1 \varepsilon_s M_t^2 \tag{5.145}$$

where α_1 is an arbitrary constant of $\mathcal{O}(1)$.

El Baz and Launder Modification   We will refer to the modification proposed by El Baz and Launder (1993) as the El Baz and Launder modification. El Baz and Launder (1993) proposed a modification to account for this extra rate of dissipation due to compressibility effects. They chose to model one of the constants in the modeled form of the transport equation for ε_s. El Baz and Launder (1993) chose to modify the constant $C_{\varepsilon 2}$ to match the observed decay rate of compressible isotropic turbulence. Based on this observation, they modified this constant as

$$C'_{\varepsilon 2} = \frac{C_{\varepsilon 2}}{1 + 3.2 M_t^2} \quad \text{where } M_t = \frac{\sqrt{K}}{a} \tag{5.146}$$

and used $C'_{\varepsilon 2}$ instead of $C_{\varepsilon 2}$ in the modeled form of the transport equation for ε. Notice that this definition of turbulent Mach number differs from the previously used definitions.

Comparison of Dilatation Dissipation Modifications   All the modifications presented above are not very much different from one another. Recently, Blaisdell et al. (1993) compared the predictive capabilities of the Sarkar et al. (1991) model and the model proposed by Zeman (1990). From their study they concluded that the Sarkar et al. (1991) modification was marginally superior to that proposed by Zeman (1990). Furthermore, they also raised the question regarding the applicability of both modifications in computing decaying compressible turbulence, because the decay rate, they observed, was very much dependent on the initial conditions.

To compare the modifications due to Sarkar et al. (1991) and El Baz and Launder (1993), we conducted numerical simulations of decaying compressible turbulence at three values of initial turbulent Mach number, $M_{t,o}$. In the case of decaying, isotropic compressible turbulence, the equations reduce to, for the Sarkar modification (Sarkar et al. 1991):

$$\frac{dK}{dt} = -\varepsilon_s \left(1.0 + \alpha_1 M_t^2\right) \tag{5.147}$$

$$\frac{d\varepsilon_s}{dt} = -C_{\varepsilon 2} \frac{\varepsilon_s^2}{K} \tag{5.148}$$

$$\frac{d\left(M_t^2\right)}{dt} = -\frac{\varepsilon_s^2}{K} M_t^2 \left(1 + \alpha_1 M_t^2\right) \left[1 + 0.5\gamma(\gamma - 1)M_t^2\right]; \tag{5.149}$$

and for the El Baz and Launder modification (El Baz and Launder 1993):

$$\frac{dK}{dt} = -\varepsilon_s \tag{5.150}$$

$$\frac{d\varepsilon_s}{dt} = -\left(\frac{C_{\varepsilon 2}}{1 + 3.2M_t^2}\right) \frac{\varepsilon_s^2}{K} \tag{5.151}$$

$$\frac{d\left(M_t^2\right)}{dt} = -\frac{\varepsilon_s^2}{K} M_t^4 \left[1 + 0.5\gamma(\gamma - 1)M_t^2\right]. \tag{5.152}$$

The above equations describe the evolution of the turbulence field. These equations were solved using a second-order Runge–Kutta scheme for various values of the initial Mach number, $M_{t,0}$. The model coefficient $C_{\varepsilon 2}$ was chosen to be 1.83 in order to reproduce the observed decay rate in the case of high-Reynolds-number incompressible turbulence.

The decay of TKE is shown in Fig. 5.32. In the figure, K_o and $(\varepsilon_s)_o$ refer to the initial values of TKE and the solenoidal dissipation rate, respectively. As Sarkar et al. (1991) point out, these computed values should not be compared with DNS simulations but just used to evaluate the decay rate of the TKE predicted by the two different modifications. For the lower-Mach-number cases, the two modifications predict the same initial decay rate. But at a later time, the El Baz and Launder (1993) modification predicts a slightly greater reduction in the TKE. There are more noticeable differences between the modifications for the initial turbulent Mach number of 0.5, which in free shear layers corresponds to a free stream Mach number of 10.

5.8.2.2 PRESSURE DILATATION

The pressure dilatation $\overline{p'd''}$, where $d'' = \frac{\partial u_k''}{\partial x_k}$, is one of the terms that appears explicitly in the governing equations, in the case of compressible turbulent flows, due to the nondivergent, fluctuating velocity field. The pressure dilatation refers to the work done due to simultaneous fluctuations in the volume of the fluid cell corresponding to the fluctuations in pressure. It can be either positive or negative, when negative representing an extra dissipation.

Modifications have been proposed by Sarkar (1993), Zeman (1992), and El Baz and Launder (1993) to model the effect of this dilatational term.

Sarkar Modification   Sarkar (1992) conducted an analysis of the evolution of the pressure dilatation correlation in both decaying compressible turbulence and homo-

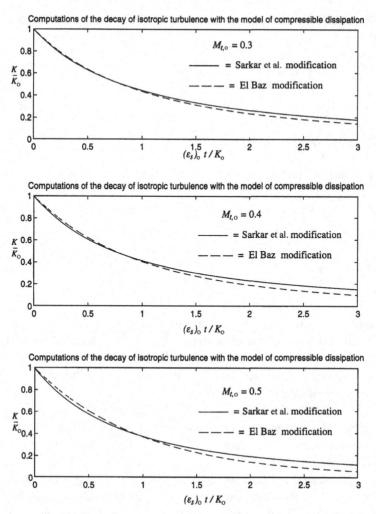

Figure 5.32 Computed decay of isotropic compressible turbulence using two different modifications for dilatational dissipation.

geneous shear turbulence. Writing a Poisson equation for the pressure fluctuations, the evolution of the incompressible part and the compressible part of the pressure fluctuations was considered. Writing the Poisson equation for the pressure fluctuations, two groups of terms can be identified. The split-up is based on the assumption that the mean density variations contribute to the evolution of the incompressible part of pressure fluctuations and the fluctuations in density and its gradients contribute to the evolution of the compressible part. Denoting the incompressible part of the pressure fluctuations as p'_I and the compressible part of the pressure fluctuations as p'_c, they considered the evolution of $\overline{p'_I d''}$ and $\overline{p'_c d''}$. These studies seem to indicate that the $\overline{p'_I d''}$ component of the pressure dilatation term was affected to a greater extent by compressibility than the $\overline{p'_c d''}$ (for both decaying and homogeneous shear turbulence). From an analysis of the $\overline{p'_I d''}$ (based on a decomposition of the pressure fluctuations into a rapid part and a slowly evolving part), they present modifications for the pressure dilatation term. The analysis did not take into account the contribution from $\overline{p'_c d''}$ and was restricted by the assumption of homogeneity in the fluctuations, that is, the spatial gradient of fluctuation correlations is equivalently zero. Borrowing ideas from the modeling of the pressure–strain correlations in incompressible turbulent flows, the model for the pressure dilatation term is derived:

$$\overline{p'd''} = -\alpha_3 P_k M_t^2 + \alpha_4 \varrho \varepsilon_s M_t^2 \tag{5.153}$$

where $\alpha_3 = 0.4$, $\alpha_4 = 0.2$, and $M_t = \frac{\sqrt{2K}}{a}$. In addition, α_1 in Eq. (5.145) is set equal to 0.5. The constants are obtained from a curve fit of the model with DNS simulations.

Zeman Modification   The model due to Zeman (1992) is based on the balance of the transport equation for the pressure fluctuation variance. That is,

$$\frac{1}{2} \frac{D \overline{p'^2}}{Dt} = -\varrho a^2 \overline{p' \frac{\partial u_i''}{\partial x_i}} - \gamma \overline{p'^2} \frac{\partial U_i}{\partial x_i} - a^2 \overline{p' u_j''} \frac{\partial \varrho}{\partial x_j}$$
$$+ \text{ higher-order terms.} \tag{5.154}$$

Making an assumption that the variance in pressure fluctuations was small, Zeman (1992) neglected the second term on the right-hand side. Assuming that the temporal variation of the pressure variance was negligible, he obtains a functional form for the pressure dilatation term:

$$\overline{p' \frac{\partial u_i''}{\partial x_i}} = F \left\{ -\frac{1}{\varrho} \overline{p' u_j''} \frac{\partial \varrho}{\partial x_j} \right\}. \tag{5.155}$$

Using the equation of state, the pressure velocity correlation can be expressed as (Lele 1994):

$$\overline{p' u_j''} = \frac{P}{\varrho} \overline{\varrho' u_j''} + \frac{P}{T} \overline{T'' u_j''}. \tag{5.156}$$

With the model for the turbulent mass flux (first term on the right side of Eq. (5.156)) and the model for the turbulent heat flux (second term on the right side of Eq. (5.156)), he derives a model for the pressure dilatation term. The steps in this derivation are:

1. From Sarkar and Lakshmanan (1991),

$$\overline{\varrho' u_j''} = -\frac{\mu_t}{\varrho \sigma_\varrho} \frac{\partial \varrho}{\partial x_j} \tag{5.157}$$

   where $\sigma_\varrho = 0.7$.
2. From the model for correlation between fluctuations in temperature and velocity,

$$\overline{T'' u_j''} = -\frac{\mu_t}{\Pr_t} \frac{1}{\varrho} \frac{\partial T}{\partial x_j} \tag{5.158}$$

   where $\Pr_t = 0.9$.

3. Substituting these two models into Eq. (5.156) and curve-fitting the model to DNS results (and invoking boundary layer assumptions), they obtain the final form of the pressure dilatation term as

$$\overline{p'd''} = g(M_t)\left(\frac{\partial \bar{\varrho}}{\partial y}\right)^2 \frac{K}{\varepsilon}\frac{a^2}{\varrho}\overline{v'^2} \tag{5.159}$$

where

$$g = 0.2\left[1 - \exp\left(\frac{-M_t^2}{0.02}\right)\right] \tag{5.160}$$

where $M_t = \frac{\sqrt{2K}}{a}$.

El Baz and Launder Modification   This modification is largely based on the modeling of the pressure–strain correlation term in incompressible flows. Identifying (in a manner similar to the analysis of Sarkar (1992)) a rapid and a slowly evolving part of the pressure fluctuations, El Baz and Launder (1993) proposed to model only the rapid evolution part (that is, contribution to the evolution of the pressure fluctuations comes only from the mean strain). El Baz and Launder (1993) use a method similar to that prescribed by Launder et al. (1975) for modeling this term (for incompressible flows). The variation for compressibility comes via the introduction of a constant which is considered to be an intrinsic function of compressibility which vanishes in the incompressible limit. From a contraction of the model for the rapid part of the pressure–strain correlation they obtain a model for the pressure dilatation term as

$$\overline{p'\frac{\partial u_i''}{\partial x_i}} = F\left[\frac{8}{3}\varrho K\frac{\partial U_k}{\partial x_k} - P_k\right] \tag{5.161}$$

where F is the constant (which is considered to be an intrinsic function of compressibility). This constant F is assumed to be a function of the turbulent Mach number, that is,

$$F = \beta M_t^2 \quad \text{where } M_t = \frac{\sqrt{K}}{a} \tag{5.162}$$

where β is an arbitrary constant whose value is prescribed to be 1.5. Again, this definition of turbulent Mach number differs from the previously used definitions.

5.8.2.3   ADDITIONAL MODIFICATIONS BY KRISHNAMURTY AND SHYY

The methodology predominantly used in computing compressible flow fields is to use Favre averages for velocity components and temperature, and Reynolds averages for pressure and density. The stress tensor and the heat flux vector are computed using Reynolds averages. The implicit assumption here is that the turbulence is homogeneous and therefore the turbulent mass flux and the fluctuating stress tensor are negligible. But, as shown by Ristorcelli (1993), this could be erroneous in the case of high supersonic and hypersonic flow fields. Particularly, Ristorcelli (1993) refers to a DNS study of hypersonic boundary layers to show that the turbulent mass flux terms are quite substantial in comparison with the mean velocity components. Therefore, to accurately model the exact form of the governing equations (except for the dissipation rate transport equation), account must be made of the turbulent mass flux term. Two new modifications have been proposed by Krishnamurty and Shyy (1996) to account for the enthalpic production term and a term to account for the effect of the baroclinic torque on the turbulent fluctuations, via a model for the turbulent mass flux. A brief description of these modifications is given below.

Estimation of Turbulent Mass Flux The total enthalpy H in its instantaneous form can be written as

$$H = h + \frac{1}{2}u_i u_i \tag{5.163}$$

where h is the static enthalpy and u_i are the components of the velocity vector. Both h and u_i are in their instantaneous form. Using Favre averages, Eq. (5.163) can be written in terms of a mean and fluctuating quantities. The equation then is written as

$$\tilde{H} + H'' = \tilde{h} + h'' + \frac{1}{2}(U_i + u_i'')(U_i + u_i''). \tag{5.164}$$

Expanding the third term on the right-hand side and rewriting the above equation, we get

$$\tilde{H} + H'' = \left[\tilde{h} + \frac{1}{2}U_i U_i + \frac{1}{2}u_i'' u_i''\right] + [h'' + u_i'' U_i]. \tag{5.165}$$

In most modeling procedures, the average value of stagnation enthalpy, $\tilde{H}$, is associated with the first group of terms on the right side of Eq. (5.165). Fluctuations in stagnation enthalpy are associated with the second group of terms on the right side of Eq. (5.165). Making an assumption of constant stagnation enthalpy, an expression can be derived to relate the fluctuations in enthalpy to the turbulence intensity. From their experimental measurements of the flow past an axisymmetric afterbody at an inflow Mach number of 2.3, Gaviglio et al. (1977) observe that the assumption of constant total enthalpy was indeed valid. Therefore, assuming constant enthalpy, the second term on the right side of Eq. (5.165) can be equated to zero. Thus,

$$h'' = -u_i'' U_i . \tag{5.166}$$

Assuming that the fluctuations are isobaric, a relationship between the fluctuations in density and temperature can be expressed as

$$\frac{\varrho'}{\varrho} \cong -\frac{T''}{T} \tag{5.167}$$

$$\frac{\varrho'}{\varrho} \cong -\frac{T''}{T} = \frac{h''}{\tilde{h}} . \tag{5.168}$$

The above equation coupled with Eq. (5.166) gives a relationship between the fluctuations in density, temperature, and velocity:

$$\frac{\varrho'}{\bar{\varrho}} = -\frac{T''}{\tilde{T}} = \frac{(\gamma - 1)}{a^2} U_i u_i'' . \tag{5.169}$$

From definition,

$$\overline{u_i''} = -\frac{\overline{\varrho' u_i''}}{\bar{\varrho}} . \tag{5.170}$$

Multiplying Eq. (5.169) throughout by u_j'' and averaging we get the functional relationship

$$\overline{u_j''} = C_1 \left\{ \left(\frac{\mu_t C_p}{\mathrm{Pr}}\right) \left(\frac{\gamma}{\bar{\varrho} a^2}\right) \frac{\partial \tilde{T}}{\partial x_j} + \left[\frac{(\gamma - 1)}{\bar{\varrho} a^2} \tilde{U}_i \overline{\varrho u_i'' u_j''}\right] \right\} \tag{5.171}$$

where C_p is the specific heat at constant pressure and C_1 is an arbitrary constant. Comparing this expression with that of Ristorcelli (1993), we get

$$C_1 = \frac{2M_t}{1 - M_t} \quad \text{where } M_t = \frac{\sqrt{2K}}{a}. \tag{5.172}$$

It is to be noted that Eq. (5.171) expresses the functional dependence of the turbulent mass flux on the thermal gradients and the intensity of the velocity fluctuations. The assumption was used as a starting point for modeling the turbulent mass flux. It does not imply that the fluctuations are isobaric, and if we make the assumption of isobaricity of the fluctuations, then the terms inside the parenthesis in Eq. (5.171) cancel each other out.

Estimation of pressure dilatation   The estimate for the pressure dilatation term follows closely the derivation for the corresponding term by Rubesin (1990). From the continuity equation we get the transport equation for the fluctuating density,

$$\frac{\partial \varrho'}{\partial t} + \frac{\partial}{\partial x_k} [\varrho' \tilde{U}_k + \varrho u_k''] = 0. \tag{5.173}$$

Multiplying by ϱ', rearranging the equations, and neglecting the triple correlation term, we get

$$\frac{\partial}{\partial t}(\varrho'^2) + \tilde{U}_k \frac{\partial}{\partial x_k}(\varrho'^2) + 2\varrho'^2 \frac{\partial \tilde{U}_k}{\partial x_k} + 2\varrho' u_k'' \frac{\partial \bar{\varrho}}{\partial x_k} + 2\varrho' \bar{\varrho} \frac{\partial u_k''}{\partial x_k} = 0. \tag{5.174}$$

Averaging and rearranging we get

$$\overline{\varrho' \frac{\partial u_k''}{\partial x_k}} = -\frac{1}{2\bar{\varrho}} \left[\frac{\partial}{\partial t}(\overline{\varrho'^2}) + \tilde{U}_k \frac{\partial}{\partial x_k}(\overline{\varrho'^2})\right] - \left(\frac{\overline{\varrho'^2}}{\bar{\varrho}^2}\right) \frac{\partial \tilde{U}_k}{\partial x_k} - \overline{u_k''} \frac{\partial \bar{\varrho}}{\partial x_k}. \tag{5.175}$$

Define a new variable β,

$$\beta = \frac{(\overline{\varrho'^2})^{\frac{1}{2}}}{\bar{\varrho}}; \quad \text{or} \quad \frac{\overline{\varrho'^2}}{\bar{\varrho}^2} = \beta^2. \tag{5.176}$$

Making the assumption that the intensity of pressure fluctuations is directly proportional to the intensity of density fluctuations, we write

$$\frac{p'}{P} \simeq \frac{\varrho'}{\bar{\varrho}}. \tag{5.177}$$

Therefore by making a change of variables to β^2, we can derive a relationship for the pressure dilatation correlation as

$$\overline{p'\frac{\partial u_k''}{\partial x_k}} = P\left\{\frac{1}{\bar{\varrho}}\frac{\partial \bar{\varrho}}{\partial x_k}\overline{u_k''} - \frac{1}{2}\left[\frac{\partial}{\partial t}(\beta^2) + \tilde{U}_k\frac{\partial}{\partial x_k}(\beta^2)\right]\right\}. \tag{5.178}$$

To estimate β^2 we use the empirical observations of Gaviglio et al. (1977) and obtain

$$\beta^2 = \frac{\overline{\varrho'^2}}{\bar{\varrho}^2} = [(\gamma-1)M^2]^2\left[\frac{\overline{u_i''u_i''}}{\tilde{U}_i\tilde{U}_i}\right]. \tag{5.179}$$

It should be noted that this derivation for the pressure dilatation correlation is strictly applicable for cases where the fluctuations cannot be considered to be isobaric.

Effect of Baroclinic Torque The modeling of the transport equation for ε_s usually follows the incompressible form and ignores the effect of the baroclinic torque. ε_s is usually defined as the correlation between the vorticity fluctuations, $\varepsilon_s = \nu\overline{\omega_i''\omega_i''}$, where ν is the kinematic viscosity. The assumption made in the proposal for the algebraic modification for the dilatational dissipation is that the solenoidal dissipation rate is relatively unaffected in the case of compressible flows. DNS studies (Sarkar et al. 1991, Zeman 1990) of decaying isotropic turbulence and homogeneous free shear layers, at low Reynolds numbers, seem to warrant this assumption. But the validity of such an assumption in the case of high-Reynolds-number flows and flow fields of increased complexity is open to question. The solenoidal part of the rate of dissipation of TKE cannot be assumed to be independent of compressibility effects. A look at the exact form of the governing equation for ε_s will reveal this.

The exact form of the governing equation for the solenoidal dissipation rate, ε_s, is written as:

$$\frac{\partial}{\partial t}(\varrho\varepsilon) + \frac{\partial}{\partial x_k}(\varrho U_k\varepsilon) = P_\varepsilon + D_\varepsilon + \Phi_\varepsilon - \nu\nabla^2\varepsilon + B_\varepsilon \tag{5.180}$$

where

$$P_\varepsilon = -2\nu\overline{\omega_p''u_k''}\frac{\partial\Omega_p}{\partial x_k} + 2\nu\overline{\omega_p''\omega_k''}\frac{\partial U_p}{\partial x_k} + 2\nu\Omega_k\left(\overline{\omega_p''\frac{\partial u_p''}{\partial x_k}}\right) \tag{5.181}$$

$$D_\varepsilon = -\overline{u_k''\frac{\partial}{\partial x_k}(\nu\omega_p''\omega_p'')} + 2\nu\overline{\omega_p''\omega_k''\frac{\partial u_p''}{\partial x_k}} \tag{5.182}$$

$$\Phi_\varepsilon = -2\nu^2\overline{\left(\frac{\partial\omega_p''}{\partial x_k}\right)\left(\frac{\partial\omega_p''}{\partial x_k}\right)} \tag{5.183}$$

$$B_\varepsilon = 2\nu\frac{\varepsilon_{pqi}}{\varrho^2}\left\{\frac{\partial\varrho}{\partial x_q}\overline{\omega_p''\frac{\partial p'}{\partial x_i}} + \frac{\partial P}{\partial x_i}\overline{\omega_p''\frac{\partial\varrho'}{\partial x_q}} + \overline{\omega_p''\frac{\partial\varrho'}{\partial x_q}\frac{\partial p'}{\partial x_i}}\right\}. \tag{5.184}$$

$$\underbrace{\phantom{\frac{\partial\varrho}{\partial x_q}}}_{\text{Term 1}} \qquad \underbrace{\phantom{\frac{\partial P}{\partial x_i}}}_{\text{Term 2}} \qquad \underbrace{\phantom{\frac{\partial\varrho'}{\partial x_q}}}_{\text{Term 3}}$$

In the equations above, P_ε represents the "production of dissipation," D_ε represents the "turbulent diffusion of dissipation," and Φ_ε represents the "destruction of dissipation." The fourth term in Eq. (5.180) represents the viscous dissipation of ε.

B_ε (Eq. (5.180)) represents the baroclinic term and arises due to differences in direction between the gradients of pressure and density, that is, the term arising due to $\left[\nabla\left(\frac{1}{\varrho}\right) \times (\nabla p)\right]$. In the case of the mean flow, the baroclinic term represents a production of vorticity due to the interaction of the pressure and density gradients. We therefore assume that this term represents a production of fluctuations in vorticity. To evaluate and to suitably account for the effect of this term, we conducted an order-of-magnitude analysis comparison of the three terms.

A characteristic length scale L_d is chosen as representing the spatial extent of the distortion, and $\Delta\varrho$ and ΔU are chosen to represent variations in ϱ and U_i, respectively. The gradient in pressure is assumed to be of the order of $(\varrho U \Delta U/L_d)$ and the fluctuations in vorticity are assumed to be of the order of ε/K, where the reciprocal represents a characteristic time scale of the turbulent fluctuations (in the eddy viscosity models). With these the order of magnitude of the terms in the expression for B_ε work out to be:

$$\text{Term 1} = \left\{\overline{\frac{\partial\varrho}{\partial x_q}\omega_p''\frac{\partial p'}{\partial x_i}}\right\} = \frac{\Delta\varrho}{L_d}\frac{\varepsilon}{K}\left(\frac{\Delta p'}{L_d}\right) \tag{5.185}$$

$$\text{Term 2} = \left\{\overline{\frac{\partial P}{\partial x_i}\omega_p''\frac{\partial\varrho'}{\partial x_q}}\right\} = \left(\frac{\varrho U \Delta U}{L_d}\right)\frac{\varepsilon}{K}\left(\frac{\Delta\varrho'}{L_d}\right) \tag{5.186}$$

$$\text{Term 3} = \left\{\overline{\omega_p''\frac{\partial\varrho'}{\partial x_q}\frac{\partial p'}{\partial x_i}}\right\} = \frac{\varepsilon}{K}\left(\frac{\Delta\varrho'}{L_d}\right)\left(\frac{\Delta p'}{L_d}\right). \tag{5.187}$$

Comparing the orders of magnitude of Terms 1 and 2 we get

$$\text{Term 1/Term 2} = \frac{\left[\frac{\Delta\varrho}{L_d}\frac{\varepsilon}{K}\frac{\Delta p'}{L_d}\right]}{\left[\frac{\varrho U \Delta U}{L_d}\frac{\varepsilon}{K}\frac{\Delta\varrho'}{L_d}\right]} = \frac{\Delta\varrho}{\varrho}\left(\frac{1}{U\Delta U}\right)\left(\frac{\Delta p'}{\Delta\varrho'}\right) < \mathcal{O}(1) \tag{5.188}$$

where it has been assumed that $(\Delta p')$ and $(\Delta\varrho')$ are of the same order. Comparing the orders of magnitude of Terms 2 and 3 we get,

$$\text{Term 2/Term 3} = \frac{\left[\frac{\varrho U \Delta U}{L_d}\frac{\varepsilon}{K}\frac{\Delta\varrho'}{L_d}\right]}{\left[\frac{\varepsilon}{K}\frac{\Delta\varrho'}{L_d}\frac{\Delta p'}{L_d}\right]} = \frac{\varrho U \Delta U}{\Delta p'} > \mathcal{O}(1) \tag{5.189}$$

where we have assumed that the fluctuations in pressure are of the order of magnitude (suggested by Lele 1994) $\frac{p'^2}{P^2} = \gamma^2 M^2 M_t^2$, where $M = \frac{Sl}{a}$ and $M_t^2 = \frac{q}{a^2}$, where S is representative of the mean flow time scale, l is a characteristic length scale of the turbulent eddies, a is the speed of sound, and $q = u_i''u_i''$. It should be pointed out that the order-of-magnitude analysis is just a first-order estimate and that the orders of magnitude used can be debated. However, due to a lack of information to provide us a better estimate of the orders of magnitude, we will use these estimates to obtain a modification for the term representing the effect of the baroclinic term.

Thus from the order-of-magnitude analysis, it turns out that the second term in Eq. (5.184) is larger than the other three terms. Comparing this term with the transport equation for TKE, it is modeled as $-C_{\varepsilon 1}\frac{\varepsilon}{K}\overline{u_i''}\frac{\partial P}{\partial x_i}$. We chose to model the second term in Eq. (5.184) as a production term because we know that for the mean flow the baroclinic term acts as a production of vorticity. The constant is based on conventional modeling of the "production of dissipation" term in the ε equation. There is no experimental or DNS data to guide our choice of this constant. Future experimental observations or DNS studies may help us in making a better estimate of the constant.

Therefore, the modeled form of the transport equations representing the evolution of K and ε are given as

$$\frac{\partial}{\partial t}(\bar{\varrho}K) + \frac{\partial}{\partial x_j}(\bar{\varrho}\tilde{U}_j K)$$

$$= P_k - \bar{\varrho}\tilde{\varepsilon} + \frac{\partial}{\partial x_j}\left[\mu + \frac{\mu_t}{\sigma_K}\left(\frac{\partial K}{\partial x_j}\right)\right] - \overline{u_i''}\frac{\partial P}{\partial x_i} + \overline{p'\frac{\partial u_i''}{\partial x_i}} \qquad (5.190)$$

and

$$\frac{\partial}{\partial t}(\bar{\varrho}\varepsilon) + \frac{\partial}{\partial x_j}(\bar{\varrho}\tilde{U}_j\varepsilon)$$

$$= C_{\varepsilon 1}\frac{\varepsilon}{K}\left(P_k - \overline{u_i''}\frac{\partial P}{\partial x_i}\right) - C_{\varepsilon 2}\bar{\varrho}\frac{\varepsilon^2}{K} + \frac{\partial}{\partial x_j}\left[\mu + \frac{\mu_t}{\sigma_\varepsilon}\left(\frac{\partial \varepsilon}{\partial x_j}\right)\right]. \qquad (5.191)$$

Comment on Additional Modifications   The merits of the modification proposed by Krishnamurty and Shyy (1996) are that it provides:

a) *A representation for the turbulent mass flux term, using experimentally validated assumptions.* The representation for the turbulent mass flux proposed by Rubesin (1990) has the inherent problem of an arbitrary parameter in terms of the polytropic coefficient. Also, Rubesin's modification (Rubesin 1990) will predict no turbulent mass flux when there is no heat flux.

b) *A representation for the enthalpic production,* which tends to be an important term when considering flows involving strong expansions and compressions. This representation, again, avoids the use of an arbitrary polytropic coefficient. Also, the entire modification (Rubesin 1990) is based on an ad hoc hypothesis.

c) *A representation for the effect of the baroclinic term on the dissipation rate of TKE.* The necessity for modeling this term in highly compressible or stratified flows is clear; in fact, this will become abundantly clear when we look at the solutions to the flow past the axisymmetric afterbody, see Krishnamurty (1996).

Computational assessment of these modifications has been conducted by Krishnamurty (1996) and Krishnamurty and Shyy (1996). It appears that the effect of these terms will not become noticeable until the turbulent Mach number becomes quite high.

5.8.2.4 ADDITIONAL ISSUES

The modifications given by Eq. (5.142) through (5.162) were intended for mixing layers. Consequently, they have been shown to be successful in predicting the reduction in the growth rates of free shear layers (El Baz and Launder 1993, Sarkar 1992, Sarkar and Lakshmanan 1991, Wilcox 1992, Zeman 1990, Zeman 1991).

From their DNS study of homogeneous shear layers, Blaisdell et al. (1993) conclude that the Sarkar et al. (1991) model gave a better estimate of ε_d than the model proposed by Zeman (1990). However, the validity of these algebraic modifications in predictions made of time-dependent flow fields is in question. For turbulent boundary layers (Wilcox 1993), they predict reduced levels of TKE, thereby aggravating the model deficiencies in predicting skin friction coefficient and wall heat transfer rates. All these modifications have not been adequately tested to infer their applicability for more complex flow fields, which is one of the issues being addressed here.

For incompressible flows, equilibrium in the logarithmic region yields the following relation between the coefficients in the dissipation rate equation:

$$\varkappa^2 = (C_{\varepsilon 2} - C_{\varepsilon 1})\sqrt{c_\mu}\sigma_\varepsilon \tag{5.192}$$

where $\varkappa$ is the von Karman constant and is 0.42. It follows from Huang et al. (1992) that if the model satisfies the above equation, then in the case of compressible flows there is a balance of terms expressing a dependence on the spatial gradients in density. None of the proposed compressibility modifications addresses this issue. This balance of terms is responsible for the deficiencies seen in the computations made of turbulent wall layers. Huang et al. (1992) also indicate that the K-ω model is dependent on the spatial gradients of density to a lesser extent than is the K-ε model. The dependence of the modeling procedure on the spatial gradients of density can be shown as follows:

We will first address the dependence of the ε equation on the spatial gradients in density and next address the same in the ω equation. The K-ω model in its current form indicates that it is dependent on the density gradients to a lesser degree. But this is mainly because of the neglect of cross-diffusion terms in the ω equation. We intend to show this in the next few paragraphs.

Consider the region in the log layer of a flat plate boundary layer. The equations for K and ε reduce to

$$-\frac{\partial}{\partial y}\left(\frac{\mu_t}{\sigma_K}\frac{\partial K}{\partial y}\right) = P_k - \varrho\varepsilon \tag{5.193}$$

$$-\frac{\partial}{\partial y}\left(\frac{\mu_t}{\sigma_\varepsilon}\frac{\partial \varepsilon}{\partial y}\right) = C_{\varepsilon 1}\varrho\frac{\varepsilon}{K}P_k - C_{\varepsilon 2}\varrho\frac{\varepsilon^2}{K} \tag{5.194}$$

respectively, where

$$\mu_t = C_\mu\frac{\varrho K^2}{\varepsilon}; \qquad P_k = -\overline{\varrho u''v''}\frac{\partial U}{\partial y} \tag{5.195}$$

and $C_\mu = 0.09, C_{\varepsilon 1} = 1.44, C_{\varepsilon 2} = 1.92, \sigma_K = 1.0$, and $\sigma_\varepsilon = 1.3$.

In the log layer we can assume an equilibrium between the rate of production of TKE and the rate of dissipation of TKE, that is, $P_k = \varepsilon$ (Huang et al. 1992).

Following Huang et al. (1992) and using the compressible form of the wall function method, the values of K, ε, and P_k can be prescribed as

$$K = \left(\frac{\tau_w}{\varrho}\right)\frac{1}{\sqrt{C_\mu}}; \qquad \varepsilon = \left(\frac{\tau_w}{\varrho}\right)^{3/2}\frac{1}{\varkappa y} = P_k. \tag{5.196}$$

Substituting these assumptions into Eq. (5.194), after mathematical manipulations we get

$$\frac{\sqrt{C_\mu}}{\varkappa^2}\sigma_\varepsilon(C_{\varepsilon 2} - C_{\varepsilon 1}) = 1 + 3\left(\frac{y}{\varrho}\right)^2\left(\frac{\partial\varrho}{\partial y}\right)^2 - \frac{3}{2}\frac{y^2}{\varrho}\left(\frac{\partial^2\varrho}{\partial y^2}\right) + \frac{y}{\varrho}\frac{\partial\varrho}{\partial y}. \tag{5.197}$$

For the K-ω model the equations reduce to

$$-\frac{\partial}{\partial y}\left(\frac{\mu_t}{\sigma_k}\frac{\partial K}{\partial y}\right) = P_k - \varrho C_\mu\omega K \qquad (K\text{ equation}) \tag{5.198}$$

$$-\frac{\partial}{\partial y}\left(\frac{\mu_t}{\sigma_\omega}\frac{\partial\omega}{\partial y}\right) = C_{\omega 1}\varrho\frac{\omega}{K}P_k - C_{\omega 2}\varrho C_\mu\omega^2 \qquad (\omega\text{ equation}) \tag{5.199}$$

and the eddy viscosity is defined as

$$\mu_t = \frac{K}{\omega} \tag{5.200}$$

where $C_\mu = 0.09$, and $C_{\omega 1}$, $C_{\omega 2}$, σ_ω are constants. The exact values of these constants are not important for our analysis. They can be found in Wilcox (1993).

The reciprocal time scale variable ω is defined as $\omega = \frac{\varepsilon}{C_\mu K}$. Conducting a similar analysis, as the one above for the K-ε model, $\omega = \left(\frac{\tau_w}{\varrho}\right)^{\frac{1}{2}}\frac{1}{\varkappa y}\frac{1}{\sqrt{C_\mu}}$ in the log region. Substituting the values for K and P_k from Eq. (5.196), and using the value of ω given above, we get

$$\frac{\sqrt{C_\mu}}{\varkappa^2}\sigma_\omega(C_{\omega 2} - C_{\omega 1}) = 1 + \frac{1}{2}\left(\frac{y}{\varrho}\right)^2\left(\frac{\partial\varrho}{\partial y}\right)^2 - \frac{1}{2}\frac{y^2}{\varrho}\left(\frac{\partial^2\varrho}{\partial y^2}\right). \tag{5.201}$$

Comparing Eqs. (5.197) and (5.201), we see that the K-ω model depends to a lesser extent on the spatial gradients in density than the K-ε model. This is the analysis used by Huang et al. (1992). However, they have not included the effect of the cross-diffusion term in their analysis. As Speziale et al. (1991) point out, the cross-diffusion term needs to be included in the ω equation to get asymptotically consistent values for K (as we approach the wall). This cross-diffusion term is written as $\frac{\mu}{K}\left(\frac{\partial K}{\partial y}\right)\left(\frac{\partial\omega}{\partial y}\right)$. If we conduct the same analysis on this cross-diffusion term we end up with

$$\frac{\mu}{K}\left(\frac{\partial K}{\partial y}\right)\left(\frac{\partial\omega}{\partial y}\right)$$

$$= \frac{(\tau_w)^{1/2}}{\varkappa}\frac{1}{\sqrt{C_\mu}}\left[\left(\frac{1}{\varrho^3 y}\right)\left(\frac{\partial\varrho}{\partial y}\right)^2 + \left(\frac{1}{\varrho^2 y^2}\right)\frac{\partial\varrho}{\partial y}\right]. \tag{5.202}$$

The cross-diffusion term in essence adds extra terms, which when taken along with the terms for the equation for ω considered by Huang et al. (1992), would lead to an equivalent dependence on the spatial gradients in density. Therefore, the K-ω model in the form given by Wilcox (1992), while dependent to a lesser extent on the spatial gradients in density, will not produce asymptotically consistent values for K in regions close to wall boundary. When the cross-diffusion term is added to the ω equation to achieve this asymptotic consistency, we have to pay the price of increased dependence on the spatial gradients in density.

5.9   Concluding Remarks

The modeling issues of the K-ε turbulence model related to the nonequilibrium, rotational, and compressibility effects are discussed in the context of the two-equation model. The nonequilibrium K-ε model, where the ε equation is modified according to the imbalance between the production term and dissipation term of the turbulent kinetic energy, produces improved results for certain complex flow computations. As evidenced by the test problems, the nonequilibrium model appears to be more responsive to the change in flow structure, often resulting in a lower maximum eddy viscosity and reduced turbulent transport. It may, however, have too strong a sensitivity in the region not far from nonequilibrium. Further optimization of $C_{\varepsilon 1}, C_{\varepsilon 2}$, and C_μ is desirable.

The K-ε turbulence model with rotational effect is investigated. The momentum equations and the Reynolds-stress equations in the rotational frame are first presented. It has been shown that the Coriolis force and the centrifugal force can explicitly affect the velocity and pressure fields through the momentum equations. The turbulent structure can be modified explicitly by the interaction between the Coriolis force and the Reynolds stresses and implicitly modified by the centrifugal force through an interaction with the velocity field. The analysis of K and ε equations shows that the rotation affects the turbulent quantities only indirectly through the Reynolds-stress term. This indirect effect from rotation can not be represented in the original K-ε model. The displaced particle stability analysis and the simplified Reynolds-stress analysis clearly identify the role of system rotation interacting with flow vorticity and affecting turbulence structure and the corresponding nondimensional parameter. Based on heuristic analyses, the rotational effect on the turbulence structure can be modeled through the modification of the source term of the ε equation.

The effect of compressibility on the turbulence structure is an important but difficult issue in turbulence modeling. Modeling issues in both the production and dissipation of turbulent kinetic energy need to be addressed to account for Mach number effects. Several proposed treatments dealing with the dilatation dissipation and the pressure dilatation correlation are discussed along with additional refinement of the turbulent mass flux, the enthalpic production, and the baroclinic term.

6 Volume-Averaged Macroscopic Transport Equations

In the previous chapter, turbulence modeling issues have been discussed in the context of engineering computations. In this chapter, we will employ the concept of volume averaging to tackle scale disparity in complex transport phenomena. Specifically, the formulation of the macroscopic transport equations will be presented. This approach has its roots in the analysis of multiphase flow, encountered in a wide variety of engineering problems involving porous media, particle suspensions, solute sedimentation, energy conversion, chemical reaction, etc. Of these applications, the study of macroscopic momentum transport in porous media has received much attention. Theoretical analysis and experimental verification in this area is abundant in the literature (Drew 1983, Ghaddar 1995, Slattery 1967, Whitaker 1967).

We will use materials solidification as an example to discuss the macroscopic transport equations of mass, momentum, energy, and species. Some issues that have not been addressed adequately in the literature (Beckermann and Viskanta 1993, Ganesan and Poirier 1990) will be pointed out. Since macroscopic transport equations are derived upon the classical microscopic transport equations, a brief discussion of the microscopic equations is presented first.

6.1 Microscopic Transport Equations

Solids, liquids, and gases are composed of distinct molecules or atoms. However, in many engineering models materials are conveniently treated as continuous media instead of individual molecules or atoms. This is because engineers are mostly concerned about the averaged features of materials, represented by such quantities as density, velocity, pressure, temperature, and so on, which vary continuously in space and time. For example, density is defined in a continuum by

$$\varrho = \lim_{\Delta V \to 0} \frac{\Delta M}{\Delta V} \tag{6.1}$$

231

where ΔV is an infinitesimal volume and ΔM is the mass contained in ΔV. Physically, ΔV cannot be allowed to shrink to zero, since if ΔV has a characteristic size so comparable to the length of the molecular free path, ΔM would vary discontinuously, depending on the number of molecules in ΔV. In order for continuum mechanics to hold, the lower limit of ΔV should be selected to some finite (but small) value ε in order to eliminate molecular effects. Transport equations based on this hypothesis are referred to as microscopic transport equations.

According to Reynolds' transport theorem (Malvern 1969), the general form of microscopic transport equations of a certain quantity Φ can be rigorously derived from conservation laws and is given by

$$\frac{\partial \Phi}{\partial t} + \nabla \cdot (\Phi V) = \nabla J + S \tag{6.2}$$

where Φ represents mass ϱ, momentum ϱV, enthalpy ϱH, or species concentration ϱC; J is a tensor of order one greater than that of Φ and accounts for the diffusive effect of momentum, heat, or species; and S is a source term. Neglecting viscous heat dissipation, compression work, volumetric energy, and species sources, which is appropriate for many practical solidification systems, the microscopic transport equations of mass, momentum, energy, and species are

$$\frac{\partial \varrho}{\partial t} + \nabla \cdot (\varrho V) = 0 \tag{6.3}$$

$$\frac{\partial \varrho V}{\partial t} + \nabla \cdot (\varrho V V) = \nabla \cdot \sigma + b \tag{6.4}$$

$$\frac{\partial \varrho H}{\partial t} + \nabla \cdot (\varrho H V) = -\nabla q \tag{6.5}$$

$$\frac{\partial \varrho C}{\partial t} + \nabla \cdot (\varrho C V) = -\nabla j \tag{6.6}$$

where σ is the stress tensor, b is the body force, q is the heat flux, and j is the species diffusion flux. The detailed expressions for these terms are supplied by constitutive equations based on the characteristics of molecular or atomic movement.

6.2 Background of Macroscopic Transport Equations

Although transport phenomena in materials solidification can be described in detail by transport equations at the microscopic level capable of handling the morphological characteristics during phase change, the direct application of the microscopic transport equations to solidification modeling is generally difficult, mainly because of the morphological complexities and the disparate scales.

6.2.1 Morphological Complexities and Disparate Scales

In the solidification process of alloys, a mushy zone forms between the solid and the liquid, where the solid and the liquid coexist as shown in Fig. 6.1. The geometry of the morphological structure between the solid and liquid interface in this

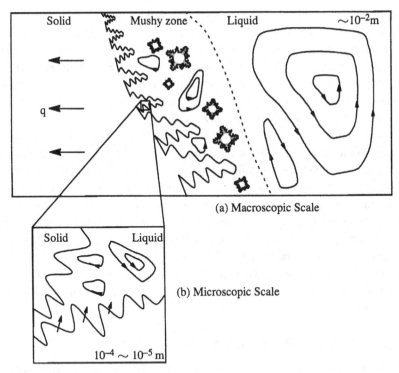

Figure 6.1 Disparate scales in solidification process, and schematic of convection patterns (in arrows).

region is very complicated and depends on the materials properties and the phase change environment such as thermal conditions and species distribution. Although analytical and numerical efforts have been made to investigate the details of the morphological structure in solidification (Langer 1980, Shyy et al. 1996a), difficulties exist in extending the computations to cover the whole mushy zone because of the disparate scales involved.

As an illustration, Fig. 6.1 shows schematically two typical scales and corresponding physical phenomena that are involved in solidification. Other scales also exist in the solidification process (Kurz and Fisher 1989, Shyy et al. 1996a,b), such as solute diffusion length scale and capillary length scale, but are not discussed here.

First, at the macroscopic scale (i.e., characteristic size of an ampoule, normally having an order of magnitude of at least 10^{-2} m for most of the materials processing systems), convective–conductive heat transfer and macroscopic segregation are present (Fig. 6.1a). Second, at the microscopic scale or morphological scale (e.g., $10^{-4}\sim 10^{-5}$ m for the typical microstructure of aluminum alloys), the interdendritic flow due to shrinkage, latent heat release due to phase change, and species partition between solid and liquid phases appear (Fig. 6.1b). It is noted that dendrites, eutectics, or combinations of these make up the microscopic morphology of any metallic microstructure after solidification (Kurz and Fisher 1989). The formation of a dendrite begins with the breakdown of an unstable planar solid/liquid interface. Perturbations are amplified until a marked difference in tip growth and depressions of the perturbed interface have occurred. The growth rate, morphology, and spacing of dendrites are all largely dependent upon the growth behavior of the tip region, where heat or solute are redistributed. Dendrites are often encountered in undercooled pure materials or in directional solidification of dilute binary alloys (Beckermann and Viskanta 1993, Langer 1980). Eutectic morphologies also exist, which are characterized by the simultaneous growth of two (or more) phases from the liquid, and exhibit a wide variety of geometrical arrangements (Kurz and Fisher 1989).

As the morphological (microscopic) scale of dendritic or eutectic evolution and the macroscopic scale of global heat and fluid flows are different by several orders of magnitude, solving an entire solidification process as a microscopic problem is a formidable task. It requires grid resolutions of at least the dendritic or eutectic arm spacings for numerical modeling so that the detailed morphologies can be captured. Such direct numerical modeling is impractical even with the computational power available today. In order to make the computational task tractable, a micro–macro model needs to be developed (Rappaz 1989, Shyy et al. 1996b). In other words, it is necessary to introduce some intermediate scale models to represent the physics of scales smaller than the computational mesh. In the following, a methodology based on the volume-averaging procedure is presented.

6.2.2   Definition of Averaging Volume

The averaging volume dV is defined such that the scale it represents is small enough to capture the global fluid motion, heat transfer, and species distribution as well as to track the liquid, solid, and mushy zones, but large enough to smooth out the details of morphological complexities and the interdendritic fluid flow, heat release, and species redistribution. In addition, based on the continuum hypothesis, the volume-averaged quantities such as density, velocity, and temperature should still be continuous functions of space and time. The morphological characteristics will be accounted for by the construction of constitutive equations, introduction of some effective coefficients of diffusivity, and their incorporation into the macroscopic equations derived by volume averaging.

As an illustration, Fig. 6.2 shows the definition of the average density in the mushy zone as a function of the characteristic size of the averaging volume (Ganesan

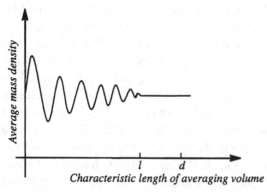

Figure 6.2   Average mass density as a function of the characteristic length of the averaging volume.

and Poirier 1990, Gray 1975). The average density is defined as the total mass of the solid and the liquid in the averaging volume divided by the averaging volume, dV. When dV is very small, the average density of the interdendritic mixture fluctuates because portions of liquid or solid phases included in dV are location-dependent. This fluctuation diminishes as the size of dV increases, and within a certain characteristic size of dV, the density or any other average quantities become independent of dV. This region is represented by the horizontal line in Fig. 6.2. Accordingly, the averaging volume dV should have a characteristic length that satisfies

$$\mathcal{O}(l) < \mathcal{O}(d) \tag{6.7}$$

where l is the microscopic characteristic length (e.g., dendritic arm spacings), and d is the characteristic length of the averaging volume. However, the characteristic size of dV should not be too large; otherwise, it becomes dependent upon the global variations of location, temperature, and composition.

The macroscopic transport equations based on the averaging volume dV have been formulated in the literature mainly from two approaches. One is the so-called volume-averaging approach (Beckermann and Viskanta 1993, Drew 1983, Ganesan and Poirier 1990, Gray 1975, Gray 1983, Slattery 1967, Whitaker 1967). In this approach, the macroscopic equations of transport are directly derived from the microscopic equations. The second is called the mixture approach (Bennon and Incropera 1987a, Poirier et al. 1991, Prescott and Incropera 1994, Voller et al. 1989). With this approach, certain mixture quantities in the mushy zone are first defined at the macroscopic scale, and then macroscopic equations are written down based on the conservation laws and mixture quantities. In the following, the salient features of these approaches will be given. Our emphasis will be primarily on the volume-averaging approach to confine the scope of presentation.

6.3   Volume-Averaging Approach

The application of the concept of volume averaging to solidification processes is relatively recent. For example, Ganesan and Poirier (1990) derived in detail the macroscopic transport equations of mass and momentum for the flow of interdendritic liquid during solidification. They introduced two permeability coefficients in order to better describe the resistance of the solid to interdendritic liquid flow in the mushy zone. Further numerical calculation of permeability for a high-volume fraction of liquid as a function of microscopic morphology, which merges with the empirical data at low volume fraction, is recently reported in Bhat et al. (1995a,b).

Beckermann and Viskanta (1993) systematically formulated the macroscopic transport equations of energy and species as well as mass and momentum. They discussed the possible mathematical constructions of the constitutive models that could incorporate the microscopic complexities into macroscopic equations. Follow-up numerical modeling based on their derived equations and the sensitivity study of solution characteristics to selections of constitutive models are also reported (Schneider and Beckermann 1995). They found that different constitutive models of microstructure predict quite different solidification characteristics in terms of solid concentration, eutectic fractions, number, length, and orientation of segregated channels, etc. They further suggested that constitutive models as a function of mushy zone microstructure should be obtained by combining the experimental measurement and numerical simulation.

Despite these efforts several important issues related to the formulation of macroscopic transport equations remain unresolved. For example, microscopic deviation terms resulting from the nonuniformity of the dendritic growth rate, temperature, and species concentration have not been addressed, although both Ganesan and Poirier (1990) and Viskanta and Beckermann (1993) mentioned that these terms are important in regions of high-volume fraction of liquid in the mushy zone. To date, no detailed models have been proposed in the literature to capture such phenomena as columnar or equiaxed growth, constitutive undercooling, and morphologies of dendritic or eutectic evolution. Another issue that needs to be clarified further is the introduction of the effective coefficients of diffusivity which are supposed to account for the reduced diffusive effects of heat and species in the interdendritic meshwork.

In this chapter, we will attempt to address the derivation of unified macroscopic transport equations, valid in solid, liquid, and mushy zones, for the modeling of materials solidification, based on the existing literature (Beckermann and Viskanta 1993, Drew 1983, Ganesan and Poirier 1990, Gray 1975, Gray 1983, Slattery 1967, Whitaker 1967). In particular, we propose a microscopic deviation model adopting the concept of the subgrid scale model in turbulent large eddy simulation and also discuss the corresponding closure problems. Concepts that are used by the averaging procedure are given first, followed by detailed derivations of the macroscopic transport equations. Then the application of this approach to mass, energy, species, and momentum equations is given. All the assumptions made during the derivations are highlighted. They are invoked only at the final stage of derivation and can be easily relaxed.

6.3.1   Definitions and Theorems

6.3.1.1   PHASE FUNCTION γ_k

The phase function γ_k is a function of space and time, equalling 1 in phase k and zero elsewhere. It has also been denoted by X_k (Beckermann and Viskanta 1993, Drew 1983) or α_k (Whitaker 1967) in the literature.

6.3.1.2   VOLUME FRACTION g_k

The volume fraction g_k is defined as

$$g_k = \frac{1}{dV} \int_{dV} \gamma_k(r, t) dv = dV_k/dV \tag{6.8}$$

where dV_k is the portion of dV that is occupied by phase k, and dv is an infinitesimal element relative to dV. Some researchers denote volume fraction as ε_k (Beckermann and Viskanta 1993, Gray 1975).

6.3.1.3 INTRINSIC VOLUME-AVERAGED (OR PORE) QUANTITY $\overline{\Psi_k}^k$

The intrinsic volume-averaged quantity is defined as

$$\overline{\Psi_k}^k = \frac{1}{dV_k} \int_{dV} \Psi_k(\boldsymbol{r}, t)\gamma_k(\boldsymbol{r}, t)dV. \tag{6.9}$$

$\langle \Psi_k \rangle^k$ has also been used by others to represent this quantity. The definition of $\overline{\Psi_k}^k$ here represents the averaged value of $\Psi_k(\boldsymbol{r}, t)$ in the control volume dV_k; if $\Psi_k(\boldsymbol{r}, t)$ is uniformly distributed in dV_k, then $\overline{\Psi_k}^k = \Psi(\boldsymbol{r}, t)$. (The arguments $\boldsymbol{r}$ and t will be dropped subsequently for conciseness.)

6.3.1.4 VOLUME-AVERAGED (OR SUPERFICIAL) QUANTITY $\overline{\Psi_k}$

The volume-averaged quantity is defined as

$$\overline{\Psi_k} = \frac{1}{dV} \int_{dV} \Psi_k\gamma_k dV. \tag{6.10}$$

It is the averaged value of quantity Ψ_k in phase k over the entire averaging volume dV. By comparing Eqs. (6.9) and (6.10), we obtain

$$g_k\overline{\Psi_k} = \overline{\Psi_k}^k. \tag{6.11}$$

6.3.1.5 FLUCTUATING COMPONENT $\hat{\Psi}_k$

The fluctuation component $\hat{\Psi}_k$ represents the deviation of Ψ_k from the intrinsic volume average, and is given by

$$\hat{\Psi}_k = \left(\Psi_k - \overline{\Psi_k}^k\right)\gamma_k. \tag{6.12}$$

In phase k, it is zero only when Ψ_k is uniformly distributed (i.e., $\Psi_k = \overline{\Psi_k}^k$). According to this definition, we have the following identity for changing the average of two products to the product of averaged quantities (Gray 1975):

$$\overline{\Psi_k\Phi_k} = g_k\overline{\Psi_k}^k\overline{\Phi_k}^k + \overline{\hat{\Psi}_k\hat{\Phi}_k} \tag{6.13}$$

where

$$\overline{\hat{\Psi}_k\hat{\Phi}_k} = \frac{1}{dV} \int_{dV} \hat{\Psi}_k\hat{\Phi}_k dv.$$

Care needs to be exercised to handle this component. For example, the first term on the right side of Eq. (6.13) has been expressed as either $\overline{\Psi_k\Phi_k}$ or $\overline{\Psi_k}^k\overline{\Phi_k}^k$ (without g_k) (Beckermann and Viskanta 1993, Ganesan and Poirier 1990) Gray (1975) gives a very clear derivation of this identity.

6.3.1.6 VOLUME-AVERAGING THEOREMS

The volume-averaging theorems listed below have been derived rigorously by several workers (Gray 1975, Howes and Whitaker 1985).

Theorem 1   *Relates the average of the time derivative to the time derivative of the average:*

$$\frac{1}{dV}\int_{dV}\left(\frac{\partial \Psi_k}{\partial t}\right)\gamma_k dv = \frac{\partial}{\partial t}\left[\frac{1}{dV}\int_{dV}\Psi_k\gamma_k dv\right] - \frac{1}{dV}\int_{dA_k}\Psi_k \boldsymbol{w}\cdot\boldsymbol{n}_k da$$

or

$$\overline{\frac{\partial \Psi_k}{\partial t}} = \frac{\partial \overline{\Psi_k}}{\partial t} - \frac{1}{dV}\int_{dA_k}\Psi_k \boldsymbol{w}\cdot\boldsymbol{n}_k da \tag{6.14}$$

where dA_k is the interfacial area of phase k with other phases, $\boldsymbol{n}_k$ is the outward unit normal of the infinitesimal element of area da of phase k, and $\boldsymbol{w}$ is the velocity of the microscopic interface.

Theorem 2   *Relates the average of the spatial derivative to the spatial derivative of the average:*

$$\frac{1}{dV}\int_{dV}(\nabla\Psi_k)\gamma_k dv = \nabla\left[\frac{1}{dV}\int_{dV}\Psi_k\gamma_k dv\right] + \frac{1}{dV}\int_{dA_k}\Psi_k\boldsymbol{n}_k da$$

or

$$\overline{\nabla\Psi_k} = \nabla\overline{\Psi_k} + \frac{1}{dV}\int_{dA_k}\Psi_k\boldsymbol{n}_k da. \tag{6.15}$$

A variation of this theorem is given by (Gray 1975) as

$$\overline{\nabla\Psi_k} = g_k\nabla\overline{\Psi_k}^k + \frac{1}{dV}\int_{dV}\hat{\Psi}_k\boldsymbol{n}_k da. \tag{6.16}$$

6.3.2   The Derivation Procedure

The derivation summarized here basically follows most of the volume-averaging procedure reported in the literature (Beckermann and Viskanta 1993, Drew 1983, Ganesan and Poirier 1990, Gray 1975, Gray 1983, Slattery 1967, Whitaker 1967). Applying the general microscopic transport equation (6.2) to phase k, we have

$$\frac{\partial \Phi_k}{\partial t} + \nabla\cdot(\Phi_k\boldsymbol{V}_k) = \nabla\cdot\boldsymbol{J}_k + S_k. \tag{6.17}$$

Multiplying each side of Eq. (6.17) by γ_k, averaged over dV, we obtain

$$\underbrace{\frac{1}{dV}\int_{dV}\frac{\partial \Phi_k}{\partial t}\gamma_k\, dv}_{\text{1st}} + \underbrace{\frac{1}{dV}\int_{dV}\nabla\cdot(\Phi_k\boldsymbol{V}_k)\gamma_k\, dv}_{\text{2nd}}$$

$$= \underbrace{\frac{1}{dV}\int_{dV}\nabla\cdot\boldsymbol{J}_k\gamma_k\, dv}_{\text{3rd}} + \underbrace{\frac{1}{dV}\int_{dV}S_k\,\gamma_k dv}_{\text{4th}}. \tag{6.18}$$

Applying Theorem 1, shown in Eq. (6.14), to the first term, we obtain

$$\frac{1}{dV}\int_{dV}\left(\frac{\partial \Phi_k}{\partial t}\right)\gamma_k dv = \frac{\partial \overline{\Phi_k}}{\partial t} + \frac{1}{dV}\int_{dA_k}\Phi_k w_k \cdot n_k da. \tag{6.19}$$

Applying Theorem 2, shown in Eq. (6.15), to the second, third, and fourth terms, we get

$$\frac{1}{dV}\int_{dV}(\nabla \Phi_k V_k)\gamma_k dv = \nabla \cdot \overline{\phi_k V_k} + \frac{1}{dV}\int_{dA_k}\Phi_k V_k \cdot n_k da \tag{6.20}$$

$$\frac{1}{dV}\int_{dV}\nabla \cdot J_k \gamma_k dv = \nabla \cdot \overline{J}_k + \frac{1}{dV}\int_{dA_k}J_k \cdot n_k da \tag{6.21}$$

$$\frac{1}{dV}\int_{dV}S_k \gamma_k dv = \overline{S}_k. \tag{6.22}$$

Furthermore, by applying the identity of Eq. (6.13), $\nabla \cdot \overline{\Phi_k V_k}$ in Eq. (6.20) becomes

$$\nabla \cdot \overline{\Phi_k V_k} = \nabla \cdot g_k \overline{\Phi_k}^k \overline{V_k}^k + \nabla \cdot \overline{\hat{\Phi}_k \hat{V}_k}. \tag{6.23}$$

Substituting Eqs. (6.19)–(6.23) into Eq. (6.18), we get

$$\frac{\partial \overline{\Phi_k}}{\partial t} + \nabla \cdot g_k \overline{\Phi_k}^k \overline{V_k}^k = \nabla \cdot \overline{J}_k + \overline{S}_k$$

$$+ \nabla \cdot \underbrace{\frac{1}{dV}\int_{dV}(-\hat{\Phi}_k \hat{V}_k)dv}_{I_k^D} + \underbrace{\frac{1}{dV}\int_{dA_k}J_k \cdot n_k da}_{I_k^J}$$

$$+ \underbrace{\frac{1}{dV}\int_{dA_k}\Phi_k(w_k - V_k) \cdot n_k da}_{I_k^Q}. \tag{6.24}$$

This is the exact form of the macroscopic transport equations. Compared with the exact microscopic equations of transport (i.e., Eq. (6.17)), three extra terms – I_k^D, I_k^J, and I_k^Q – appear from this volume-averaging procedure. We call I_k^D the microscopic deviation term and I_k^J and I_k^Q the interfacial terms, because I_k^D involves volume integration and I_k^J and I_k^Q involve surface integration. In order to solve Eq. (6.24), these terms have to be modeled, that is, the so-called macroscopic constitutive equations must be constructed, as discussed below.

6.3.3   The Treatment of Microscopic Deviation and Interfacial Terms

6.3.3.1   MICROSCOPIC DEVIATION TERM I_k^D

This term appears because of the deviation of field quantities $\Phi_k(r, t)$ and $V_k(r, t)$ from their intrinsically averaged counterparts. From the definition of Eq. (6.13), I_k^D is zero only when $\Phi_k(r, t)$ or $V_k(r, t)$ are uniformly distributed in dV_k, that is, $\hat{\Phi}_k$ or $\hat{V}_k$ is zero. However, at the microscopic (or morphological) scale characterizing the mushy zone, there always exist species, temperature, and velocity gradients in the liquid, due to solute rejection, heat release, and no slip velocity conditions; hence, $\Phi_k(r, t)$ and $V_k(r, t)$ are not uniform within the averaging volume. Despite this fact, almost all solidification models reported in the literature neglect the microscopic deviation term (Bennon and Incropera 1987b, Ganesan and Poirier 1990, Prescott and Incropera 1994, Schneider and Beckermann 1995, Voller et al. 1989). By neglecting this term, the real mechanisms of columnar or equiaxed growth cannot be captured. Therefore, a microscopic deviation model needs to be developed to relate the deviation of velocity, temperature, and species concentration from their averaged counterparts to the corresponding gradients. Analogy can be found in the subgrid scale (SGS) model in large eddy simulation (LES).

In large eddy simulation, a subgrid-scale model representing the action of scales smaller than the computational mesh of large scale is introduced in order to increase the Reynolds number in the simulations (Lesieur 1993, Moin et al. 1991). The large-scale field is computed directly from the solution of the local volume-averaged Navier-Stokes equations, and the subgrid scale stresses are modeled. The subgrid scale model then represents the effect of the small scales on the large scale motions. This concept of LES can help alleviate the severe Reynolds number restriction in direct numerical simulation (DNS) by modeling the subgrid behavior. Among the many SGS models existing in the literature, the simplest one is the Smagorinsky model (Lesieur, 1993, Moin et al. 1991). If we denote the extra terms that arise from the volume-averaging procedure of the inertial term in the Navier-Stokes equation as τ'_{ij}, which appears because of the fluctuation of turbulent velocity from its volume-averaged counterpart, then this term can be expressed similarly in terms of the strain rates of the flow field through an eddy viscosity coefficient, that is,

$$\tau'_{ij} = 2KS_{ij} \tag{6.25}$$

where S_{ij} is the volume-averaged strain rate tensor, expressed as

$$S_{ij} = \frac{1}{2}\left(\frac{\partial \overline{V_i}}{\partial x_j} + \frac{\partial \overline{V_j}}{\partial x_i}\right) \tag{6.26}$$

and

$$K = (C_s \Delta)^2 (2S_{ij}S_{ij})^{\frac{1}{2}}. \tag{6.27}$$

A basic ingredient that characterizes the subgrid scale model used in large eddy simulation is that the characteristic length scale is usually given by the grid size (Germano 1992). In the above expression, Δ represents a grid spacing and C_s is an adjustable constant. C_s can be obtained by either direct numerical simulation of isotropic turbulence, homogeneous shear flow, etc., or conducting corresponding experiments (Moin et al. 1991). It might be interpreted as the ratio of the characteristic subgrid size to the computational grid size.

The concept of averaging volume we have used for the formulation of macroscopic transport equations is analogous to the concept of subgrid scale. Since the microscopic deviation term I_k^D also arises from the deviation of microscopic quantities $\Phi_k(r, t)$ and $V_k(r, t)$ from their averaged counterparts in dV_k, it is then also plausible for us to relate this term to the gradients of $\overline{\Phi}_k(r, t)$ and $\overline{V}_k(r, t)$. Therefore, we can adopt a formulation similar to that of the Smagorinsky model, in which the integral portion of I_k^D is expressed as

$$\frac{1}{dV} \int_{dV} (-\hat{\Phi}_k \hat{V}_k) dv = 2K \nabla \overline{\Phi}_k^k. \tag{6.28}$$

Thus

$$I_k^D = \nabla \cdot \frac{1}{dV} \int_{dV} (-\hat{\Phi}_k \hat{V}_k) dv = \nabla \cdot \left(2K \nabla \overline{\Phi}_k^k \right). \tag{6.29}$$

In this expression, $\nabla \overline{\Phi}_k^k$ has been extended to represent not only the strain rate as in LES, but also the gradient of temperature and species concentration. K is given by the same format as in Eq. (6.27),

$$K = (\alpha_k l)^2 (2S_{ij} S_{ij})^{\frac{1}{2}} \tag{6.30}$$

where α_k is also an adjustable constant of phase k that should be decided based on experimental evidence. It should be noted that since the capillary effect has not been explicitly accounted for in the averaging process, its implication will need to be reflected by the adjustable coefficient. Accordingly, the value of this coefficient will change based on the variations in microscopic morphologies. In addition, l here does not represent the computational mesh size, but the morphological length scale, which, according to the established theory (Langer 1980, Shyy 1994), can be taken as the square root of the product of the capillary and conduction scales for dendritic structures.

The formulation of Eq. (6.29) can be seen to reflect some basic features of interdendritic actions. For example, in the mushy zone where the liquid volume fraction is close to unity, dendrites or eutectics grow as a result of gradients in velocity, temperature, and species concentration. According to Eq. (6.29), the microscopic deviation effect in this region is therefore significant. Hence, the inclusion of a microscopic deviation term is expected to be able to capture in more detail the dendritic or eutectic evolution in this region. In the mushy zone where the liquid volume fraction is approaching zero, that is, near the dendrite root region, however, as the interdendritic fluid motion is nearly stagnant, the microscopic deviation term will gradually diminish as the solid region is approached.

It should be mentioned here that, although the treatment of the microscopic deviation term is virtually absent in the literature concerning the modeling of solidification phenomena, it has been discussed elsewhere. For example, Whitaker (1967) considered the microscopic deviation term in the macroscopic equation of species transport as a function of variables such as the averaged species concentration and the gradients of concentration and velocity. Although it provided a basis for the correlation of experimental data and illustrated the connection between the microscopic and macroscopic equations, his definition of the fluctuating component was ambiguous and led to a nonzero value of microscopic deviation term even when Φ_k is uniform in dV_k (Gray 1975). With the present definition of fluctuating components and the proposed model for the microscopic deviation term, these problems will not appear.

6.3.3.2 INTERFACIAL TERMS I_k^J AND I_k^Q

It can be seen from Eq. (6.24) that these two terms arise from the interfacial integration over dA_k. I_k^Q accounts for the interfacial transfer due to phase change. Physically, it represents advection of interfacial mass, momentum, heat, or species of phase k due to the relative motion of the solid/liquid interface. I_k^J accounts for the effect of interfacial stress, heat, and species transfer. It represents the transport phenomena between the phases within dV by diffusion and is related to the gradients of microscopic velocity, temperature, and species concentration on each side of the solid/liquid interface dA_k (Beckermann and Viskanta 1993). Depending on different modeling efforts, these terms are treated differently, as outlined below.

The Two-Phase Model   In this model, the transport equation (6.24) is applied to each phase and is solved separately (Ni and Beckermann 1991). The equations are coupled through the interfacial conditions estimated by I_k^J and I_k^Q, which are expressed as a function of the interfacial area concentration $S_v(S_v = dA_k/dV)$. The variable S_v has been adopted by many researchers in the multiphase research community to account for the morphological complexities of porous media or solidification dendritic meshwork. Gray (1975) formed the interfacial species transfer due to the concentration gradient as a product of mass transfer coefficient and a concentration driving force. Beckermann and Viskanta (1993) represented the interfacial transfer due to phase change $\left(I_k^Q\right)$ and the interfacial transfer due to variable gradient $\left(I_k^J\right)$ as the product of the interfacial area concentration S_v and a mean interfacial flux.

The main advantage of the two-phase model, as argued by Ni and Beckermann (1991), is that it can account for nonequilibrium effects during solidification, such as rapid solidification, floating and settling of small equiaxed crystals, nucleation, etc. However, numerical computations with such a model require explicit treatment of the boundaries between highly irregular domains. In addition, a detailed and accurate formulation of these interfacial transfer terms has to be provided.

The One-Phase Model   In this model, since the averaged macroscopic equations from different phases are added within the averaging volume dV, detailed modeling of the interfacial transfer terms I_k^J and I_k^Q can be avoided. The heat or mass lost from interface by one phase is gained by other phases, i.e.,

$$\sum I_k^Q = 0 \quad \text{and} \quad \sum I_k^J = 0. \tag{6.31}$$

Hence, the explicit treatment of interfacial flux is not necessary here. However, this one-phase model has the disadvantage that relationships between the macroscopic variables of each phase have to be provided. In the following, the one-phase model will be presented, including the averaged macroscopic equations of mass, heat, species, and momentum in the context of alloy solidification.

6.4 Macroscopic Transport Equations via the Volume-Averaging Approach

To aid the discussion, assumptions in the modeling of binary solidification processes are first made. In these assumptions, μ is the viscosity, k is the thermal conductivity, and D is the coefficient of binary species diffusivity. In the following derivations, the momentum equation will be featured last because it has been studied widely in the research community and is also the most controversial.

Assumption 1   *Only two phases, liquid and solid, are considered, that is, $k = l$ or s.*

Assumption 2   *Variations of material properties in dV_k are neglected, although globally they may vary, that is,*

$$\overline{\varrho_k}^k = \varrho_k = \overline{\varrho_k}/g_k, \qquad \overline{\varrho_k \Phi_k} = \varrho_k \overline{\Phi_k}$$

$$\overline{\mu_k}^k = \mu_k, \qquad \overline{k_k}^k = k_k, \qquad \overline{D_k}^k = D_k.$$

6.4.1   Macroscopic Equation of Mass Conservation

In this case, $\Phi = \varrho$, $\boldsymbol{J} = \boldsymbol{0}$, and $S = 0$. According to Eq. (6.24) the liquid and solid mass equations yield, respectively,

$$\frac{\partial \overline{\varrho_l}}{\partial t} + \nabla \cdot g_l \varrho_l \overline{\boldsymbol{V}_l}^l = I_l^Q \tag{6.32}$$

$$\frac{\partial \overline{\varrho_s}}{\partial t} + \nabla \cdot g_s \varrho_s \overline{\boldsymbol{V}_s}^s = I_s^Q \tag{6.33}$$

where

$$I_l^Q = \frac{1}{dV} \int_{dA_k} (\boldsymbol{w}_l - \boldsymbol{V}_l) \cdot \boldsymbol{n}_l da \qquad \text{and} \qquad I_s^Q = \frac{1}{dV} \int_{dA_k} (\boldsymbol{w}_s - \boldsymbol{V}_s) \cdot \boldsymbol{n}_s da.$$

Here the microscopic deviation term does not appear due to Assumption 2 (i.e., $\hat{\varrho}_l = \hat{\varrho}_s = 0$). Now adding Eqs. (6.32) and (6.33) and considering the interfacial mass flux balance, that is, that the mass lost by one phase is gained by the other phase, or $I_l^Q + I_s^Q = 0$, we get

$$\frac{\partial \left(\overline{\varrho_l} + \overline{\varrho_s} \right)}{\partial t} + \nabla \cdot \left(g_l \varrho_l \overline{\boldsymbol{V}_l}^l + g_s \varrho_s \overline{\boldsymbol{V}_s}^s \right) = 0. \tag{6.34}$$

6.4.2   Macroscopic Equation of Energy Conservation

In this case, $\Phi = \varrho H$, where H represents the total enthalpy, and $S = 0$. Furthermore, we utilize Fourier's law, neglecting other heat diffusion fluxes such as Soret and Dufour effects (Bird et al. 1960) as they are less important in the conventional solidification process:

Assumption 3   *Fourier's law for heat flux, that is, $\boldsymbol{J} = -k \nabla T$.*

Then for liquid and solid phases, the averaged energy equations yield:

$$\frac{\partial \varrho_l \overline{H_l}}{\partial t} + \nabla \cdot g_l \varrho_l \overline{H_l^l V_l^l} = \nabla \cdot \overline{(k_l \nabla T_l)} + I_l^D + I_l^Q + I_l^J \tag{6.35}$$

$$\frac{\partial \varrho_s \overline{H_s}}{\partial t} + \nabla \cdot g_s \varrho_s \overline{H_s^s V_s^s} = \nabla \cdot \overline{(k_s \nabla T_s)} + I_s^D + I_s^Q + I_s^J. \tag{6.36}$$

The physical meanings of the terms on the left side of Eqs. (6.35) and (6.36) are obvious, as they are the same as their microscopic counterparts. The terms on the right side of Eqs. (6.35) and (6.36), however, need further discussion. Equation (6.35) will be used for this purpose.

6.4.2.1 $\nabla \cdot \overline{k_l \nabla T_l}$ TERM

By applying Theorem 2 (Eq. (6.16)) and using Assumption 2, $\overline{k_l \nabla T_l}$ can be further expanded as

$$\overline{k_l \nabla T_l} = g_l k_l \nabla \overline{T_l^l} + k_l \frac{1}{dV} \int_{dA_k} \hat{T}_l \boldsymbol{n}_l \, da \tag{6.37}$$

where the term involving the area integration is called the tortuosity factor (Gray 1975) because it accounts for a decreased diffusion rate due to the geometry of the dendrites. In addition, since the thermal diffusion effect should be zero when the temperature in the liquid is uniform, this implies that this term is related to the temperature gradient. Based on this argument, an effective conductivity is introduced and is related to the macroscopic variable as (Beckermann and Viskanta 1993, Whitaker 1967)

$$\overline{k_l \nabla T_l} = g_l k_l^* \nabla \overline{T_l^l} \tag{6.38}$$

where the contribution of the area integration is accounted for with the newly defined effective conductivity k_l^*. To close this formulation, k_l^* as a function of the microscopic morphology needs to be specified. A commonly used variable to account for the morphological tortuosity is the interfacial area concentration S_v, and k_l^* is written as a function of S_v. An alternative variable to account for the tortuosity is the dendritic arm number N. As the number of dendritic arms increases, the value of S_v increases and hence the structure is more tortuous. The exact functional form of k_l^* is based on empirical studies. Beckermann and Viskanta (1993) have reviewed several empirical relations proposed in the literature, in the form of $k_l^* = k_l^*(S_v)$, where S_v is expressed as a function of the liquid volume fraction g_l. To illustrate such an approach, let us use the following example.

It is known that during solidification, S_v in the mushy region increases from zero near the dendrite root, eventually reaches a maximum value, and then decreases again to zero near the primary dendrite tip region where the fluid fraction is approaching 1. According to these characteristics, an empirical relation suggested by Speich and Fisher (1966) is

$$S_v = c g_l (1 - g_l) \tag{6.39}$$

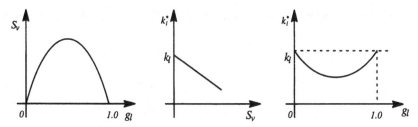

Figure 6.3 Schematic of the effective conductivity as a function of microscopic morphology.

where c is a constant. Furthermore, as the effective conductivity of $k_l^*(S_v)$ should reduce to its microscopic counterpart as it approaches the liquid or the solid zone, the following relation can be considered:

$$k_l^*(S_v) = k_l(1 - S_v/c).$$ (6.40)

Hence

$$k_l^*(S_v) = k_l\left(1 - g_l + g_l^2\right).$$ (6.41)

These relations are shown in Fig. 6.3.

The above example is just a schematic illustration of the closure and the construction of the effective coefficient as a function of microscopic complexity. In reality, the relation might not be a simple parabolic or linear function and should be determined from experimental measurements or direct numerical modeling of dendrite morphology (Bhat et al. 1995a,b). This is an area that has recently received considerable attention in the solidification community; however, further effort is required.

6.4.2.2 I_l^D TERM

From the microscopic deviation model of Eq. (6.29) in Section 6.3.3, I_l^D can be written as

$$I_l^D = \nabla \cdot (2\varrho C p K)\nabla \overline{T_l}^l$$

and

$$K = (\alpha_l l)^2 (2S_{ij}S_{ij})^{\frac{1}{2}}$$

where α_l is a constant that should be decided from experimental or theoretical analysis. Furthermore, let us define the microscopic deviation conductivity as $k_l^D = 2\varrho C p K$, then Eq. (6.35) becomes

$$\frac{\partial \varrho_l \overline{H_l}}{\partial t} + \nabla \cdot g_l \varrho_l \overline{H_l}^l \overline{V_l}^l = \nabla \cdot \left(g_l k_l^* + g_l k_l^D\right)\nabla \overline{T_l}^l + I_l^Q + I_l^J.$$ (6.42)

It is noted that k_l^* accounts for the reduced diffusion effect due to tortuosity, while k_l^D accounts for the increased effect of diffusion due to the gradient of temperature. The deviation of k_l^* from its microscopic counterpart is significant when dendritic morphology is complex, as shown in Fig. 6.3, while k_l^D is significant when the volume fraction of liquid approaches unity.

In the same manner, Eq. (6.36) can also be written as

$$\frac{\partial \varrho_s \overline{H_s}}{\partial t} + \nabla \cdot g_s \varrho_s \overline{H_s}^s \overline{V_s}^s = \nabla \cdot \left(g_s k_s^* + g_s k_s^D\right) \nabla \overline{T_s}^s + I_s^Q + I_s^J. \tag{6.43}$$

6.4.2.3 I_l^J AND I_l^Q TERMS

Since the one-phase model is applied here, these interfacial terms need not be modeled because the interfacial heat flux and heat transfer between phases during solidification are balanced with each other, that is,

$$\left(I_l^Q + I_s^Q\right) = \left(I_l^J + I_s^J\right) = 0.$$

So adding Eqs. (6.42) and (6.43), the final form of the macroscopic transport equation of energy conservation is

$$\frac{\partial\left(\varrho_l \overline{H_l} + \varrho_s \overline{H_s}\right)}{\partial t} + \nabla \cdot \left(g_l \varrho_l \overline{H_l}^l \overline{V_l}^l + g_s \varrho_s \overline{H_s}^s \overline{V_s}^s\right)$$
$$= \nabla \cdot \left[g_l \left(k_l^* + k_l^D\right) \nabla \overline{T_l}^l + g_s \left(k_s^* + k_s^D\right) \nabla \overline{T_s}^s\right]. \tag{6.44}$$

6.4.3 Macroscopic Equation of Species Conservation

In this case, $\Phi = \varrho C$, in which C is the species concentration in weight percentage and $S = 0$. Furthermore, we utilize Fick's law for species diffusion flux.

Assumption 4   *Fick's law for binary species flux, $\mathbf{J} = -\varrho D \nabla C$.*

D is the binary diffusion coefficient. For liquid and solid phases, the averaged species equations yield, respectively,

$$\frac{\partial \varrho_l \overline{C_l}}{\partial t} + \nabla \cdot g_l \varrho_l \overline{C_l}^l \overline{V_l}^l = \nabla \cdot \overline{(\varrho_l D_l \nabla C_l)} + I_l^D + I_l^Q + I_l^J \tag{6.45}$$

$$\frac{\partial \varrho_s \overline{C_s}}{\partial t} + \nabla \cdot g_s \varrho_s \overline{C_s}^s \overline{V_s}^s = \nabla \cdot \overline{(\varrho_s D_s \nabla C_s)} + I_s^D + I_s^Q + I_s^J. \tag{6.46}$$

Following the same argument as in the derivation of the energy equation, the first term on the right side of Eq. (6.45) can be written in terms of the effective diffusion coefficient as

$$\overline{\varrho D_l \nabla C_l} = g_l \varrho_l D_l^* \nabla \overline{C_l}^l \tag{6.47}$$

where the area integral effect has been brought into the newly defined effective diffusivity D_l^*, which may be written as a function of area concentration S_v. The detailed expression for D_l^*, which accounts for the reduced diffusivity due to increased tortuosity, can be determined by an argument similar to that presented in Section 6.4.2 for the treatment of k_l^*.

In addition, from Eq. (6.29), the liquid microscopic deviation term can be written as

$$I_l^D = \nabla \cdot (2\varrho K)\nabla\overline{C_l}^l.$$

Defining the microscopic deviation diffusion coefficient as $D_l^D = 2\varrho K$, Eq. (6.45) becomes

$$\frac{\partial \varrho_l \overline{C_l}}{\partial t} + \nabla \cdot g_l \varrho_l \overline{C_l}^l \overline{V_l}^l = \nabla \cdot \left(g_l \varrho_l D_l^* + g_l \varrho_l D_l^D\right)\nabla_l\overline{C_l}^l + I_l^Q + I_l^J. \quad (6.48)$$

Equation (6.46) can also be rewritten in the same manner, as

$$\frac{\partial \varrho_s \overline{C_s}}{\partial t} + \nabla \cdot g_s \varrho_s \overline{C_s}^s \overline{V_s}^s = \nabla \cdot \left(g_s \varrho_s D_s^* + g_s \varrho_s D_s^D\right)\nabla\overline{C_s}^s + I_s^Q + I_s^J. \quad (6.49)$$

Adding Eqs. (6.48) and (6.49) together and taking into account that the interfacial species flux and species transfer due to solidification balance each other, that is,

$$\left(I_l^Q + I_s^Q\right) = \left(I_l^J + I_s^J\right) = 0,$$

the final form of the macroscopic mixture equation of species conservation is

$$\frac{\partial\left(\varrho_l\overline{C_l} + \varrho_s\overline{C_s}\right)}{\partial t} + \nabla \cdot \left(g_l\varrho_l\overline{C_l}^l\overline{V_l}^l + g_s\varrho_s\overline{C_s}^s\overline{V_s}^s\right)$$
$$= \nabla \cdot \left[g_l\varrho_l\left(D_l^* + D_l^D\right)\nabla\overline{C_l}^l + g_s\varrho_s\left(D_s^* + D_s^D\right)\nabla\overline{C_s}^s\right]. \quad (6.50)$$

6.4.4 Macroscopic Equation of Momentum Conservation

In this case, $\Phi = \varrho V$, in which V is velocity vector and $S = b$ is the body force vector. Furthermore, we assume a Newtonian fluid and hence the viscous stress is proportional to the ratio of deformation:

Assumption 5 *Newtonian Fluid Flow, that is,* $J = -\nabla P + \mu[\nabla V + (\nabla V)^t]$, *where* $(\nabla V)^t$ *is the transposed tensor of* ∇V.

Then for liquid and solid phases, the averaged momentum equations yield:

$$\frac{\partial \varrho_l \overline{V_l}}{\partial t} + \nabla \cdot g_l \varrho_l \overline{V_l}^l \overline{V_l}^l$$
$$= -\nabla\overline{P_l} + \nabla \cdot \left[\overline{\mu_l\nabla V_l} + \overline{\mu_l(\nabla V_l)^t}\right] + I_l^D + I_l^Q + I_l^J + \overline{b_l} \quad (6.51)$$

$$\frac{\partial \varrho_s \overline{V_s}}{\partial t} + \nabla \cdot g_s\varrho_s\overline{V_s}^s\overline{V_s}^s$$
$$= -\nabla\overline{P_s} + \nabla \cdot \left[\overline{\mu_s\nabla V_s} + \overline{\mu_s(\nabla V_s)^t}\right] + I_s^D + I_s^Q + I_s^J + \overline{b_s}. \quad (6.52)$$

Following the same argument as in the derivation of energy and species equations, the second term on the right side of Eq. (6.51) can be written in terms of the effective viscosity as

$$\overline{\mu_l\nabla V_l} + \overline{\mu_l(\nabla V_l)^t} = \mu_l^*\left[\nabla\overline{V_l}^l + \left(\nabla\overline{V_l}^l\right)^t\right]. \quad (6.53)$$

With this assumption, combined with the model we proposed earlier for the microscopic deviation term shown in Eq. (6.29), that is,

$$I_l^D = \nabla \cdot (2\varrho K)\left[\nabla \overline{V}_l^{\,l} + (\nabla \overline{V}_l^{\,l})'\right],$$

Eq. (6.51) becomes

$$\frac{\partial \varrho_l \overline{V}_l}{\partial t} + \nabla \cdot g_l \varrho_l \overline{V}_l^{\,l} \overline{V}_l^{\,l} = -\nabla \overline{P}_l + \nabla \cdot \left(g_l \mu_l^* + g_l \mu_l^D\right)$$
$$\times \left[\nabla \overline{V}_l^{\,l} + (\nabla \overline{V}_l^{\,l})'\right] + I_l^Q + I_l^J \tag{6.54}$$

where the microscopic deviation viscosity is defined as $\mu_l^D = 2\varrho K$. Equation (6.52) can also be rewritten in the same manner as

$$\frac{\partial \varrho_s \overline{V}_s}{\partial t} + \nabla \cdot g_s \varrho_s \overline{V}_s^{\,s} \overline{V}_s^{\,s} = -\nabla \overline{P}_s + \nabla \cdot \left(g_s \mu_s^* + g_s \mu_s^D\right)$$
$$\times \left[\nabla \overline{V}_s^{\,s} + (\nabla \overline{V}_s^{\,s})'\right] + I_s^Q + I_s^J. \tag{6.55}$$

Equations (6.54) and (6.55) may be added according to the fact that the interfacial momentum fluxes due to solidification balance each other, that is,

$$\left(I_l^Q + I_s^Q\right) = 0.$$

However, the interfacial stress in this case is not balanced; across the interface, there is a jump condition due to surface tension (Drew 1983),

$$\left(I_l^J + I_s^J\right) = \sigma \varkappa$$

where σ is the surface tension, assumed to be constant, and $\varkappa$ is the mean curvature of the interface. Many previous derivations do not explicitly include this term (Beckermann and Viskanta 1993, Ganesan and Poirier 1990), since this effect is small in alloy solidification and hence often negligible. This term is considered here for completeness. The final form of the macroscopic transport equation of momentum conservation then yields

$$\frac{\partial \left(\varrho_l \overline{V}_l + \varrho_s \overline{V}_s\right)}{\partial t} + \nabla \cdot \left(g_l \varrho_l \overline{V}_l^{\,l} \overline{V}_l^{\,l} + g_s \varrho_s \overline{V}_s^{\,s} \overline{V}_s^{\,s}\right)$$
$$= -\nabla\left(\overline{P}_l + \overline{P}_s\right) + \nabla \cdot \left\{g_l\left(\mu_l^* + \mu_l^D\right)\left[\nabla \overline{V}_l^{\,l} + (\overline{V}_l^{\,l})'\right]\right.$$
$$\left. + g_s\left(\mu_s^* + \mu_s^D\right)\nabla\left[\overline{V}_s^{\,s} + (\overline{V}_s^{\,s})'\right]\right\} + b + \sigma \varkappa. \tag{6.56}$$

6.4.5 Comment on the Formulation of Effective Coefficients

In Sections 6.4.2–6.4.4, we used Eq. (6.16) of Theorem 2 to expand the terms that are related to the viscosity, conductivity, and species diffusivity. This method was also adopted by Beckermann and Viskanta (1993) and Gray (1975). This procedure leads to the introduction of effective coefficients to handle the corresponding morphological complexity. However, there is another method of derivation for treating these terms. This method is summarized below.

Using the macroscopic energy equations (6.35) and (6.36) as an example, the first terms on their right sides can be expanded according to the first equation of Theorem 2, Eq. (6.15). This method is adopted by Whitaker (1967) and Poirier (1990). We then obtain the following:

$$\overline{k_l \nabla T_l} = k_l \nabla g_l \overline{T_l}^l + k_l \frac{1}{dV} \int_{dA_k} T_l n_l \, da \tag{6.57}$$

and

$$\overline{k_s \nabla T_s} = k_s \nabla g_s \overline{T_s}^s + k_s \frac{1}{dV} \int_{dA_k} T_s n_s \, da. \tag{6.58}$$

Considering a continuous distribution of temperature over the liquid–solid phase boundary (i.e., $T_l = T_s$) and the relation between the unit normal (i.e., $n_l = -n_s$) the two integral terms from Eqs. (6.57) and (6.58) cancel each other when they are added, resulting in

$$\overline{k_l \nabla T_l} + \overline{k_s \nabla T_s} = k_l \nabla g_l \overline{T_l}^l + k_s \nabla g_s \overline{T_s}^s. \tag{6.59}$$

In the above, no effective conductivity is introduced, but g_l and g_s appear within the spatial derivatives. The previous method with effective conductivity yields

$$\overline{k_l \nabla T_l} + \overline{k_s \nabla T_s} = k_l^* g_l \nabla \overline{T_l}^l + k_s^* g_s \nabla \overline{T_s}^s \tag{6.60}$$

with g_l and g_s appearing outside the spatial derivatives. Equation (6.59) can be expanded with respect to the spatial derivative to yield

$$\overline{k_l \nabla T_l} + \overline{k_s \nabla T_s} = k_l \left(g_l \nabla \overline{T_l}^l + \overline{T_l}^l \nabla g_l \right) + k_s \left(g_s \nabla \overline{T_s}^s + \overline{T_s}^s \nabla g_s \right). \tag{6.61}$$

Comparing with Eq. (6.60), it can be seen that

$$k_l^* = k_l \left\{ 1 + \left(\frac{\nabla g_l}{g_l} \right) \left[\frac{\overline{T_l}^l}{\nabla \overline{T_l}^l} \right] \right\}$$

and

$$k_s^* = k_s \left\{ 1 + \left(\frac{\nabla g_s}{g_s} \right) \left(\frac{\overline{T_s}^s}{\nabla \overline{T_s}^s} \right) \right\}. \tag{6.62}$$

The same derivation holds true for the treatment of the diffusivity terms in species and momentum transport equations. However, in the species transport process, in general there is a jump condition of species concentration over the solid/liquid interface, that is, $C_l \neq C_s$. The corresponding surface integrations shown in Eq. (6.57) or (6.58) will not cancel each other; thus the effective diffusivity of species as a function of microscopic complexity still has to be introduced.

6.4.6 Comments on the Momentum Equation

Investigations on the formulation of the macroscopic momentum equation are abundant in the literature because of its wide application in porous media problems and two-phase flows (Drew 1983). Most formulations eventually reduce the liquid momentum equation to the form of Darcy's law or may include an additional term to account for inertial effects. Then the solution of the momentum equation is in essence transferred into a problem of deciding the permeability that is required by Darcy's law.

While the measurement of permeability for flows with low liquid volume fraction or isotropic porous structure is well established, the study of the permeability of flows through anisotropic porous media or flows with high liquid volume fraction is difficult experimentally. Numerical or theoretical efforts have been combined with experimental evidence to tackle this problem. For example, Ochoa-Tapia and Whitaker (1995a,b) developed a method to continuously connect the velocity boundary between a porous media and a homogeneous fluid. Ghaddar (1995) developed a parallel computational method to compute the permeability of a randomly distributed array. Bhat et al. (1995a,b) numerically computed the permeability at high volume fraction of a liquid with results that merge with the empirical data at the lower volume fraction.

In materials solidification, Darcy's law is applicable only when the solid phase is stationary and the fluid moves with a sufficiently low velocity. In traditional porous media flows, the porous media is assumed to be a fixed structure. During the solidification process, however, the morphology of dendrites varies in space and time; thus the description of permeability in solidification modeling is more challenging. It is also noticed that if solid particles are dispersed into the liquid and move with it, then Darcy's law no longer applies. In this case, one has to directly apply the general form of the macroscopic equation of momentum, Eq. (6.56). Voller et al. (1989) gave an illustration of this application. In the following we present a derivation based on Darcy's treatment, starting from the liquid momentum equation (6.51), which can also be written as

$$\frac{\partial}{\partial t}\left(\varrho_l g_l \overline{V_l}^l\right) + \nabla \cdot \left(\varrho_l g_l \overline{V_l}^l \overline{V_l}^l\right) = -\nabla\left(g_l \overline{P_l}^l\right) + \nabla \cdot \left[\overline{\mu_l \nabla V_l + \mu_l \left(\nabla V_l\right)^t}\right]$$
$$+ I_l^D + I_l^Q + I_l^J + g_l \varrho_l \mathbf{g} \qquad (6.63)$$

where I_l^D represents the microscopic deviation momentum flux, I_l^J represents the interfacial stresses, and I_l^Q represents the interfacial momentum transfer due to phase change.

As only the liquid phase is considered here, all these terms need to be modeled in detail as discussed in Beckermann and Viskanta (1993), Drew (1983), Ganesan and Poirier (1990), and Slattery (1967). Discrepancies in the interpretation of different terms may arise from different derivations. Nevertheless, they can all eventually be reduced to the same form, that is, Darcy's law, after making appropriate assumptions. For example, if I_l^D is interpreted as the interfacial stress of a combination of "viscous drag" and "form drag," and assumed to be dependent on the relative mass-averaged velocity between the liquid and the solid, it can be written as (Ganesan and Poirier 1990, Whitaker 1967)

$$I_l^J = -\mu\Gamma(V_l - V_s)$$

where Γ is a resistance coefficient which is a function of the morphology of the solid in the mushy zone.

Furthermore, assuming that the momentum exchange caused by volume change during phase change is negligible (i.e., $I_l^Q = 0$) and neglecting the momentum microscopic deviation term (i.e., $I_l^D = 0$), Eq. (6.63) yields

$$\frac{\partial}{\partial t}\left(\varrho_l g_l \overline{V_l}^l\right) + \nabla \cdot \left(\varrho_l g_l \overline{V_l}^l \overline{V_l}^l\right)$$
$$= -\nabla\left(g_l \overline{P_l}^l\right) + g_l \varrho_l \mathbf{g} + \nabla \cdot \left[\overline{\mu \nabla\left(g_l V_l^l\right)} + \overline{\mu \nabla\left(g_l V_l^l\right)^l}\right]$$
$$- \mu \Gamma \cdot \left(\overline{V_l}^l - \overline{V_s}^s\right). \tag{6.64}$$

In addition, by assuming (a) steady and slow flow (i.e., $\overline{V_l}^l \to 0$) and hence that the inertial term can be neglected; (b) that the volume function of liquid g_l is uniform and constant; and (c) that the forces of the liquid–liquid interactions are negligibly small, Eq. (6.64) then becomes

$$\overline{V_l}^l = -\frac{g_l^2 \Gamma}{\mu}\left[\nabla \overline{P_l}^l - \varrho_l \mathbf{g}\right] \tag{6.65}$$

which is Darcy's law with an additional body force for approximating the velocity in the mushy zone.

6.5 Mixture Approach

From the above derivation, it can be seen that by the volume-averaging approach, the macroscopic equations are rigorously derived from the exact microscopic equations. If appropriate relations between the macro–micro scales are given, the phase change process can be accurately modeled. In the mixture approach, as the validity of certain continuum relations on a macroscopic scale has to be first assumed, no direct connection has been established in the literature with regard to the microscopic characteristics. This approach applies the conservation law of a certain quantity Φ_k directly within the averaging volume dV and expresses the balance of the net outflow with convection, diffusion, and volumetric production:

$$\frac{\partial}{\partial t}\int_{dV}\varrho_k \Phi_k dV_k + \int_{dA}\left(\varrho_k V_k \Phi_k\right) \cdot \mathbf{n} dA_k = \int_{dA} -\mathbf{J}_k \cdot \mathbf{n} dA_k + \int_{dV} S_k dV_k \tag{6.66}$$

where $\mathbf{n}$ is the outward unit normal to the surface, $\mathbf{J}_k$ is a diffusive flux vector, and S_k is a volumetric source term (Incropera and Bennon 1987a). When g_k is assumed continuous, the integrands of the above equations are continuous and differentiable functions, and the integral theorem of Leibnitz and Gauss can be applied. Furthermore, as the averaging control volume is arbitrary, the above equation can then be expressed in differential form as

$$\frac{\partial}{\partial t}\left(\varrho_k g_k \Phi_k\right) + \nabla \cdot \left(\varrho_k g_k V_k \Phi_k\right) = -\nabla \cdot \left(g_k \mathbf{J}_k\right) + g_k S_k \tag{6.67}$$

where Φ_k equals 1; V_k, H_k, and C_k for mass, momentum, energy, and species equations, respectively. By applying this equation to different phases in the averaging control volume dV and then adding them, the so-called macroscopic continuum mixture equations by the mixture approach result.

6.6 Application to the Columnar Solidification of Binary Alloys

In this section, the macroscopic transport equations derived above are applied to a special case of columnar solidification of binary liquid-solid alloys. In order to accomplish this, first simplified governing equations are obtained from the general macroscopic transport theory based on assumptions related to the characteristics of columnar solidification of binary mixture. Then, corresponding numerical techniques for the solutions of governing equations and implementations of the phase diagram are developed to a point that computer code can be developed based on the materials presented.

6.6.1 The Simplification of Transport Equations

Several additional assumptions have been made to simplify the governing equations for the columnar solidification of binary alloys. They are listed in Table 6.1. Assumption 6 is made because in columnar solidification, the solid phase is a rigid region that is fixed to a cooled wall, and therefore the velocity in the solid region is zero. Assumption 7 can be justified by the high thermal diffusivity of metal alloys and the presence of creeping flow in the columnar mushy zone (Beckermann and Viskanta 1993). Assumption 8 is typically good for the liquid within the interdendritic meshwork, but it produces errors at the dendritic tips due to solute undercooling

Table 6.1 *Assumptions for the Modeling of Columnar Solidification of Binary Alloys*

Assumption 6

$$\overline{V}_s^s = \overline{V}_s = V_s = 0.$$

Assumption 7   All phases in the averaging volume are in thermal equilibrium, that is,

$$\overline{T}_s^s = \overline{T}_l^l = T.$$

Assumption 8   The liquid in the averaged volume is solutally well mixed, that is,

$$\overline{C}_l^l = C_l$$

and, at the liquid–solid interface, the relation of solidus and liquidus concentration is given by the phase diagram.

Assumption 9   The species flux in the solid region is neglected, that is,

$$D_s = 0.$$

(Flood and Hunt 1987, Huppert 1990). The main drawback of this assumption is that a kinetic law for the movement of the dendritic tips can not be introduced (Rappaz 1989). But this effect can now be accounted for with our consideration of microscopic deviation terms. Assumption 9 applies to almost all solidification processes because the species (solute) diffusion in the solid is much smaller than that in the liquid (i.e., $D_s < D_l$). However, this assumption does not necessarily imply that the species (solute) distribution in the solid of the averaged volume is uniform. In fact, species (solute) concentration in the solid phase is usually assumed in the literature with either complete or no diffusion (Level Rule or Scheil) models (Kurz and Fisher 1989), which will be explained in detail later.

With these assumptions, the macroscopic transport equations of mass, heat, and species can then be simplified from Eqs. (6.34), (6.44), and (6.50), respectively, and yield

$$\frac{\partial(g_l\varrho_l + g_s\varrho_s)}{\partial t} + \nabla \cdot \left(g_l\varrho_l\overline{V}_l^l\right) = 0 \tag{6.68}$$

$$\frac{\partial\left(g_l\varrho_l\overline{H}_l^l + g_s\varrho_s\overline{H}_s^s\right)}{\partial t} + \nabla \cdot \left(g_l\varrho_l\overline{H}_l^l\overline{V}_l^l\right) = \nabla \cdot \left[g_l\left(k_l^* + k_l^D\right) + g_sk_s^*\right]\nabla T \tag{6.69}$$

$$\frac{\partial\left(g_l\varrho_l\overline{C}_l^l + g_s\varrho_s\overline{C}_s^s\right)}{\partial t} + \nabla \cdot \left(g_l\varrho_l\overline{C}_l^l\overline{V}_l^l\right) = \nabla \cdot \left[g_l\varrho_l\left(D_l^* + D_l^D\right)\nabla\overline{C}_l^l\right]. \tag{6.70}$$

Further simplification of momentum equation (6.56) is rather difficult because the effective viscosity in the mushy region can not be estimated directly. For columnar solidification, with the solid region fixed, we will directly adopt Darcy's approximation. The momentum equation can then be obtained from Eq. (6.64) as

$$\frac{\partial}{\partial t}\left(\varrho_l g_l\overline{V}_l^l\right) + \nabla \cdot \left(\varrho_l g_l\overline{V}_l^l\overline{V}_l^l\right) = -\nabla\left(g_l\overline{P}_l^l\right) + g_l\varrho_l g + \nabla \cdot \overline{\left[\mu\nabla\left(g_l\overline{V}_l^l\right)\right.}$$
$$\left. + \mu\nabla\left(g_l\overline{V}_l^l\right)^T\right] - \mu\Gamma\overline{V}_l^l \tag{6.71}$$

where Γ is the resistance coefficient. Adopting the Kozeny-Carman equation for flow in a porous media (Bennon and Incropera 1987b), we get

$$\Gamma = K_0\frac{(1 - g_l)^2}{g_l^3} \tag{6.72}$$

where K_0 is a permeability constant depending on the morphology of the mushy zone during phase change.

Furthermore, we define that

$$\varrho = \varrho_s g_s + \varrho_l g_l \qquad \varrho f_s = \varrho_s g_s \qquad \varrho f_l = \varrho_l g_l \tag{6.73}$$

and

$$V = f_l\overline{V}_l^l \qquad \varrho_l g_l\overline{V}_l^l = \varrho f_l\overline{V}_l^l = \varrho V \tag{6.74}$$

where ϱ is the averaged density, V is the averaged velocity in the averaging volume dV, f_s and f_l are mass fractions of solid and liquid respectively, and $f_l + f_s = 1$ for binary alloys with no pores. Then Eq. (6.68) of mass conservation becomes

$$\frac{\partial \varrho}{\partial t} + \nabla \cdot (\varrho V) = 0 \tag{6.75}$$

and Eq. (6.71) for momentum conservation becomes

$$\frac{\partial}{\partial t}(\varrho V) + \nabla \cdot (\varrho g V V)$$

$$= -\nabla(P) + \varrho g + \nabla \cdot \left[\mu \nabla V + \mu \nabla V^T\right] - \mu K_0 \frac{(1-g_l)^2}{g_l^3} V. \tag{6.76}$$

This equation reveals that, in the pure solid, when $g_l = 0$ and $\Gamma \to \infty$, $V = 0$; in the pure liquid, $g_l = 1$ and $\Gamma \to 0$, and the equation reduces to the classical Navier-Stokes equations; and in the mushy zone, it incorporates the Darcy's law for porous media flow.

For the energy conservation equation, we define

$$\Delta H = \overline{H_l}^l - \overline{H_s}^s \tag{6.77}$$

where $\overline{H_l}^l$ is the averaged total enthalpy of the interdendritic liquid. $\overline{H_s}^s$ is the averaged total enthalpy of the solid and ΔH is an effective latent heat, which varies during the solidification of alloys (Poirier et al. 1988). Substituting $\overline{H_l}^l$ by $\overline{H_s}^s + \Delta H$ into the time-dependent term of Eq. (6.69) of energy conservation and using the definition of Eqs. (6.73) and (6.74), Eq. (6.69) can then be cast in the form

$$\frac{\partial}{\partial t}\left(\varrho \overline{H_s}^s + \varrho f_l \Delta H\right) + \nabla \cdot \left[\varrho V\left(\overline{H_l}^s + \Delta H\right)\right]$$

$$= \nabla \cdot \left[\left(g_l k_l^* + g_s k_s^*\right) + g_l k_l^D\right]\nabla T. \tag{6.78}$$

Since $\overline{H_s}^s = CpT$, where Cp is a function of both temperature and composition (Poirier and Nandapurkar 1988), the energy equation can then be further cast as

$$\frac{\partial}{\partial t}(\varrho CpT) + \nabla \cdot (\varrho CpTV) = \nabla \cdot \left[\left(g_l k_l^* + g_s k_s^*\right) + g_l k_l^D\right]\nabla T$$

$$- \left[\frac{\partial}{\partial t}(\varrho f_l \Delta H) + \nabla \cdot (\varrho V \Delta H)\right]. \tag{6.79}$$

For the solute (species) conservation equation, the extent of diffusion in the local solid of a control volume dV must be estimated in order to account for the partitioning of the solute between the local dendritic solid and the interdendritic liquid (Poirier et al. 1991). Two models, those of complete diffusion and of no diffusion, also referred in the literature as Level Rule model and Scheil model respectively (Kurz and Fisher 1989), are commonly used. The complete-diffusion model assumes that the solute concentrations in both the local solid portion (C_s) and in the local liquid portion (C_l) within an averaging volume dV are completely uniform. On the other hand, the no-diffusion model assumes that no diffusion takes place in the solid while the solute concentration in the local liquid (C_l) is uniform.

Now let us introduce

$$\overline{C} = f_l \overline{C_l}^l + f_s \overline{C_s}^s \tag{6.80}$$

or

$$f_l = \frac{\overline{C} - \overline{C_s}^s}{\overline{C_l}^l - \overline{C_s}^s} \tag{6.81}$$

where $\overline{C}$ is the averaged weight percentage of solute concentration in the averaged volume dV.

According to the definition of Eq. (6.80), Eq. (6.70) then becomes

$$\frac{\partial}{\partial t}(\varrho \overline{C}) + \nabla \cdot (\varrho \overline{C_l}^l V) = \nabla \cdot \left[\varrho f_l (D_l^* + D_l^D) \nabla \overline{C_l}^l \right]. \tag{6.82}$$

Replacing $\overline{C_l}^l$ by $\overline{C_l}^l + \overline{C} - \overline{C}$, Eq. (6.82) can further be written as

$$\frac{\partial}{\partial t}(\varrho \overline{C}) + \nabla \cdot (\varrho \overline{C} V) = \nabla \cdot \left[\varrho f_l (D_l^* + D_l^D) \nabla \overline{C} \right] + \left\{ \nabla \cdot \left[\varrho (\overline{C} - \overline{C_l}^l) V \right] \right.$$
$$\left. + \nabla \cdot \left[\varrho f_l (D_l^* + D_l^D) \nabla (\overline{C_l}^l - \overline{C}) \right] \right\}. \tag{6.83}$$

This form of solute equation has the advantage of allowing implementation of different microscopic diffusion models, avoiding the necessity of changing the form of governing equations every time the microscopic diffusion model is changed. For example, Voller et al. (1989) and Poirier et al. (1991) have to change their solute equations when their diffusion models are changed. In order to do this, they first represent $\overline{C_l}^l$ in terms of $\overline{C}$ according to the definition of Eq. (6.80), substitute it back into Eq. (6.82), and then apply mathematical operations such as Newton-Leibnz's rule to regroup the equation. This way, different diffusion models yield different forms of governing equations, which makes numerical implementation rather difficult. With the present way of deriving the solute equation, however, any type of diffusion model can be accounted for by Eq. (6.83). The effect of diffusion models comes into play via the different constructions of Eq. (6.81), which will be detailed later.

Equations (6.75), (6.76), (6.79), and (6.83) are then the final forms of our simplified governing equations for the modeling of columnar solidification of binary alloys. Fluid fraction f_l can be determined from Eq. (6.81), where the phase diagram comes into play. However, to close the problem, the constitutive equations of the microscopic deviation and effective coefficients included in these equations also have to be supplemented as a function of microscopic transport characteristics. According to previous discussions, the microscopic deviation coefficients of diffusivity are constructed as

$$k_l^D = 2\varrho C p K \quad \text{and} \quad D_l^D = 2\varrho K \tag{6.84}$$

where K is given by Eq. (6.27).

For the purpose of showing the basic approach of incorporating the microscopic complexities into the macroscopic transport equations, the formulations of effective coefficients of diffusivity adopted here are simple empirical models as discussed previously (Speich and Fisher 1966). They are given in the following:

$$k_l^* = k_l\left(1 - g_l + g_l^2\right), \qquad k_s^* = k_s\left(1 - g_s + g_s^2\right),$$

and

$$D_l^* = D_l\left(1 - g_l + g_l^2\right). \tag{6.85}$$

It should be noted that the accurate construction of these constitutive equations of microscopic deviation or effective coefficients requires extensive experimental investigations or direct numerical simulations of morphological evolution (Rappaz 1989).

6.6.2   Implementation of the Numerical Procedure

6.6.2.1   SOLUTION PROCEDURE

The calculation of the coupled effect of mass, momentum, heat, and solute transport is a transient process. The following steps are employed in the current study to accomplish this process.

(1) Give the initial conditions at $t = 0$, that is, the initial values of $V, T, \overline{C}, f_l,$ and $\overline{C}_l^{\,l}$.
(2) Go to $t = t + \Delta t$.
(3) Solve V, T, and $\overline{C}$ according to Eqs. (6.75), (6.76), (6.79) and (6.83).
(4) Update f_l according to the newly obtained T and $\overline{C}$ via Eq. (6.81).
(5) With the newly updated f_l and $\overline{C}_l^{\,l}$, recalculate the source term of Eq. (6.83), go back to step (3), and repeat the procedure. If the solution is converged, then go to step (6).
(6) Go back to step (2) for the next time step.

In this procedure, step 4 is where the phase diagram comes into play and where the energy and solute equations are coupled. Different microscopic diffusion models will lead to different expressions of f_l. This issue is discussed in more detail next.

6.6.2.2   IMPLEMENTATION OF THE DIFFUSION MODEL

Most of the numerical models reported in the literature consider only the complete diffusion model (Beckermann and Viskanta 1993, Bennon and Incropera 1987a, Prakash and Voller 1989, Prescott and Incropera 1994, Viskanta 1990). Voller et al. (1989) derived the solute equation assuming that no diffusion takes place in the local solid. However, their numerical updating method for f_l seems to still adopt the one based on the complete diffusion model and is inconsistent with the governing equation. Poirier et al. (1991) took full account of both models and derived the corresponding governing equations, but their calculation is quite limited and they presented

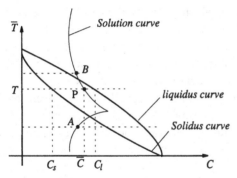

Figure 6.4 Schematic of the calculation of fluid fraction and mushy region.

only qualitative results. They assume that the cooling speed is known *a priori* such that they can define a new coordinate system z' that moves with the solidus interface. The boundary condition at $z' = 0$ is then given by the eutectic temperature and the eutectic solute composition. In addition, their computation is only one dimensional. Since these calculations are based on the solute transport equations derived via different models and an inadequate numerical resolution, it is difficult to separate physical and numerical uncertainties.

Here, for the complete diffusion model, it is assumed that the solute concentrations of the local solid (C_s) and of the local liquid (C_l) in a control volume dV are completely uniform:

$$\overline{C} = f_l C_l + f_s C_s. \tag{6.86}$$

Then combining Eq. (6.81) and (6.86), we can directly obtain

$$f_l = \frac{\overline{C} - C_s}{C_l - C_s} \tag{6.87}$$

where C_s and C_l can be decided by the phase diagram as shown schematically in Fig. 6.4 when temperature T is given. The mushy zone can then be identified in the following way. If $\overline{C} < C_s$ and thus $f_l < 0$, for example at point A, then we specify that $f_l = 0$ and know that A is in the solid region; if $\overline{C} > C_l$ and thus $f_l > 1$, for example at point B, then we let $f_l = 1$ and know that B is in the liquid zone; otherwise, if $0 < f_l < 1$, the corresponding point P must be located in the mushy zone. With this approach of updating f_l, no assumption is necessary for the liquidus and solidus curves to be straight lines, nor is it necessary that the constant partition coefficient k_p be specified. As long as the liquidus or solidus curves are given in the form of a set of data or certain curve fitting functions, C_s and C_l can be obtained for a given temperature.

For the no-diffusion model, it is assumed that the solute concentration of the local solid (C_s) is not diffusive, while the solute concentration in the local liquid (C_l) is uniform. Then it follows that

$$\overline{C} = \int_0^{f_s} C_s(\eta) d\eta + f_l C_l \tag{6.88}$$

where $C_s(\eta)$ is an unknown function, in which η represents the solidus volume fraction. In this case f_l cannot be obtained directly from Eq. (6.81). However, we can always approximate the function $C_s(\eta)$ by curve fitting. For example, if we want to update f_l of a designated averaging volume, we can record the solute concentration at the instant when the material in that volume starts to solidify, that is, $C_s(0)$ at $f_s = 0$ from the previous calculations. At the current time instant, the solidus concentration at the solid/liquid interface, which is in equilibrium with the liquidus concentration C_l, is given by $C_s^*(f_s)$. Then by assuming a linear variation of $C_s(\eta)$ as a function of η, it can be shown that

$$\int_0^{f_s} C_s(\eta) d\eta = \frac{[C_s(0) + C_s(f_s^*)]}{2}. \tag{6.89}$$

So we can still update f_l by the expression

$$f_l = \frac{\overline{C} - C_s^*}{C_l - C_s^*}$$

which is in the same form as Eq. (6.87), except that

$$C_s^* = \frac{[C_s(0) + C_s^*(f_s)]}{2}, \tag{6.90}$$

which is no longer the real equilibrium solidus concentration that corresponds to the equilibrium temperature T. We call it the effective solidus concentration, which accounts for the nonuniform distribution of solidus concentration in the local control volume.

This approach of updating f_l is flexible and easy to modify. For example, if we want to obtain a more accurate approximation of $C_s(\eta)$ as a function of solidus fraction η, we just need to store one more point calculated at the previous time instant ($C_s(f_{s1})$ at $f_s = f_{s1}$). By assuming a quadratic function of $C_s(\eta)$, C_s^* can thus be expressed in terms of $C_s(0)$, $C_s(f_{s1})$, and $C_s^*(f_s)$. In addition, different liquidus diffusion models, which are necessary near the dendritic tip region where liquidus solute gradients play an important role, can also be accounted for by the current approach.

6.7 Concluding Remarks

In this chapter, the macroscopic transport equations are systematically derived from the microscopic continuum equations via a volume-averaging approach. These equations can be applied to solve a variety of multiphase problems including alloy solidification, in which the coupled effects of fluid flow, heat transfer, as well as species transport are present. The formulation of microscopic deviation terms, which arise from the averaging procedure and have been largely neglected in the literature, have been addressed via an approach analogous to turbulence modeling. Using the mathematical formulations of macroscopic transport equations presented in this chapter, coupling between complex morphologies at the microscopic level and transport phenomena at the macroscopic level can be established via appropriate constructions of effective coefficients of diffusivity and microscopic deviation terms.

We have also applied the macroscopic transport equations to the columnar solidification of binary alloys. A convenient form of solute transport equation has been derived for the easy update of fluid fraction via the introduction of an effective solidus concentration. In the case of complete diffusion (Level Rule model), this effective solidus concentration is simply equal to the equilibrium solidus temperature that corresponds to the local temperature; in the case of zero solid diffusion (Scheil model), this effective solidus concentration is some weighted value between the local equilibrium solidus concentrations and the solidus concentration recorded from previous time instants.

7 Practical Applications

Based on the techniques presented in previous chapters, many practically relevant problems can be handled effectively. In this chapter, selected examples are presented to illustrate the effectiveness of the computational and modeling techniques detailed earlier in this book, such as the multiblock method, multilevel modeling, turbulence closure, and volume averaging for phase change problems. The cases presented are diverse in nature, including the turbulent fluid flow inside hydraulic power plants, the double-diffusive flow in a thermohaline stratification, and the vertical Bridgman growth of NiAl crystals. Both the physical relevance of these problems and the computational issues associated with these cases are discussed. An attempt is made to describe the physics associated with these problems in some detail, but the intent is more to provide a flavor of the dynamic and geometric complexity of some representative practical problems of engineering interest that can be handled by the techniques presented in this book.

7.1 Flow in a Hydraulic Turbine

The first case involves turbulent flows in a hydraulic turbine. The geometries used in the computations to be presented are that of components of a hydraulic turbine including the distributor and the draft tube.

7.1.1 Distributor Flow Computations

In the following, a discussion of the distributor flow computations will be presented to shed more light on the application of the multiblock technique. Two distributor configurations are considered.

7.1.1.1 CASE A: 5 BLADES

In this configuration, a single periodic sector consists of two fixed stay vanes (first row of airfoils) and three pivoting wicket gates (second row of airfoils). Figure 7.1

260

a periodic sector within which computations are conducted

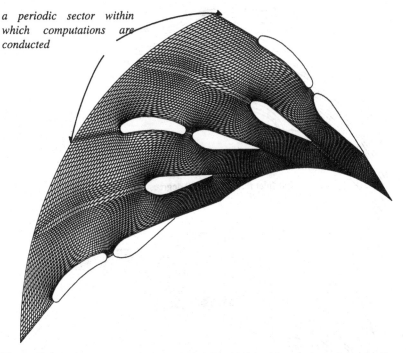

Figure 7.1 Distributor A: The two block grid with 86 × 23 nodes in each block. Shown here are two consecutive periodic sectors.

shows the flow domain and the grid of two consecutive periodic sectors. In the actual computation, only one periodic sector is computed. For a single passage, to accomodate the three wicket gates, the flow domain consists of two blocks (86 × 23 and 86 × 23 nodes), each of which is between a pair of wicket gates. The grid interfaces are continuous in this case. For the upper and lower boundaries, the periodic boundaries are alternated with the solid walls representing the stay vane and the wicket gate. It is noted that, due to the nonparallel arrangement between two consecutive blades, the periodic condition needs to be applied with the consideration of coordinate rotation.

The present two-dimensional flow analysis is based on the full Reynolds-averaged Navier-Stokes equations. The Reynolds number based on the inlet flow condition is 10^6. The original K-ε two-equation turbulence model is adopted here as the closure form. As to the discretization, second-order central difference schemes are applied to all derivatives except the convection terms, which are treated by the second-order upwind scheme (Shyy 1994, Shyy et al. 1992a, Thakur and Shyy 1993). At the inlet of the flow domain, a uniform flow distribution is assigned with a given flow angle, which is measured relative to the tangent of the circle defining the locations of the leading edge of the stay vanes. For the present case, the flow analyses are performed at the inlet flow angles of 20 and 30 degrees. At the exit, zero first-order derivative along the streamwise direction is taken for all dependent variables, except for static pressure which is regulated by the exit mass flux condition. At the nodal position next to the solid wall, the wall function treatment is applied. The solution is considered

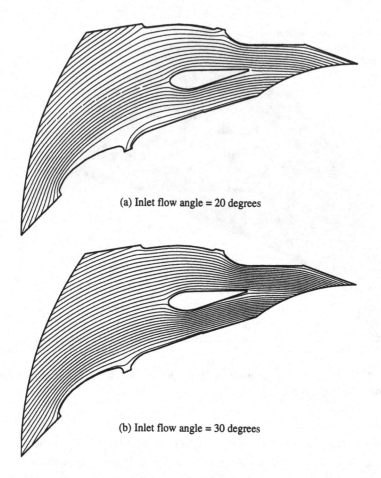

(a) Inlet flow angle = 20 degrees

(b) Inlet flow angle = 30 degrees

Figure 7.2   Streamline patterns for two inlet flow angles within a periodic sector.

to be converged when the sum of the momentum and mass residuals, normalized by the inlet momentum and mass fluxes respectively, is below 10^{-4}. The streamlines are presented in Fig. 7.2. At a 20-degree inlet flow angle, the flow separates at the upper surface of stay vane; while at a 30-degree inlet flow angle, no separation happens. Figure 7.3 shows the corresponding pressure distributions. In both cases, substantial fluid flow occurs across the periodic boundary, requiring a careful treatment there.

7.1.1.2   CASE B: 4 BLADES

In this configuration, a single periodic sector consists of two fixed stay vanes and two wicket gates. Figure 7.4 shows the flow domain and the grid (142×23 nodes for a single sector). Again, two sectors are stacked together to help visualize the periodic boundaries. For the present domain, a single block grid with continuous periodic boundary condition is found to be fairly skewed throughout a majority of the domain, because grid lines along the bottom boundary proceed directly to their corresponding points on the top boundary. Instead, a single block grid with the discontinuous grid

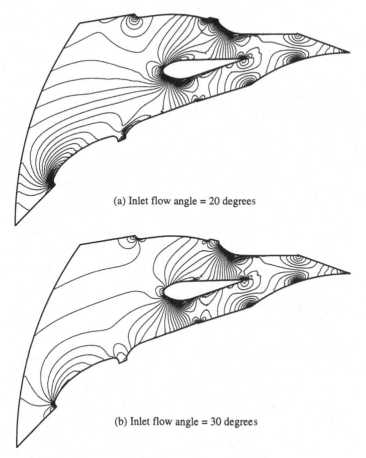

(a) Inlet flow angle = 20 degrees

(b) Inlet flow angle = 30 degrees

Figure 7.3 Pressure fields for two inlet flow angles within a periodic sector.

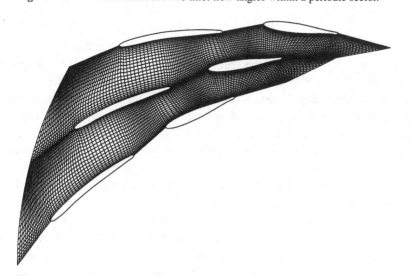

Figure 7.4 Distributor B: Single block grid with discontinuous periodic boundaries. Two periodic sectors are shown here.

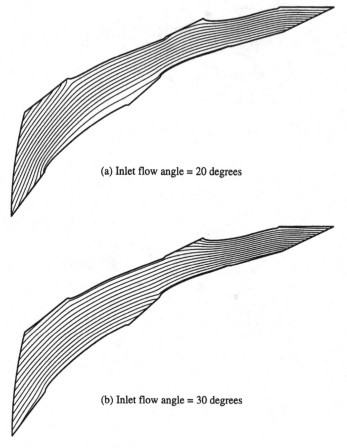

(a) Inlet flow angle = 20 degrees

(b) Inlet flow angle = 30 degrees

Figure 7.5   Streamline patterns for two inlet flow angles within a periodic sector.

distribution across the periodic boundary is adopted, as illustrated in Fig. 7.4. At the first periodic boundary pair, the grid spacings on the two sides of the interface are of the ratio 2:1, with the upper boundary acting as the fine grid interface. At the second periodic boundary the grid is continuous, and at the third periodic boundary, the grid spacings on the two sides of the interface are of a ratio of 1:2 with the lower boundary acting as the fine grid interface. The periodic boundaries are treated as nonoverlapping discontinuous grid interfaces, where the continuity of dependent variables and the local conservation of mass flux are maintained by employing a piecewise-linear interpolation with conservative correction. It can be seen that with the introduction of the discontinuous grid interface, the grid skewness is greatly reduced for this type of domain consisting of a staggered array of airfoils. In this case, the incoming flow condition is similar to that in case A. The flow analyses are also performed for the inlet flow angles of 20 and 30 degrees. When convergence is reached, both the normalized momentum and mass residuals are below 10^{-4}, showing that the discontinuous grid interface treatment is successful. The corresponding streamline and pressure distributions are shown in Figs. 7.5 and 7.6, respectively.

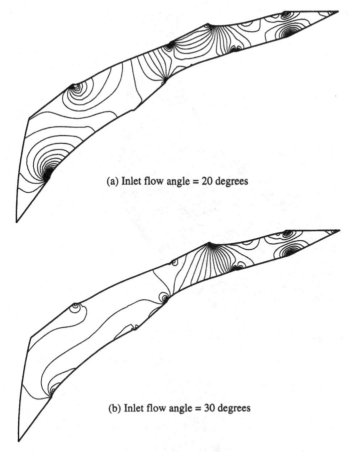

(a) Inlet flow angle = 20 degrees

(b) Inlet flow angle = 30 degrees

Figure 7.6   Pressure fields for two inlet flow angles within a periodic sector.

7.1.2   Flow in the Draft Tube

In this section we present the flow in the draft tube, shown in Fig. 7.7. It is used as a means to recover the pressure potential energy of the flow leaving a hydraulic turbine. The flow entering the draft tube in its true environment is turbulent and highly swirling due to the presence of the turbine directly upstream. The main purpose of this example is to illustrate the capability of the multiblock technique detailed in Chapter 4 for handling three-dimensional, turbulent, swirling flows using grids with discontinuous interfaces between the multiple blocks in the flow domain. For the case under consideration here, $Re = 1 \times 10^6$. Two blocks are used to discretize the domain. The upstream block has $35 \times 21 \times 21$ nodes while the downstream block has $17 \times 21 \times 41$ nodes, resulting in a 2:1 discontinuous interface in the cross-stream direction (seen in Fig. 7.7). The inlet swirling velocity profile is prescribed from experimental data. Figure 7.8 shows a three-dimensional view of the configuration with a number of computed stream ribbons which have been obtained by seeding particles at the draft tube inlet and integrating to obtain the trajectory through the domain. Figure 7.9 shows a two-dimensional view from above the draft

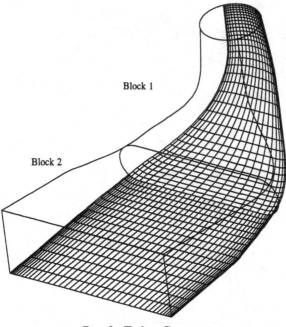

Draft Tube Geometry

Figure 7.7   Draft tube with two-block discontinuous grid along the cross-stream direction.

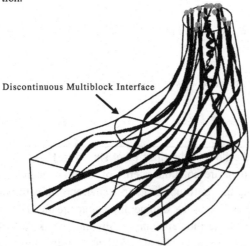

Turbulent Swirling Flow in Draft Tube

Figure 7.8   Three-dimensional view of turbulent swirling flow in the draft tube.

tube, showing the streamlines of the swirling flow. It can be observed that the flow passes smoothly through the multiblock interface without distortion, illustrating the efficacy of the conservative multiblock interface treatment for complex turbulent flows.

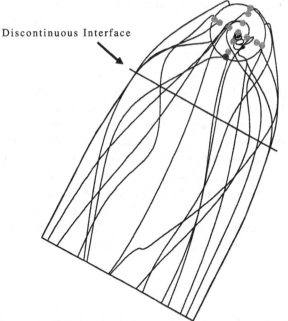

Discontinuous Interface

Turbulent Swirling Flow in Draft Tube

Figure 7.9 Two-dimensional top view of the draft tube showing streamlines passing through the discontinuous grid interface.

7.2   Thermohaline Stratification

7.2.1   Introduction

In this section we present yet another application of engineering interest, which involves a particular class of double-diffusive flows in which an initially quiescent fluid with linear solute and temperature stratifications is heated from a sidewall. Interest in this particular flow has stemmed primarily from observations of stably stratified convective layers in natural reservoirs, such as lakes and oceans (Hoare 1966, Huppert and Turner 1980), as well as human-made systems, including solar ponds (Akbarzadeh and Manins 1988, Sherman and Imberger 1991). This example is chosen to serve as a demonstration of the effectiveness of the various numerical techniques presented in this book, especially the multiblock method with selective grid refinement, for yet another class of flows exhibiting complex physics. We first discuss the physical aspects of the flow along with experimental observations, followed by the results of the numerical simulation.

Double-diffusive convection encompasses the spectrum of buoyancy-induced fluid motions which occur when two or more components having different molecular diffusivities and making competing contributions to the fluid density gradients exist simultaneously. Since the molecular diffusivities of the components can often differ by an order of magnitude or more (such as when temperature and a salt are the two components of interest), the motions encountered in double-diffusive systems can be quite varied and complex, even for simple physical systems, as demonstrated by Bergman and Ungan (1986), Kamakura and Ozoe (1993), and Shyy (1994), among others. The nature of double-diffusive flows and many reported findings can be found in the literature, including Huppert and Turner (1981) and Turner (1974, 1979, 1985).

The first experimental investigations of the lateral heating of a stably stratified fluid were performed by Chen et al. (1971) and Thorpe et al. (1969). Both studies considered a fluid of constant temperature with a vertical salinity gradient subjected to heating from a sidewall held at a fixed temperature above that of the initial temperature of the interior fluid. These investigations clarified the nature of the flow at the sidewall as a stability phenomenon by showing that under certain supercritical heating conditions, convection cells appear simultaneously along the entire heated portion of the wall. In addition, the nature of the formation and propagation of the intrusions was observed, where it was noted that the initial convection cells merged to form larger intrusions as they propagated laterally into the bulk of the domain. Chen et al. (1971) proposed a vertical length scale for the initial intrusions, h, given by

$$h = \frac{\beta_T \Delta T}{\beta_S (\partial S / \partial y)_i} \tag{7.1}$$

which characterizes the height to which a heated fluid element at the wall would rise in the initial density gradient. In the above expression, β_T represents the coefficient of thermal expansion, β_S represents the coefficient of expansion due to salinity, ΔT is the difference between the heated sidewall temperature and the initial temperature of the interior fluid, and $(\partial S / \partial y)_i$ is the initial vertical solute gradient, as designated by the subscript i. Based on this length scale, a critical Rayleigh number, defined by

$$Ra_c = \frac{g \beta_T \Delta T h^3}{\nu \alpha_T} \tag{7.2}$$

of $1.5 \times 10^4 \pm 2.5 \times 10^3$ was established, below which lateral intrusions would not form. In Eq. (7.2), ν and α_T are respectively the kinematic viscosity and thermal diffusivity of the working fluid.

Early investigations involving sidewall heating focused on the formation of intrusions due to heating from a constant temperature wall; however, in many circumstances, a constant lateral heat flux at the wall is the appropriate boundary condition. By performing a nondimensional analysis of the governing equations and assuming that the length scales in both the horizontal and vertical directions are identical at initiation, Narusawa and Suzukawa (1981) showed that in addition to the Prandtl number ($\Pi_1 = \nu / \alpha_T$) and Lewis number ($\Pi_2 = \alpha_T / \alpha_S$, where α_S is the mass diffusivity of the solute), a third nondimensional parameter is given by the following:

$$\Pi_3 = \frac{-\beta_T (q/k)}{\beta_S (\partial S / \partial y)_i} \tag{7.3}$$

where q represents the constant applied lateral heat flux at the wall, k is the thermal conductivity of the fluid, and other quantities are as previously defined in Eq. (7.1). Since they considered a constant temperature fluid with only an initial linear solute stratification, this parameter can be physically interpreted as the ratio of the horizontal density gradient at the vertical heated sidewall to the initial vertical density gradient. For a constant sidewall heat flux, this third parameter was clearly identified as the appropriate stability parameter.

An experimental study of double-diffusive systems containing initial linear vertical stratifications of both temperature and salinity has been carried out by Schladow et al. (1992). In their experiments, they used two nondimensional quantities to characterize the flows, one the Rayleigh (or buoyancy) ratio, given by

$$R_\varrho = \frac{\beta_S (\partial S/\partial y)_i}{\beta_T (\partial T/\partial y)_i} \tag{7.4}$$

which is formed from the vertical saline and thermal Rayleigh numbers defined by the initial salinity and temperature gradients (again denoted by the subscript i), both of which are negative here, and the other a lateral ratio, defined by

$$R_1 = \frac{-\beta_T (q/k)}{(-\beta_T \partial T/\partial y + \beta_S \partial S/\partial y)_i}. \tag{7.5}$$

The Rayleigh ratio, R_ϱ, provides a measure of the gravitational stability of the system, larger values indicating gravitational stability. The lateral ratio, R_1, was proposed as a modification to the nondimensional parameter Π_3 used by Narusawa and Suzukawa (1981) which explicitly takes into account the two-component vertical stratification. The quantity R_1 provides a measure of the forcing at the heated sidewall and can also be interpreted as a ratio of the lateral density gradient at the sidewall to the initial vertical density gradient. Based on these two parameters, a series of experiments was conducted using a NaCl-based salt solution, aimed at delineating the physical characteristics of the intrusions that developed under various states of gravitational and lateral stability.

Three classes of intrusions, denoted as I, II, and III, were clearly identified. Class I intrusions, which developed under conditions of relatively high gravitational and lateral stability ($R_\varrho > 5, R_1 \approx 1$) were characterized by relatively quiescent, well-defined cells less than 10 mm thick. The internal structure of these intrusions consisted of a highly stable vertical temperature stratification and a well-mixed salinity field. The intrusions propagated at speeds less than 10 cm/h during the period in which the heat flux at the wall was applied, and following removal of the heat flux, they were observed to quickly diffuse away. Similar characteristics were observed by Narusawa and Suzukawa (1981) for cases where the lateral stability parameter $\Pi_3 \leq 1$. In their case $R_\varrho = \infty$, since they considered a constant temperature fluid. Intrusions categorized in class II developed under conditions of relatively low gravitational and lateral stability ($2.3 \lesssim R_\varrho \lesssim 5, R_1 \gg 1$). These intrusions were observed to be larger (~ 10–40 mm), more dynamic, and less defined than class I intrusions. In contrast to class I intrusions, the vertical salinity stratification was found to be slightly unstable; however, the temperature field was still highly stable. In addition, intrusions categorized in class II propagated at speeds much higher than class I intrusions (~ 10–30 cm/h). Like class I intrusions, they also diffused away immediately following removal of the sidewall heating.

Class III intrusions developed under the conditions of lowest gravitational and lateral stability ($R_\varrho \approx 2, R_1 \gg 1$) and were found to be the most dynamic of the three classes, propagating at speeds greater than 25 cm/h. Due to stronger convective motions within these intrusions, which reduce the time scales of the fluid flow, both the temperature and salinity fields were found to be well mixed. While the physical appearances of class II and class III intrusions were observed to be very similar, the defining characteristic of class III intrusions was their ability to continue propagating long after the removal of the sidewall heating. In this situation, sidewall heating serves only as a triggering mechanism for the instability. A weak destabilizing mechanism along the vertical direction, associated with a blockage effect ahead of the intrusions, which reduces the local value of R_ϱ below a critical value, has been proposed by Schladow et al. (1992) to account for this behavior.

Regarding the intrusion merging process, it was noted that the physical mechanism for class I intrusions appears to be fundamentally different than that observed for classes II and III. Following the formation of the initial intrusions for a class I flow, merging seems to take place via a process whereby weaker intrusions are forced back to the heated wall by a blockage effect from the intrusions above and below, with no initial exchange of fluid between neighboring intrusions. Conversely, the merging process for class II and class III flows occurs via specific events, in which jets of fluid near the heated wall, originating from some intrusions, penetrate into neighboring intrusions from below, causing a breakdown of the interface separating the intrusions and leading to merger. Observations of the merging processes similar to that discussed above for class II and III flows have also been made by Tanny and Tsinober (1988).

While many experimental investigations of the development and propagation of intrusions due to sidewall heating have been performed, few detailed numerical simulations have appeared in the literature to date, most likely due to the excessive computational resources required to capture these flows. The physical time scales of the fine-scale fluid motion are usually $\mathcal{O}(0.1 \text{ s})$, while the time required to achieve full formation of the intrusions and to observe the merging processes is $\mathcal{O}(0 \text{ min})$ or more for slowly propagating intrusions. Besides the need for considering multiple convection cells, the disparity of the Prandtl number (Pr) and Schmidt number (Sc) generally encountered means that high grid resolutions are required to accurately capture the sharp velocity, temperature, and salinity gradients located in the interfaces between the intrusions. In terms of overall computational requirements, the problem is a challenging one.

One of the first numerical simulations involving a stably stratified fluid subjected to lateral heating was performed by Wirtz and Liu (1975). They investigated the flow in a narrow slot with an initial linear solute stratification subjected to a constant temperature sidewall heating. Details of the initial formation, growth, and merging of convective cells that were obtained were consistent with previous experimental observations and linear stability analysis. Due to the lack of computational resources at the time, only extremely coarse grids could be used, and thus no detailed analysis of the interaction between multiple convection cells could be considered. Other works have appeared in the literature since then, most notably those of Heinrich (1984),

Kamakura and Ozoe (1993), and Lee and Hyun (1991). In each of these works, emphasis was placed on the basic qualitative details of the cell formation and the global features of the time-evolving flow field. The grid resolutions employed were still relatively coarse considering the large Schmidt and Prandtl number flows being simulated. Thus, it is doubtful that the thin salinity and thermal layers separating intrusions were faithfully captured. In order to obtain quantitative comparison with experiment for quantities such as the intrusion front propagation speed, which by its nature is dependent on the interaction of neighboring intrusions, it is essential that the interfaces between intrusions be accurately handled. In addition, to capture the features of class II flows (including the details of interface breakdown), which exhibit much stronger convective motions than class I flows, high grid resolutions must be used throughout the entire near-wall region, since the sharp salinity and temperature interfaces can become highly contorted and deviate substantially from a generally horizontal alignment. It is noted that the flows considered in the above works correspond closely (in terms of internal structure and relative convection strength) to class I flows, so the convective motions are relatively benign compared to the class II flows to be simulated here.

In this section, detailed numerical simulations of two of the experimental cases run by Schladow et al. (1992) are presented, based on a work by Wright and Shyy (1996). With a multiblock composite grid approach, the grid resolution considerations mentioned above can be handled very effectively due to the localized nature of the intrusion development. The composite grid internal boundary treatment is based upon a locally – and thus globally – conservative treatment of the mass flux and momentum flux tangential to the boundary. Global conservation of the momentum flux normal to the internal boundary is used to obtain the arbitrary constants of the pressure field within the various blocks of the grid, which arise due to the decoupling of the pressure fields (Shyy 1994).

In order to compute natural convection flows, modifications to the existing methodology have been made to handle the buoyancy term which now appears in the v-momentum equation. For the present flows of interest, the governing equations for temperature and salinity must also be solved and an appropriate internal boundary treatment devised for communicating the temperature and salinity field information between the blocks of the composite grid. In this regard, it is shown that the internal boundary treatment, whereby the fluxes at the internal boundary are obtained entirely from the neighboring block in a locally conservative manner, allows an arbitrary jump in the temperature and salinity field across the interface which is nonphysical. To circumvent this problem, an interface scheme based on a linear interpolation with conservative correction is employed, which allows the temperature and salinity fields to maintain the proper continuity conditions across the interface.

7.2.2 Grid Refinement Strategy

Due to the localized nature of the growth and development of the intrusions at the heated sidewall, use of the composite grid method with a mesh refinement strategy based on the local addition and deletion of grid lines provides an efficient

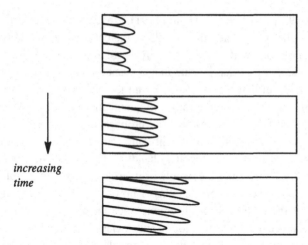

increasing time

Figure 7.10 Time evolution of typical class I intrusions.

and accurate solution procedure (for other conditions fixed, such as the convection scheme, etc.). Figure 7.10 shows a typical time sequence for the development of class I intrusions over a portion of the heated wall. At initiation, convection cells appear nearly simultaneously at the wall and quickly merge to form the initial intrusions. The intrusions are characterized by a relatively high-speed clockwise rotation, in which heated fluid flows away from the wall along the top of the intrusions, cools as it moves along, and then returns along the bottom of the intrusions. This process is evidenced in the characteristic downward-sloping appearance of the intrusions. Typical fluid speeds are $\mathcal{O}(1 \text{ mm/s})$, while the propagation speed of the front is only $\mathcal{O}(10 \text{ cm/h})$. Since the intrusions are abutting, sharp velocity gradients exist along the interfaces between the intrusions. In addition, the intrusions are usually well mixed in terms of temperature and salinity and are separated at the interfaces by sharp temperature and salinity gradients. In regions away from the intrusions, flow velocities are much smaller, and the temperature and salinity profiles remain very close to the initial stratifications.

From this physical description, it is clear that a very fine grid resolution is required in the region near the intrusions, while a much coarser grid is sufficient away from the intrusions. A single grid, clustered in the region near the intrusions, might be sufficient; however, due to the preferred directionality of the flow, in which a very fine vertical spacing must be used to capture the nearly horizontal interfaces between the intrusions, this choice is not optimal, since the fine vertical resolution in the near-wall region will also be imposed on the far-field flow. An adaptive grid strategy using a single grid within a curvilinear coordinate framework could be used to optimize the vertical resolution in the near-wall region; however, many grid points would still be wasted in the bulk of the domain, since the vertical resolution requirements in these two regions are truly disparate.

In Wright and Shyy (1996), a two-block composite grid is used to track the evolution of the sidewall intrusions. One block, with a very fine horizontal and vertical resolution, is placed in the near-wall region, extending from the wall to a location just

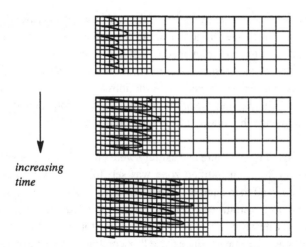

Figure 7.11 Composite grid tracking of sidewall intrusions.

ahead of the moving intrusion front. A second block with a relatively coarse horizontal and vertical resolution is used in the bulk of the domain. These two blocks share a vertical overlap region one coarse cell thick (which corresponds to several fine-grid cells). As the intrusion front moves into the bulk fluid, grid lines are added to the fine grid and removed from the coarse grid, so that the find-grid/coarse-grid interface remains ahead of the intrusion front. A visual representation of this process for the intrusion development sequence of Fig. 7.10 is shown in Fig. 7.11. By keeping the fine-grid/coarse-grid interface sufficiently far ahead of the intrusion front, we can ensure that the intergrid transfer of information required for providing starting values for the newly introduced fine grid nodes can be done in an accurate manner. Since coarse-grid lines are being deleted, no new information is required for the coarse grid upon re-gridding near the front. In the following section, we give a detailed presentation of the simulations which have been performed.

7.2.3 Numerical Simulation Results

7.2.3.1 BASICS

Two numerical simulations were performed, one corresponding to the development of class I intrusions and the other to class II intrusions. For both simulations, a 20-cm-high by 20-cm-wide domain was used. In the experiments performed by Schladow et al. (1992), the tank used was 50 cm high and 400 cm wide. This height is sufficient to accommodate approximately 30 class I intrusions and 10 class II intrusions along the heated left wall, based on the final intrusion thickness after completion of the merging process, and taking into account the unstratified fluid layers employed at the top and bottom of the tank. Similarly, a domain of 20 cm in height should allow approximately 10 class I intrusions and 5 class II intrusions to develop along the heated wall, which should be sufficient to allow comparison with the experimental results for the key points mentioned previously. Concerning the apparent disparity in the tank widths, it is noted that for the duration of the experiments, completely through the final merger process (with the exception of the self-propagation cases for class III flows), the intrusion fronts penetrated no more than about 40 cm into the tank, and thus a large portion of the tank remained essentially quiescent, except for the small disturbances created by the presence of the intrusions. Since the numerical simulations performed here will be terminated when the intrusion fronts are in the approximate vicinity of the vertical centerline of the domain, the right solid boundary should not significantly influence the motion of the intrusions developing from the left heated wall.

The initial stratifications of temperature and salinity were prescribed as follows. At the top of the domain, a 3 cm unstratified zone was set, followed by a linear stratification region of 14 cm, and then another 3 cm unstratified zone at the bottom. In the experimental simulations, unstratified layers were also employed at the top and bottom of the tank to provide a nearly constant zero flux condition for temperature and salinity for the duration of the experiment. Since these flux conditions can be exactly enforced in the numerical simulations, the unstratified zones used here serve only as a buffer region, effectively isolating the intrusion front from the effects of the upper and lower walls. A schematic of the domain used for the numerical simulations, showing the boundary conditions and initial temperature and salinity stratifications is shown in Fig. 7.12. The prescribed sidewall heating fluxes, specified initial temperature, and salinity stratifications and the corresponding Rayleigh ratio, lateral stability ratio, and intrusion classification for the numerical simulations are summarized in Table 7.1. These simulations correspond exactly (in terms of the values of R_ϱ and R_1) to the two experimental cases reported in detail. Since the mean stratification temperature and salinity of the NaCl-based solution at the start of the experimental simulations were unknown, the numerical simulations were assumed to occur at a starting mean stratification temperature of 20°C and a mean salinity stratification of 30 kg m^{-3}, for which the physical parameters of water (such as the kinematic viscosity, thermal diffusivity, and mass diffusivity), all assumed constant, were taken from Akbarzadeh and Manins (1988). Thus, small differences in both the thermal and solutal Rayleigh numbers may exist between the experimental and numerical simulations for each case,

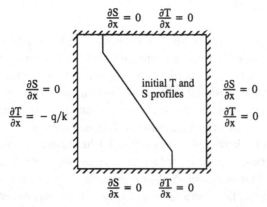

Figure 7.12   Schematic of physical domain, boundary conditions, and initial T and S profiles for numerical simulations.

Table 7.1 *Parameter Specifications for Numerical Simulations.*

Intrusion Class	Side Wall Heat Flux (W/m²)	Initial dT/dy (K/m)	Initial dS/dy (kg/m⁴)	Rayleigh Ratio R_ϱ (Eq. 7.4)	Lateral Ratio R_1 (Eq. 7.5)
I	91.67	−25.37	−129.58	8.3	0.9
II	91.61	− 7.01	− 13.80	3.2	10.8

although the Rayleigh ratios R_ϱ and lateral stability ratios R_1 are the same. With the fluid properties above, $Pr = \frac{\nu}{\alpha_T} = 8$ and $Sc = \frac{\nu}{\alpha_S} = 675$.

The grid used for the simulations, as mentioned previously, is composed of two blocks, a fine block in the vicinity of the heated wall and a coarse block away from the wall. In the initial grid, the fine block consists of 131×501 nodes, while the coarse block has 162×251 nodes. In the fine grid block, the grid lines have been clustered toward the heated wall, resulting in a grid spacing of $\Delta x = 0.012$ cm (0.06% of overall domain width) at the wall ($x = 0.0$) and increasing linearly to $\Delta x = 0.05$ cm at $x = 4.0$ cm. The vertical grid spacing in the fine block is uniform, with $\Delta y = 0.04$ cm. In the coarse block, the grid lines are uniformly spaced in both directions, resulting in grid spacings of $\Delta x = 0.1$ cm and $\Delta y = 0.08$ cm. Since the last two vertical columns of the fine grid have a grid spacing of $\Delta x = 0.05$ cm, the overlap region of the fine and coarse grids consists of exactly two fine grid cells and one coarse grid cell.

As the solution advances in time, the intrusion front will eventually approach the fine/coarse grid interface of the initial grid, and re-meshing will be required. Re-meshing is carried out here when the fastest-moving intrusion approaches a distance of about 1 cm from the interface. Since the intrusion fronts for class I and class II flows propagate at different speeds, the times at which re-meshing occurs will be different. When re-meshing is required, a 4.0 cm extension is added to the fine grid, while 4.0 cm of the coarse grid is removed, producing the new mesh. The grid spacings in the newly added portion of the fine mesh are taken to be uniform, with values of $\Delta x = 0.05$ cm and $\Delta y = 0.04$ cm. This process results in a new mesh composed of a fine block with 211×501 nodes and a coarse block with 122×251 nodes. Values for the solution at the newly introduced fine grid nodes are taken from the coarse block via linear interpolation before the coarse block is restructured. As the solutions progress, a second re-meshing may also be required, in which case a similar procedure as that described above for the first re-meshing is invoked. Figure 7.13 shows the sequence of grids employed in the simulations.

For both simulations, a time step $\Delta t = 0.2$ s was used. Since the maximum fluid velocities are $\mathcal{O}$ (1 mm/s), this value was heuristically chosen so that between successive time steps, the fastest fluid particles will have traversed at most one grid cell. To verify this choice, a single-block grid simulation corresponding to the class I simulation described above was performed with time steps of $\Delta t = 0.1$ s and $\Delta t = 0.2$ s. The grid consisted of 301×301 uniformly spaced nodes. At a physical

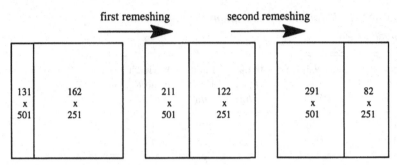

Figure 7.13 Sequence of grids employed in numerical simulations.

time of 2 minutes, vertical profiles of the field variables were taken at a location near the heated wall. Comparison of the profiles for the two time steps indicated that the solution appeared to be time step–independent near $\Delta t = 0.2$ s. In this regard, it is noted that the time step employed by Schladow et al. (1992) in their numerical simulation, $\Delta t = 0.0125$ s, seems excessively small, since the minimum grid spacing used there was 0.03 cm, which is very close to that employed by Wright and Shyy (1996). In the following, we first present results for the class I simulation, followed by results for the class II simulation.

7.2.3.2 RESULTS FOR CLASS I SIMULATION

A sequence of contour plots showing the development of class I intrusions from the beginning of the numerical simulation to the termination time of 36 minutes is shown in Fig. 7.14. Contours of stream function, temperature, and salinity have been plotted at each chosen time. In each of the contour plots, only the portion of the domain corresponding to the initial stratified region (3 cm $\leq$ y $\leq$ 17 cm) has been included. The initial intrusions form along the heated wall at a time of about 10 minutes, and by about 28 minutes they have spanned the entire stratified region along the wall. In agreement with the experimental simulation, the intrusions form exclusively from the bottom of the domain to the top (with the exception of some small circulation regions near the top). The characteristic downward-sloping appearance of the intrusions, observed in all experiments for these flows, are also observed here. From the stream function plots it is apparent that class I intrusions are characterized by relatively quiescent motions, producing a well-defined layered structure with distinct intrusions separated by thin, nearly horizontal interfaces.

Vertical profiles of the temperature and salinity for the full domain height, taken at a distance of 3.0 cm from the heated wall (15% of width from wall), correspond-ing to the times of Fig. 7.14, are given in Fig. 7.15. From these plots, the step-like nature of the intrusions is clearly evident, where it is observed that a highly sta-ble temperature profile and a well-mixed salinity profile have developed in each of the intrusions. Both of these characteristics were observed in the correspond-ing experimental case for this class. Results for the overall size of the intrusions and the front propagation speed also show favorable agreement with the values observed experimentally. For the numerical simulation, the average height of the

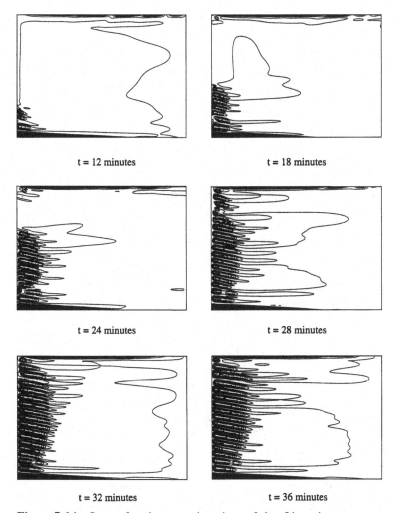

t = 12 minutes   t = 18 minutes

t = 24 minutes   t = 28 minutes

t = 32 minutes   t = 36 minutes

Figure 7.14   Stream functions at various times of class I intrusions.

intrusions measured at the heated wall at a time of 36 minutes is approximately 8 mm. The corresponding experimental value, taken from photographic images at a similar time, is about 10 mm. Using the sequence of stream function contour plots shown in Fig. 7.14, the intrusion front speed is estimated to be approximately 7 cm/h. In this computation, only the ten intrusions located about the center of the stratified region were used, since the development of the intrusions near the upper and lower portions may have been hindered by the fairly strong recirculation regions that exist in the unstratified layers at the top and bottom of the domain. In this regard, the attempt to isolate the intrusion front from the effects of the solid walls by including unstratified layers above and below may have resulted in a greater disturbance to the upper and lower intrusions than would have existed without these layers. No value for the average speed of the intrusions for the corresponding experimental case was given, but the value computed above does agree favorably with the speed of 10 cm/h or less than that was generally observed for class I intrusions.

Regarding the merger process, some evidence of the blockage mechanism described earlier, and observed clearly in the experiments, can be seen in the final four frames of the stream function contour sequence shown in Fig. 7.14 for the fourth

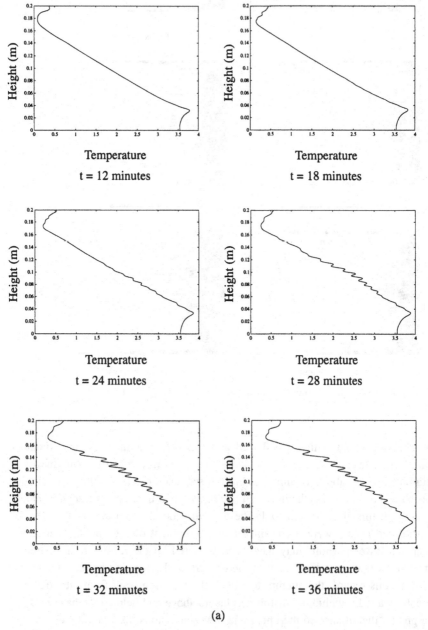

(a)

Figure 7.15 Temperature and salinity profiles taken at a distance of 3.0 cm from the heated wall at various times. (a) Temperature; (b) salinity.

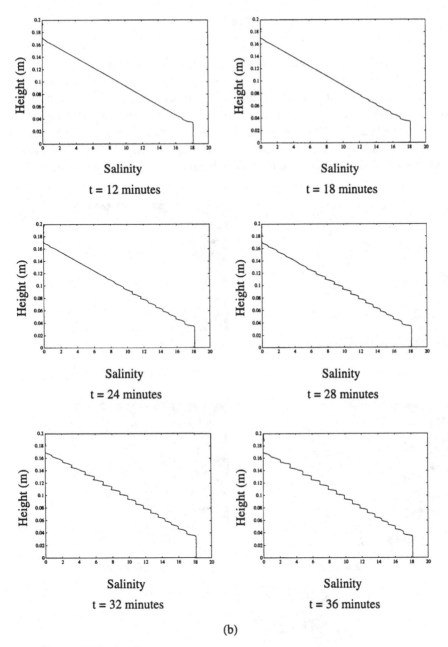

Figure 7.15 *(cont.)*

intrusion from the bottom. From a time of 24 minutes to 36 minutes, a progressive weakening of this intrusion is clearly evident as it becomes overtaken by its upper and lower neighbors. Consistent with the description of the merger mechanism, no exchange of fluid between neighboring intrusions appears to initiate the process of merger, since the interfaces between the intrusion and its upper and lower neighbors are clearly seen to be maintained during the entire process.

7.2.3.3   RESULTS FOR CLASS II SIMULATION

A sequence of contour plots for the development of the class II intrusions in increments of 4 minutes from the beginning of the simulation to the termination time of 20 minutes is shown in Fig. 7.16. Again, only the portion of the domain corresponding to the initial stratified region has been included. The initial intrusions form along the heated wall at a time of about 2 minutes, eventually spanning the length of the heated wall in the stratified region by a time of 5 minutes. As time elapses, the characteristic structure of the flow becomes evident, where, again, fairly well-mixed regions are separated by distinct interfaces in which sharp gradients of

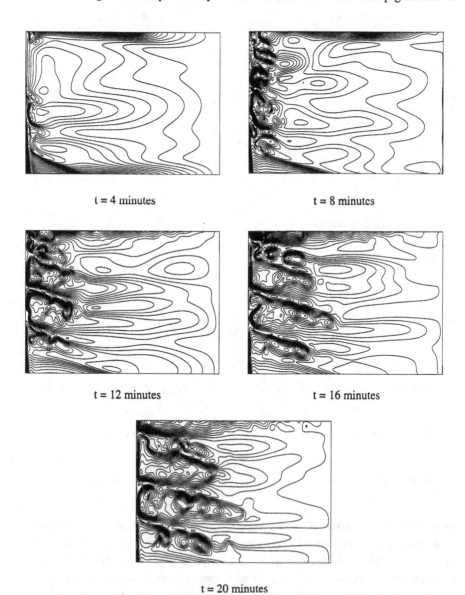

t = 4 minutes   t = 8 minutes

t = 12 minutes   t = 16 minutes

t = 20 minutes

Figure 7.16   Stream functions at various times of class II intrusions.

temperature and salinity exist. It is apparent from the salinity contours that class II intrusions are much more dynamic and contain a much finer salinity structure due to the highly convective nature of the flow within the intrusions.

Figure 7.17 displays vertical profiles of the temperature and salinity (again taken 3.0 cm from the heated wall) for the full domain height corresponding to the times shown in Fig. 7.16. Again, the step-like nature of the intrusions is evident, where it is observed that a highly stable temperature profile has developed in each of the

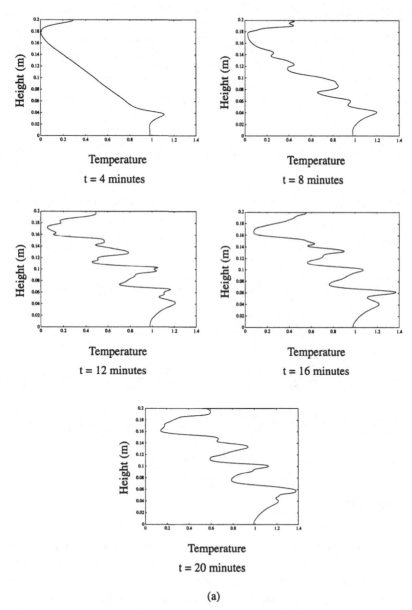

(a)

Figure 7.17 Temperature and salinity profiles taken at a distance of 3.0 cm from the heated wall at various times. (a) Temperature; (b) salinity.

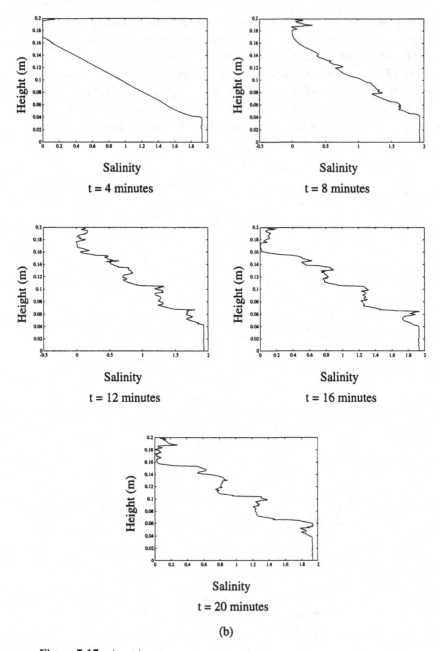

Figure 7.17   *(cont.)*

intrusions. In contrast to the class I intrusions, in which the salinity was well mixed, the salinity distribution in the class II intrusions is observed to be slightly unstable. These characteristics for the internal temperature and salinity profiles were also observed in the corresponding experimental case. Regarding the apparent "jagged" behavior of the salinity profiles compared to the temperature profiles, it is noted that since the Schmidt number is very large ($Sc = 675$), the fine-scale salinity structure of the intrusions is essentially being convected with diffusion having a role to play only within small length scales.

Results for the overall size and front propagation speed again show favorable agreement with the values observed for typical class II intrusions. Based on the general observations of the experiments, the final height for typical class II intrusions was found to fall in the range of 10–40 mm. From the salinity and temperature profile plots for the experimental case corresponding to the current numerical simulation, taken 3 cm from the heated wall at a time of 19 minutes, the average height of the intrusions is estimated to be about 40 mm. Based on a similar analysis of the temperature and salinity profiles for the numerical simulation, an average intrusion height of about 35 mm is obtained. Using the intrusion salinity sequence shown in Fig. 7.17b, the intrusion front speed was computed to be about 27 cm/h. In this computation, only the two center intrusions were used. This value also agrees very well with the range of 10–30 cm/h observed in the experiments for class II flows. Again, no estimate of the front propagation speed for the specific experimental case corresponding to the numerical simulation was given.

Finally, in terms of the merger process, it is clear that, unlike class I flows, class II flows are dominated by dynamic vertical motions in the near-wall region. These motions are evidenced in the time-history sequence by the sharp salinity gradient regions that are pulled up along the wall and are subsequently folded back into the interior of the intrusions when they reach their apex. Although details of the initial merger process are not captured in the plots, a vivid example of the merger process can be seen in the three final frames of the time sequence for the two top intrusions. Here, vertical motions from the lower intrusion have penetrated into the intrusion above. A similar merger mechanism can be seen in photographs taken by Tanny and Tsinober (1988). From the salinity contour sequence, this merging process appears to be the result of an interface breakdown scenario rather than that of interface migration (both described by Linden (1976)). In any case, this merging event is in contrast to the merger process observed in the previous section for class I intrusions, which was initiated by stronger intrusions blocking weaker ones from propagating and forcing them back to the heated wall. For both merger processes, fluid is eventually exchanged during merger; however, with class II flows the kinetic energy of the fluid in the near-wall region serves as the mechanism for merger.

7.3 Vertical Bridgman Crystal Growth

In this section, we will present the results of a simulation of the vertical Bridgman system for the single crystal growth of NiAl. A two-level simulation has been performed to simultaneously achieve accuracy as well as computational economy. To validate the current numerical model, experimental measurements of temperature profiles at designated locations in a vertical Bridgman growth system are compared with the predictions. To help optimize the thermal conditions for growing single crystal NiAl with desired quality, parametric variations will also be considered in numerical simulations. Furthermore, the exact melting temperature of NiAl, which does not seem to have been ascertained with certainty in the literature, will be assessed by combining the computational and experimental information.

7.3.1 Background

β-NiAl, having a composition of 50 at.% Ni, 50 at.% Al, with a CsCl, B2 crystal structure, is an intermetallic that is currently being investigated as a promising high-temperature structural material for application in the next generation of aircraft engines and structural components. NiAl is especially attractive because of its low density, high thermal conductivity, high melting temperature, and superior isothermal and cyclic oxidation resistance (Darolia 1991 Sen and Stefanescu 1991). Its high melting temperature (above 1640°C) allows it to sustain a high operating temperature and increases gas turbine efficiency; its low density of 5.95 g/cm^3 (approximately two-thirds the density of state-of-the-art nickel-based superalloys) helps to reduce the turbine rotor weight (blades and disk) by as much as 40%; its high thermal conductivity (4 to 8 times that of nickel-based superalloys) allows the temperature distribution in a NiAl turbine blade to be much more uniform, and the life limiting "hot spot" temperature is reduced by as much as 50°C; finally, its excellent oxidation resistance makes it widely useful for coating on virtually all high-pressure turbine blades and vanes of aircraft engines.

The physical and mechanical properties of NiAl depend strongly on both its composition and temperature (Darolia 1991, Levit et al. 1996, Noebe et al. 1993, Sen and Stefanescu 1991, Takasugi 1992, Vedula et al. 1985). For example, it has been found that single crystal NiAl with improved purity (Levit et al. 1996) or with small (<1%) additions of iron, gallium, or molybdenum (Darolia 1991) yields substantial increases in ductility at room temperature. As far as the temperature effect is concerned, it has been widely recognized that NiAl, like most other intermetallics, has the drawbacks of poor ductility at room temperature and low strength at elevated temperatures, in spite of its many other advantages (Darolia 1991, Noebe et al. 1993, Sen and Stefanescu 1991). These problems need to be resolved before NiAl can be useful in structural applications as a high-temperature material.

The vertical Bridgman growth system is one of the directional solidification configurations that are useful for producing large-size single crystals from the melt. The advantage of the Bridgman system is that the temperature gradients can be controlled, making it possible to achieve unidirectional solidification.

For the case presented here, initially the raw material of NiAl is vacuum-induction melted and chill cast into a rod with a desired size. The rod is then removed from the chill mold, placed in a high-density refractory (e.g., alumina) ampoule, which is situated on the top of a water-cooled ram (e.g., copper), remelted, and then directionally solidified by slowly withdrawing the ampoule/ram assembly from the hot zone. In most cases, a single crystal seed is placed between the chill and the feedstock in order to preselect the orientation of the single crystal NiAl. Several physical mechanisms can be identified in this vertical Bridgman system and are of key importance to the processing conditions. These include the phase change dynamics between the NiAl melt/solid interface, heat conduction across the ampoule wall and the copper ram, convection in the melt and in the encapsulated argon gas, and radiation between the outer ampoule wall and the heater (T4) surrounding the ampoule. In the following, experimental and computational results will be presented to assess the current predictive capabilities.

7.3.2 Experimental Procedure

7.3.2.1 SYSTEM SETUP

As shown in Fig. 7.18, the feedstock of NiAl was prepared with a 5-mm-diameter hole in the center to allow for installation of a thermocouple. Four type C tungsten-

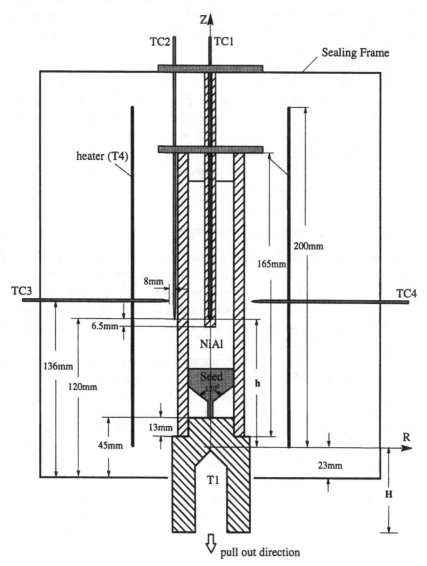

Figure 7.18 Setup of themocouples for the experimental measurement of temperature in Bridgman system.

rhenium thermocouples with 3 mm diameter were installed for measuring the temperature profiles and the furnace hot zone temperature during the crystal growth process. Figure 7.18 shows the detailed arrangement of these thermocouples. Thermocouple 1 (TC1) was positioned along the centerline of the NiAl feedstock. To protect this thermocouple from reacting with the NiAl during the melting process, it was enclosed by an alumina tube that has 3 mm inner diameter and 5 mm outer diameter with a tip length of 6.5 mm. This arrangement also allows thermocouple 1 to translate vertically along the NiAl centerline. Thermocouple 2 (TC2) was positioned along the outer ampoule wall and was also allowed to translate vertically. In addition, thermocouples 3 and 4 (TC3 and TC4) were fixed horizontally at about one-half height of the heater. They were used to help monitor the thermal condition in the furnace hot zone during the entire crystal growth process. All thermocouples were calibrated before installation. However, it has to be mentioned that the confidence limit for a type C thermocouple is 1%.

The relative positions of the various components in the system were measured with respect to the baseline of the sealing frame. The relative positions of the ampoule and thermocouples are taken with respect to the bottom of the heater, shown in Fig. 7.18 by the horizontal radial axis R. The ampoule position is denoted by H, which is initially set as 43 mm. As the ampoule is pulled down, this number increases. h denotes the tip positions of thermocouple 1 or 2. Because of the interior complexity and small scale of the current system, the measurement of positions is estimated to have a confidence limit of ± 1 mm.

7.3.2.2   TEMPERATURE MEASUREMENT

In the course of making temperature measurements, the system was first heated up for about three and one-half hours until TC3 and TC4 reached 1740°C. Then the system was held for another 30–45 minutes to allow the NiAl feedstock to completely melt. Next, the ampoule/ram assembly was withdrawn from the furnace and stopped at designated positions to allow TC1 and TC2 to take measurements at different positions, as described next.

After the NiAl feedstock was fully melted, the alumina tube with thermocouple TC1 was moved down through the melt until it stopped and touched the melt/solid interface. Hence, the interface position for the initial ampoule position of $H = 43$ mm could be located. Thereafter, the alumina tube remained at this position, whereas TC1 was allowed to translate vertically. Once the temperature readings from TC1 and TC2 were stable, TC1 and TC2 were moved to a new height and the corresponding temperatures were obtained. For the convenience of measurement, TC1 and TC2 were always moved in tandem.

After measurement at various h positions were made, the ampoule/ram assembly was withdrawn and stopped at a designated new ampoule position (e.g., $H = 59$ mm), where temperatures were once again measured. This process was repeated until the desired number of ampoule positions were measured. Temperature profiles, along the NiAl centerline and the ampoule outer wall, were measured for a total of six ampoule positions: $H = 43$ mm, 59 mm, 67 mm, 77 mm, 87 mm, and 97 mm.

Table 7.2 *Temperature Versus Thermocouple Positions (h) for Different Ampoule Positions (H)*

H (mm)	Thermocouple 1 (Located along the NiAl centerline)					
	h(mm)	82	88	98	108	118
43	T(C)	1737	1752	1766	1774	1778
	h(mm)	63	78	88	98	108
59	T(C)	1733.5	1762	1773.5	1779	1781.5
	h(mm)	56	68	83	93	103
67	T(C)	1733	1758	1772.5	1782	1784
	h(mm)	54	58	68	78	88
77	T(C)	1751.5	1757.5	1771.5	1780	1785
	h(mm)	54	63	73	83	
87	T(C)	1762	1773	1780	1780	
	h(mm)	54	58	68	73	
97	T(C)	1767	1770	1777.5	1780	

H (mm)	Thermocouple 2 (Located along the ampoule outer wall)						
	h(mm)	82	88	98	108	118	
43	T(C)	1780	1786	1791	1794.5	1795.5	
	h(mm)	54	63	78	88	98	108
59	T(C)	1761	1773	1786.5	1791	1792.5	1793
	h(mm)	56	68	83	93	103	
67	T(C)	1772.5	1783	1792	1793.5	1794.5	
	h(mm)	54	58	68	78	88	
77	T(C)	1778.5	1781.5	1788	1792.5	1794	
	h(mm)	54	63	73	83		
87	T(C)	1778.5	1787	1790	1793		
	h(mm)	54	58	68	73	83	93
97	T(C)	1781.5	1783.5	1788.5	1790.1	1793	1793.5

Table 7.2 and Fig. 7.19 summarize the data collected. The number of data points collected decreases as the ampoule is pulled out (H increases), since the thermocouples traverse a smaller length. In addition, the interface position was obtained only for the first ampoule position of $H = 43$ mm. The measured interface position, with respect to the bottom line of the heater (R axis), is 75 mm. This number is checked once again after the experiment when the solidified NiAl ingot was broken and the position of the tip trace of the alumina tube was measured. The results showed that the confidence limit for the interface measurement is about ± 2 mm. During the entire experimental process, a stable furnace temperature of $1740 \pm 1°C$ (monitored by TC3/TC4) was maintained by manually adjusting the input power about a baseline value of 7 kW.

7.3.2.3 EXPERIMENTAL OBSERVATIONS

From Table 7.2 and Fig. 7.19, it can be seen that the temperature measured along the outer ampoule wall can be as high as 1796°C, which is higher than the value of

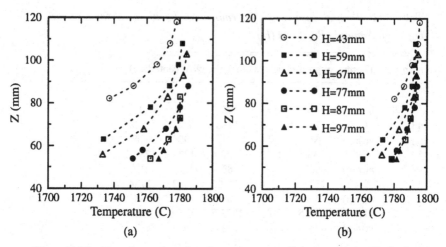

Figure 7.19 Experimental results of temperature profiles along (a) the NiAl centerline and (b) the ampoule outer wall for different ampoule position H.

1740°C measured by TC3/TC4. Even at the same height, the temperature value at the ampoule outer wall is about 40°C higher than that of TC3/TC4, although their horizontal gap is only about 8 mm, as shown in Fig. 7.18. It is also observed that when TC3 or TC4 touches the ampoule outer wall, it yields the same temperature as measured by TC2. However, when TC3 or TC4 is pulled horizontally away from the ampoule, its temperature reading drops significantly. Figure 7.20 shows this situation more clearly in a top view of the system. It can be seen that the tips of TC3 and TC4 both face directly toward the ampoule wall instead of the heater, hence they receive less radiation from the heater than TC2, resulting in a lower temperature. In addition, it is also observed from Fig. 7.19 that the temperature gradient in the axial direction is higher at the lower portion of the ampoule than at the higher portion of the ampoule. As the ampoule is pulled out (H increases), temperature profiles along both the NiAl centerline and the ampoule outer wall become more unified.

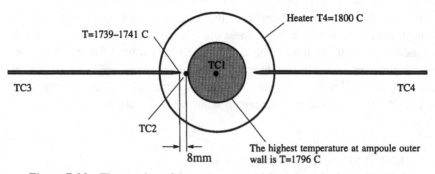

Figure 7.20 The top view of the arrangement of thermocouples in the NiAl furnace.

7.3.3 Comparison with Numerical Results

7.3.3.1 MODELING CONSIDERATIONS

To compare the modeling results with the experimental measurements within a reasonable scope, several specific customizations have been made to meet the operating conditions adopted in the experiment. First, as explained previously, the geometry of the experimental system had to be changed to accommodate the thermocouples. The inserted alumina tube at the NiAl centerline, for protecting TC1 from corrosion by the NiAl melt, has a diameter about one sixth of the ampoule inner diameter of 29 mm. Hence, the interference caused by the alumina tube had to be considered.

The second change is in the boundary condition specification. Since the highest temperature along the ampoule outer wall is about 1796°C, the heater temperature $T4$ was selected as 1800°C, based upon previous numerical results which showed a 3–5°C difference in temperature between the heater and the alumina wall (Ouyang and Shyy 1996). The cooling water temperature ranged from 5–10°C (city water). When the cooling water reaches the copper ram, it will be warmed up somewhat, so we use the upper limit of $T1 = 10°C$ as the cooling boundary condition. For the rest of the boundaries in the furnace, temperature boundary conditions are difficult to specify directly. Estimation can be made, though, according to the experimental characteristics. As mentioned earlier, during the entire crystal growth process, the furnace hot zone temperature was maintained at a fixed value of 1740°C by manually adjusting the stable power input. In other words, a condition of constant heat flux into the system is maintained. So the thermal boundary conditions of constant heat flux for the remaining furnace boundaries are assigned (i.e., $\partial T^2/\partial n^2 = 0$).

The third consideration is the accurate specification of material properties as a function of temperature. This is a very important factor in obtaining accurate numerical results. As shown in Table 7.2 and Fig. 7.19, the majority of the alumina ampoule is exposed to a temperature range above 1700°C, which exceeds the upper threshold 1127°C (1400°K) of conductivity as a function of temperature, $k(T)$, as shown in Fig. 7.21 (Perry et al. 1984). Therefore, the conductivity of alumina above $T = 1127°C$ has been estimated by an extrapolation of the gradient of $k(T)$ at $T = 1127°C$. Governing equations and other numerical treatments can be found in Ouyang and Shyy (1996) and in Shyy et al. (1996a).

7.3.3.2 MODELING RESULTS

In our computation, all six ampoule positions measured in the experiment are simulated and the computational results are compared with experimental data as follows.

7.3.3.2.1 *Comparison of Temperature Profiles* Figures 7.22 and 7.23 show that the temperature distribution along the NiAl centerline and the ampoule outer wall obtained from the computational simulation matches the experimental measurements quite well overall. Satisfactory agreement is observed for all six ampoule positions,

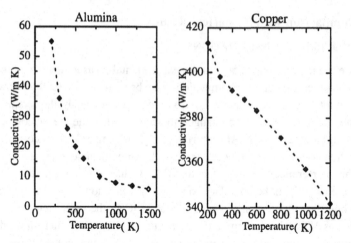

Figure 7.21   Dependency of thermal conductivities of alumina and copper on temperature (Perry et al. 1984).

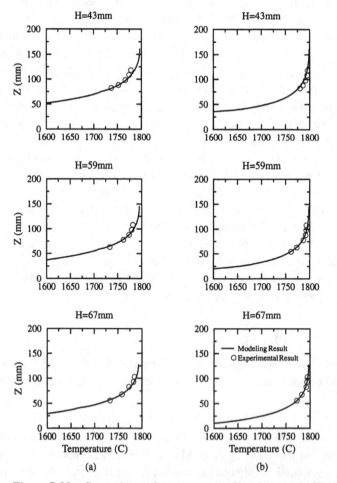

Figure 7.22   Comparisons of temperature profiles along (a) the NiAl centerline and (b) the ampoule outer wall between the modeling result and the experimental data for the three higher ampoule positions of $H = 43$ mm, $H = 59$ mm, and $H = 67$ mm.

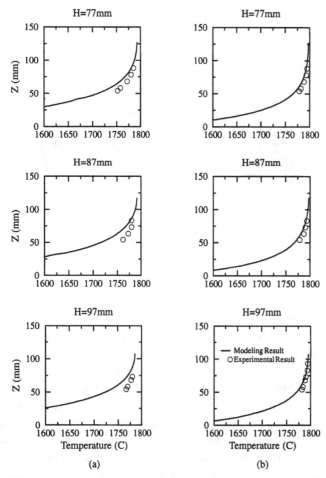

Figure 7.23 Comparisons of temperature profiles along (a) the NiAl centerline and (b) the ampoule outer wall between the modeling result and the experimental data for the three lower ampoule positions of $H = 77$ mm, $H = 87$ mm, and $H = 97$ mm.

especially for the temperature along the outer alumina wall. This implies that the simplified radiation model used for computing the heat flux between the alumina outer wall and the heater is sufficiently accurate, and that the boundary condition of constant heat flux at the heater bottom is reasonable. Along the ampoule centerline, agreement between experimental data and computational results is good for the three higher ampoule positions of $H = 43$ mm, $H = 59$ mm, and $H = 67$ mm, but worsens for the three lower ampoule positions. This trend is explained in more detail below.

Figure 7.24 shows the stream functions and isotherm distributions for two selected ampoule positions, one at a higher location ($H = 43$ mm) and the other at a lower location ($H = 87$ mm). For the higher ampoule position, Fig. 7.24a shows that the ampoule base region is contained inside the furnace hot zone. Since the treatment of radiation heat transfer and the boundary conditions appear to be reasonable there, the

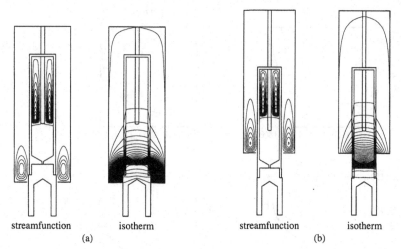

streamfunction   isotherm   streamfunction   isotherm
              (a)   (b)

Figure 7.24   Solution characteristics of NiAl furnace for selected ampoule positions of (a) $H = 43$ mm and (b) $H = 87$ mm.

computed thermal distribution of this region is accurate. Accordingly, the predicted temperature profiles along the NiAl centerline matches very well with experimental results. At the lower ampoule position, Fig. 7.24b shows clearly that the ampoule base region has been pulled out of the furnace hot zone. In this case, since the boundary condition at the outer wall of this region is still specified as a constant heat flux, this is no longer accurate. Furthermore, as the thermal distribution inside the ampoule is very sensitive to the isotherm change near the ampoule base, it is not surprising to observe in Fig. 7.23 that the deviations between numerical results and experimental data along the NiAl centerline exist. Improvement in modeling accuracy can be obtained by specifying realistic boundary conditions at the outer ampoule wall as it is pulled out of the furnace. The temperature distribution along the ampoule outer wall in the furnace hot zone, though, is not influenced by this effect.

7.3.3.2.2   Melting Temperature of β-NiAl   The melting temperature of NiAl at stoichiometric composition is reported in the literature as $T_{melt} = 1638°C$ (Noebe et al. 1993, Vedula et al. 1985). However, the accuracy of this value seems questionable. Based on the current experimental data and corresponding modeling results, we have conducted an evaluation of the NiAl melting temperature of β-NiAl.

For the ampoule position of $H = 43$ mm, we measured the interface position by moving the alumina tube through the NiAl melt until it touched the solid/melt interface, which was found to be $h_{interf} = 75$ mm. It is noted that, since the tip of TC1 is separated from the interface by the closed end of the alumina tube (with a thickness of 6.5 mm shown in Fig. 7.18), the temperature reading of TC1 is not the melt/solid interface temperature. However, the detailed temperature profile along the entire ampoule centerline can be obtained from our numerical results. Combining these two results, the interface temperature of NiAl can be determined by interpolation, as shown in Fig. 7.25, yielding $T_{interf} = 1716°C$. This result is surprising, as this value

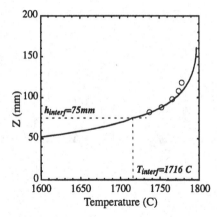

Figure 7.25 The estimation of melt/solid interface temperature of β-NiAl.

is at least 70°C higher than the melting temperature of NiAl widely reported in the literature. Waltson and Darolia (1995) have also reported recently an observation of NiAl melting temperature as high as 1682°C. Although T_{melt} could not be determined directly in our computation, the above estimation has its base. First, the current numerical model correlates well with the experimental data; hence the temperature profile we used for interpolating the interface temperature is likely to be accurate. Second, thermocouples have been calibrated before the experiment. Although it has to be kept in mind that a type C thermocouple does have a confidence limit up to 1% and possibly brings in about 18°C difference in the current temperature measurement, this will not influence our current comparison between experimental results and numerical predictions. Third, the interface position is measured twice – during the experiment and after the experiment – with a confidence limit of ± 2 mm. Considering all possible errors that might arise during the measurements of temperature and position, the melt/solid interface temperature we obtained for the current NiAl sample at stoichiometric composition from our computation is still much higher than 1638°C.

7.3.4 Analysis of Processing Conditions

β-NiAl with a high purity and an improved mechanical property of high tensile elongation at room temperature has been produced routinely with the current design of the vertical Bridgman system (Levit et al. 1996). Therefore, it is important to describe the detailed characteristics of the processing conditions that are suitable for growing high-quality single crystal NiAl.

Satisfactory comparison between experimental and numerical results is obtained for various ampoule positions with the alumina tube and thermocouple inserted. Next, we will further investigate the solidification characteristics of the actual growth process without the alumina tube.

Figure 7.26 shows the solution characteristics of the stream functions and isotherms obtained for different H values. It can be observed that the stream functions in the melt share a very similar pattern for all positions, and the maximum velocity

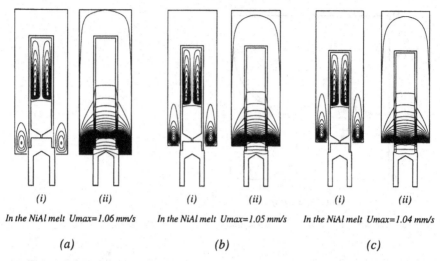

(i)	(ii)	(i)	(ii)	(i)	(ii)
In the NiAl melt Umax=1.06 mm/s		*In the NiAl melt Umax=1.05 mm/s*		*In the NiAl melt Umax=1.04 mm/s*	
(a)		(b)		(c)	

Figure 7.26 Solution characteristics: (i) stream functions and (ii) isotherms in an actual vertical Bridgman growth system for NiAl crystal for different ampoule positions. (a) $H = 43$ mm, (b) $H = 59$ mm, and (c) $H = 67$ mm. (Stream functions inside encapsulated argon gas and inside the NiAl melt are not drawn with the same scales.)

varies only slightly from 1.06 mm/s to 1.04 mm/s. The maximum velocity in the encapsulated argon gas is orders of magnitude smaller than that in the NiAl melt. It appears that convection effects in the current crystal growth system are not strong. This is a desirable condition, for otherwise isotherms can be greatly distorted by convection, which will cause nonuniform thermal conditions across the interface. The situation will, of course, change if the physical dimension of the Bridgman system increases for large scale production.

It has been recognized in the literature that for crystal growth from the melt in the vertical Bridgman growth system, a slightly convex (toward the melt) liquid–solid interface is favorable for grain selection and helps to prevent spontaneous nucleation at the ampoule wall. In addition, high temperature gradients across the interface and uniform distributions of thermal conditions along the interface are desirable for obtaining high purity crystals (Brown 1988, Kurz and Fisher 1989, Sen and Stefanescu 1991). In the following, these parameters as related to the current system are further discussed.

Figure 7.27 shows the interface shape and the vertical temperature gradient across the interface for different H values. It is observed that the interface shapes obtained from all three ampoule positions share a very similar convex pattern and develop parallel to each other. The difference between the centerline and the inner ampoule wall locations is about 2 mm. A similar feature is also observed for the temperature gradient across the interface as shown in Fig. 7.27b. For different ampoule positions, the gradients are all maintained in the range of 3.16–3.35°C/mm. However, the gradient distribution along the interface is not uniform. The temperature gradient at the interface centerline is about 5% higher than that at the inner ampoule wall. This deviation is caused by the presence of convection effects in the melt. This phenomenon

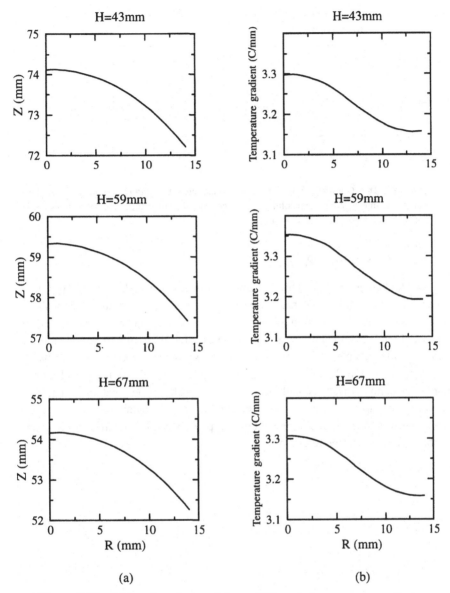

Figure 7.27 (a) Interface characteristics and (b) vertical temperature gradient across the interface for three ampoule positions of $H = 43$ mm, $H = 59$ mm, and $H = 67$ mm.

indicates that a good furnace design should minimize the convection effect in order to maintain a uniform processing condition. Figure 7.28 explains this situation more clearly, in which the interface characteristics with the convection effect and without the convection effect are compared. It is observed that convection causes the interface position to be moved up about 0.5%, which is quite small. However, it causes the temperature gradient along the interface to deviate quite noticeably along the interface (5%), as is shown in Fig. 7.28b.

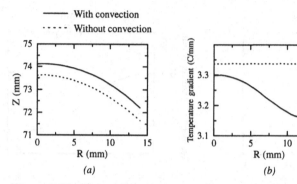

Figure 7.28 Comparison of (a) interface characteristics and (b) temperature gradient across the interface between solutions with convection and without convection effects for the ampoule position of $H = 43$ mm.

7.4   Concluding Remarks

In this chapter, an attempt has been made to provide the reader with a flavor of the broad spectrum of problems that can be effectively handled with the numerical and modeling techniques discussed in the previous chapters of this book. The problems presented include turbulent flows with a high degree of swirl and recirculation in geometrically complex domains, double-diffusive flows involving very disparate length scales in the flow domain, and crystal growth involving multilevel modeling. The examples discussed in this chapter illustrate that a combination of numerical strategies to improve accuracy and efficiency of the computational algorithm and modeling strategies to reflect the underlining physics can be effective in predicting problems of engineering relevance.

References

Akbarzadeh, A. and Manins, P. 1988. Convective Layers Generated by Side Walls in Solar Ponds, *Solar Energy*, **41**(6), 521–529.

Akselvoll, K. and Moin, P. 1995. Large Eddy Simulation of Turbulent Confined Co-Annular Jets and Turbulent Flow over a Backward Facing Step, Report No. TF-63, Thermosciences Division, Dept. of Mechanical Engineering, Stanford University.

Alef, M. 1994. Implementation of a Multigrid Algorithm on Suprenum and Other Systems, *Parallel Comput.*, **20**(10), 1547–1557.

Almeida, G.P., Durao, D.F.G., and Heitor, M.V. 1993. Wake Flow Behind Two-Dimensional Model Hills, *Exp. Thermal Fluid Sci.*, **7**, 87.

Altas, I. and Burrage, K. 1994. A High-Accuracy Defect-Correction Multigrid Method for the Steady Incompressible Navier-Stokes Equations, *J. Comput. Phys.*, **114**(20), 227–233.

Aubry, N., Holmes, P., Lumley, J.L., and Stone, E. 1988. The Dynamics of Coherent Structures in the Wall Region of a Turbulent Boundary Layer, *J. Fluid Mech.*, **192**, 115–173.

Averbuch, A., Gabber E., Gordissky B., and Medan, Y. 1990. A Parallel FFT on an MIMD Machine, *Parallel Comput.*, **15**, 61–74.

Bal, H.E., Steiner, J.G., and Tanenbaum, A.S. 1989. Programming Languages for Distributed Computing Systems, *ACM Computing Surveys*, **21**(3), 261–322, September.

Baldwin, B.S. and Lomax, H. 1978. Thin-Layer Approximation and Algebraic Model for Separated Turbulent Flows, *AIAA Paper* No. 78-257.

Bardina, J., Ferziger, J.H., and Rogallo, R.S. 1985. Effect of Rotation on Isotropic Turbulence: Computation and Modeling, *J. Fluid Mech.*, **154**, 321–336.

Beckermann, C. and Viskanta, R. 1988. Double-Diffusive Convection due to Melting, *Int. J. Heat Mass Transf.*, **31**, 2077–208.

Beckermann, C. and Viskanta, R. 1993. Mathematical Modeling of Transport Phenomena During Alloy Solidification, *Appl. Mech. Rev.*, **46**(1), 1–27.

Ben-Ari, M. 1982. *Principles of Concurrent Programming*. Prentice-Hall, Englewood Cliffs, NJ.

Benek, J.A., Sterger, J.J., and Dougherty, F.C. 1983. A Flexible Grid Embedding Technique with Application to the Euler Equations, *AIAA-83-1944-CP*.

297

Bennon, W.D. and Incropera, F.P. 1987a. A Continuum Model for Momentum, Heat and
    Species Transport in Binary Solid–Liquid Phase Change Systems – I. Model
    Formulation, *Int. J. Heat Mass Transfer*, **30**(10), 2161–2170.

Bennon, W.D. and Incropera, F.P. 1987b. A Continuum Model for Momentum, Heat and
    Species Transport in Binary Solid–Liquid Phase Change Systems – II. Application to
    Solidification in a Rectangular Cavity, *Int. J. Heat Mass Transfer*, **30**(10), 2171–2187.

Berger, M.J. 1987. On Conservation at Grid Interfaces, *SIAM J. Numer. Anal.*, **24**(5),
    967–984.

Bergman, T.L. and Ungan, A. 1986. Experimental and Numerical Investigation of
    Double-Diffusive Convection Induced by a Discrete Heat Source, *Int. J. Heat Mass
    Transfer*, **29**(11), 1695–1709.

Bertsekas, D.P. and Tsitsiklis, J.N. 1989. *Parallel and Distributed Computation: Numerical
    Methods*. Prentice-Hall, Englewood Cliffs, NJ.

Bhat, M.S., Poirier, D.R., and Heinrich, J.C. 1995a. Permeability for Cross Flow Through
    Columnar-Dendritic Alloys, *Metall Trans. B*, **26B**, 1049–1056.

Bhat, M.S., Poirier, D.R., and Heinrich, J.C. 1995b. A Permeability Length Scale for Cross
    Flow Through Model Structures, *Metall. Trans. B*, **26B**, 1091–1092.

Bird, R.B., Steward, W.E., and Lightfoot, E.N. 1960. *Transport Phenomena*, John Wiley,
    New York, NY.

Blaisdell, G.A., Mansour, N.N., and Reynolds, W.C. 1993. Compressibility Effects on the
    Growth and Structure of Homogeneous Turbulent Shear Flow, *J. Fluid Mech.*, **256**,
    443–485.

Blosch, E.L. 1994. *Pressure-Based Methods on Single-Instruction-Stream/
    Multiple-Data-Stream Computers*, Ph.D. dissertation, University of Florida, Department
    of Aerospace Engineering, Mechanics and Engineering Science, December.

Blosch, E.L. and Shyy, W. 1994. Sequential Pressure-Based Navier-Stokes Algorithms on
    SIMD Computers – Computational Issues, *Numerical Heat Transfer Part B*, **26**(2),
    115–132.

Blosch, E.L. and Shyy, W. 1996. Scalability and Performance of Data-Parallel
    Pressure-Based Multigrid Methods for Viscous Flows, *J. Comput. Phys.*, **125**, 338–353.

Blosch, E., Shyy, W., and Smith, R.W. 1993. The Role of Mass Conservation in
    Pressure-Based Algorithms, *Numer. Heat Transf.*, **24**, 415–430.

Bodin, F., Erthel, J., and Priol, T. 1993. Parallel Sparse Matrix by Vector Multiplication Using
    a Shared Virtual Memory Environment. In *Proceedings of the Sixth SIAM Conference on
    Parallel Processing for Scientific Computing*, 421–428.

Braaten, M.E. and Connell, S.D. 1996. Three-Dimensional Unstructured Adaptive Multigrid
    Scheme for the Navier-Stokes Equations, *AIAA J.*, **34**, 281–290.

Braaten, M.E. and Shyy, W. 1986. A Study of Recirculating Flow Computation Using
    Body-Fitted Coordinates: Consistency Aspect and Mesh Skewness, *Numer. Heat
    Transf.*, **9**, 559–574.

Braaten, M.E. and Shyy, W. 1987. Study of Pressure-Correction Methods with Multigrid for
    Viscous Flow Calculations in Nonorthogonal Curvilinear Coordinates, *Numer. Heat
    Transf.*, **11**, 417–442.

Bradshaw, P. 1969. The Analogy between Streamline Curvature and Buoyancy in Turbulent
    Shear Flow, *J. Fluid Mech.*, **36**, 177–191.

Brandt, A. 1977. Multi-Level Adaptive Solutions to Boundary-Value Problems, *Math.
    Comput.*, **31**, 333–390.

Brandt, A. 1984. *1984 Multigrid Guide with Application to Fluid Dynamics*. Lecture Notes in

Computational Fluid Dynamics, von Karman Institute for Fluid Dynamics, Rhode-Saint-Genese, Belgium. Available from Department of Computer Science, University of Colorado, Denver, CO.

Brandt, A. and Ta'asan, S. 1984. Multigrid Solution to Quasi-Elliptic Schemes. In E.M. Murman and S.S. Abarbanel, editors, *Progress and Supercomputing in Computational Fluid Dynamics, Proceedings of U.S.-Israel Workshop*, 235–255, Birkhauser, Boston.

Brandt, A. and Yavneh, I. 1992. On Multigrid Solution of High-Reynolds Incompressible Entering Flows, *J. Comput. Phys.*, **101**, 151–164.

Briggs, W. 1987. *A Multigrid Tutorial*. SIAM, Philadelphia.

Briggs, W. 1993. Wavelets and Multigrid, *SIAM J. Sci. Comput.*, **14**(2), 506–510.

Briggs, W. and McCormick, S.F. 1987. Introduction. In S.F. McCormick, editor, *Multigrid Methods*, chapter 1, SIAM, Philadelphia.

Brown, R.A. 1988. Theory of Transport Processes in Single Crystal Growth from the Melt, *AIChE J.*, **34**, 881–911.

Bruneau, C.-H. and Jouron, C. 1990. An Efficient Scheme for Solving Steady Incompressible Navier-Stokes Equations, *J. Comput. Phys.*, **89**, 389–413.

Canright, D. and Davis, S.H. 1991. Buoyancy Effects of a Growing, Isolated Dendrite, *J. Crystal Growth*, **114**, 173–189.

Carriero, N. and Gelernter, D. 1990. *How to Write Parallel Programs*, MIT Press, Cambridge, MA.

Cebeci, T. and Smith, A.M.O. 1974. *Analysis of Turbulent Boundary Layers*, Academic Press, New York.

Chan, T.F. and Tuminaro, R.S. 1988. A Survey of Parallel Multigrid Algorithms. In A.K. Noor, editor, *Parallel Computations and Their Impact on Mechanics*, **AMD-86**, 155–170, ASME, New York.

Chen, C.F., Briggs, D.G., and Wirtz, R.A. 1971. Stability of Thermal Convection in a Salinity Gradient Due to Lateral Heating, *Int. J. Heat Mass Transfer*, **14**, 57–65.

Chen, Y.S. and Kim, S.W. 1987. Computation of Turbulent Flows Using an Extended K-ε Turbulence Closure Model, *NASA CR*-179204.

Chesshire, G. and Henshaw, W.D. 1990. Composite Overlapping Meshes for the Solution of Partial Differential Equations, *J. Comp. Phys.*, **90**, 1–64.

Chesshire, G. and Henshaw, W.D. 1994. A Scheme for Conservative Interpolation on Overlapping Grids, *SIAM J. Sci. Comput.*, **15**(4), 819–845.

Chorin, A.J. 1967. Numerical Solution of the Navier-Stokes Equations, *Math. Comput.*, **22**(106), 745–762.

Christenson, M.S. and Incropera, F.P. 1989. Solidification of an Aqueous Ammonium Chloride Solution in a Rectangular Cavity – I. Experimental Study, *Int. J. Heat Mass Transfer*, **32**(1), 47–68.

Christenson, M.S., Bennon, W.D., and Incropera, F.P. 1989. Solidification of an Aqueous Ammonium Chloride Solution in a Rectangular Cavity – II. Comparison of Predicted and Measured Results, *Int. J. Heat Mass Transfer*, **32**(1), 69–79.

Chyu, W.J., Rimlinger, M.J., and Shih, T.I.-P. 1995. Control of Shock-Wave/Boundary-Layer Interactions by Bleed, *AIAA J.*, **33**, 1239–1247.

Coirier, W.J. and Powell, K.G. 1996. Solution-Adaptive Cartesian Cell Approach for Viscous and Inviscid Flows, *AIAA J.*, **34**, 938–945.

Crowley, A.B. 1983. Mathematical Modeling of Heat Flow in Czochralski Crystal Growth, *IMA J. Appl. Math.*, **30**, 173–189.

Dantzig, J.A. 1989. Modeling Liquid-Solid Phase Change with Melt Convection, *Int. J. Numer. Meths. Engrg.*, **28**, 1769–1785.

Darolia, R. 1991. NiAl Alloys for High-Temperature Structural Applications, *JOM* **March**, 44–49.

DeGregoria, A.J. and Schwartz, L.W. 1986. A Boundary-Integral Method for Two-Phase Displacement in Hele-Shaw Cells, *J. Fluid Mech.*, **164**, 383–400.

Dendy, J.E. 1982. Black Box Multigrid, *J. Comput. Phys.*, **48**, 366–386.

Dendy, J.E., Ida, M.P., and Rutledge, J.M. 1992. A Semicoarsening Multigrid Algorithm for SIMD Machines, *SIAM J. Sci. Stat. Comput.*, **13**(6), 1460–1469.

Douglas, C. and Douglas, J. 1993. A Unified Convergence Theory for Abstract Multigrid or Multilevel Algorithms, Serial and Parallel, *SIAM J. Numer. Anal.*, **30**(1), 136–158.

Douglas, C. 1994. Parallel Multilevel and Multigrid Methods, Obtained from *casper.cs.yale.edu* via anonymous FTP.

Drew, D.A. 1983. Mathematical Modeling of Two-Phase Flow, *Ann. Rev. Fluid Mech.*, 1983, **15**, 261–291.

Duff, S., Erisman, M., and Reid, J.K. 1990. *Direct Methods for Sparse Matrices*, Oxford University Press, Oxford, UK.

Dwyer, H.A. 1984. Grid Adaption for Problems in Fluid Dynamics, *AIAA J.*, **22**, 1705–1712.

Egolf, T.A. 1992. Computational Performance of CFD Codes on the Connection Machine. In H.D. Simon, editor, *Parallel Computational Fluid Dynamics: Implementations and Results*, 271–280, MIT Press, Cambridge, MA.

El Baz, A.M. and Launder, B.E. 1993. Second-Moment Modeling of Compressible Mixing Layers, *Engineering Turbulence Modeling and Experiments*, edited by Rodi, W. and Martelli, F., Elsevier Science Publishers B.V., 63–72.

Ferng, W., Petition, S.G., and Wu, K. 1993. Basic Sparse Matrix Computation on Data-Parallel Computers. In *Proceedings of the Sixth SIAM Conference on Parallel Processing for Scientific Computing*, 462–466.

Ferziger, J.H. and Peric, M. 1993. Computational Methods for Incompressible Flow. In M. Lesieur and J. Zinn-Justin, editors, *Proceedings of Session LIV of the Les Houches Conference on Computational Fluid Dynamics*. Elsevier, Amsterdam, The Netherlands.

Ferziger, J.H. and Peric, M. 1996. *Computational Methods for Fluid Dynamics*. Springer-Verlag, New York.

Fischer, P.F. and Patera, A.T. 1994. Parallel Simulation of Viscous Incompressible Flows, *Ann. Rev. Fluid Mech.*, **26**, 483–527.

Fletcher, C.A.J. 1988. *Computational Techniques for Fluid Dynamics*, Vol. 2, Springer-Verlag, New York.

Flood, S.C. and Hunt, J.D. 1987. Columnar and Equiaxed Growth, *J. Crystal Growth*, **82**, 543–551.

Flynn, M.J. 1972. Some Computer Organizations and Their Effectiveness, *IEEE Transactions on Computers*, **C-21**(9), 948–960.

Frederickson, P.O. and McBryan, O.A. 1981. Normalized Convergence Rates for the PSMG Method, *SIAM J. Sci. Stat. Comput.*, **12**, 221–229.

Galmes, J.M. and Lakshminarayana, B. 1984. Turbulence Modeling for Three-Dimensional Shear Flows over Curved Rotating Bodies, *AIAA J.*, **22**, 1420–1428.

Ganesan, S. and Poirier, D.R. 1990. Conservation of Mass and Momentum for the Flow of Interdendritic Liquid during Solidification, *Metall. Trans. B*, **21B**, 173–181.

Gannon, D. and Van Rosendale, J. 1986. On the Structure of Parallelism in a Highly Concurrent PDE Solver, *J. Parallel Distr. Comput.*, **3**, 106–135.

Garandet, J.P., Duffar, T., and Favier, J.J. 1990. On the Scaling Analysis of the Solute Boundary Layer in an Idealized Growth Configuration, *J. Crystal Growth*, **106**, 437–444.

Gaviglio, J., Dussauge, J.P., Debieve, J.F., and Favre, A. 1977. Behaviour of a Turbulent Flow, Strongly out of Equilibrium, at Supersonic speeds, *Phys. Fluids*, **20**, S179–S192.

Germano, M. 1992. Turbulence: The Filtering Approach, *J. Fluid Mech.*, **238**, 325–336.

Ghaddar, C.K. 1995. On the Permeability of Unidirectional Fibrous Media: A Parallel Computational Approach, *Phys. Fluids*, **7**(11), 2563–2586.

Ghia, U., Ghia, K.N., and Shin, C.T. 1982. High-Resolution for Incompressible Flow Using the Navier-Stokes Equations and a Multigrid Method, *J. Comp. Phys.*, **48**, 387–411.

Gleick, J. 1987. *Chaos*, Viking, New York.

Golub, G. and Ortega, J.M. 1993. *Scientific Computing: An Introduction with Parallel Computing*. Academic Press, Inc., San Diego, CA.

Gray, W.G. 1975. A Derivation of the Equations for Multi-Phase Transport, *Chem. Engng Sci.*, **30**, 229–233.

Gray, W.G. 1983. Brief Communication: Local Volume Averaging of Multiphase Systems Using a Non-constant Averaging Volume, *Int. J. Multiphase Flow*, **9**(6), 755–761.

Grebogi, C., Ott, E., and Yorke, J.A. 1987. Chaos Strange Attractors, and Fractal Basin Boundaries in Nonlinear Dynamics, *Science*, **238**, 632–638.

Griebel, M. 1993. Sparse Grid Multilevel Methods, Their Parallelization, and Their Application to CFD. In R.B. Pelz, A. Ecer, and J. Hauser, editors, *Parallel Computational Fluid Dynamics '92*, 161–174. Elsevier, Amsterdam, The Netherlands.

Gropp, W.D. and Keyes, D.E. 1992a. Domain Decomposition Methods in Computational Fluid Dynamics, *Int. J. Numer. Meths. Fluids*, **14**, 147–165.

Gropp, W.D. and Keyes, D.E. 1992b. Domain Decomposition with Local Mesh Refinement, *SIAM J. Sci. Stat. Comput.*, **13**(4), 967–993.

Gropp, W., Lusk, E., and Skjellum, A. 1994. *Using MPI: Portable Parallel Programming with the Message-Passing Interface*, Scientific and Engineering Computation Series, MIT Press, Cambridge, MA.

Gupta, A., Kumar, V., and Sameh, A.H. 1993. Performance and Scalability of Preconditioned Conjugate Gradient Methods on Parallel Computers. In *Proceedings of the Sixth SIAM Conference on Parallel Processing for Scientific Computing*, 664–674.

Gupta, S.N., Zubair, M., and Grosch, C.E. 1992. A Multigrid Algorithm for Parallel Computers: CPMG. *J. Sci. Comput.*, **7**(3), 263–279.

Gustafson, J.L. 1990. Fixed Time, Tiered Memory, and Superlinear Speedup. In *Proceedings of the Fifth Distributed Memory Computing Conference*, 1255–1260, Charleston, SC, IEEE Computer Society Press.

Gustafson, J.L., Montry, G.R., and Benner, R.E. 1988. Development of Parallel Methods for a 1024-Processor Hypercube, *SIAM J. Sci. Stat. Comput.*, **9**(4), 609–638.

Gukenheimer, J. and Holmes, P.J. 1983. *Nonlinear Oscillations, Dynamical Systems and Bifurcations of Vector Fields*, Springer, New York.

Hackbusch, W. 1980a. Survey of Convergence Proofs for Multi-Grid Iterations. In J. Frehse, D. Pallaschke, and U. Trottenberg, editors, *Special Topics of Applied Mathematics – Functional Analysis, Numerical Analysis, and Optimization*, 151–164. North-Holland, Amsterdam, The Netherlands.

Hackbusch, W. 1980b. Convergence of Multi-Grid Iterations Applied to Difference Equations, *Mathematics of Computation*, **34**(150), 425–440.

Hammond, S.W. 1992. Mapping Unstructured Grid Computations to Massively Parallel Computers, Technical report, RIACS, NASA Ames Research Center, Mountain View,

CA, Technical Report 92.14.

Hänel, D. and Schwane, R. 1989. An Implicit Flux-Vector Splitting Scheme for the Computation of Viscous Hypersonic Flow, *AIAA Paper No. 89-0274*.

Hänel, D., Schwane, R., and Seider, G. 1987. On the Accuracy of Upwind Schemes for the Solution of the Navier-Stokes Equations, *AIAA Paper No. 87-1105*.

Harten, A. 1983. High-Resolution Schemes for Hyperbolic Conservation Laws, *J. Comp. Phys.*, **49**, 357–393.

Hatcher, P.J. and Quinn, M.J. 1991. *Data-Parallel Programming on MIMD Computers*, MIT Press, Cambridge, MA.

Heath, M.T. and Raghavan, P. 1993. Distributed Solution of Sparse Linear Systems, Technical report, Department of Computer Science, University of Illinois, Urbana, IL. Technical Report 93-1793.

Heinrich, J.C. 1984. A Finite Element Model for Double Diffusive Convection, *Int. J. Numer. Methods Eng.*, **20**, 465–477.

Heinrich, J.C., Felicelli, S., Nadapukar, P., and Poirier, D.R. 1989. Thermosolutal Convection during Dendritic Solidification of Alloys, Part II: Nonlinear Convection, *Metall. Trans.*, **20B**, 883–891.

Henshaw, W.D. 1994. A Fourth-Order Accurate Method for the Incompressible Navier-Stokes Equations on Overlapping Grids, *J. Comp. Phys.*, **113**, 13–25.

Hinatsu, M. and Ferziger, J.H. 1991. Numerical Computation of Unsteady Incompressible Flow in Complex Geometry Using a Composite Multigrid Technique, *Intl. J. Numer. Meths Fluids*, **13**, 971–997.

Hinze, O. 1959. *Turbulence*, McGraw Hill, New York.

Hiranandani, S., Kennedy, K., Koelbel, C., Kremer, U., and Tseng, C. 1992. An overview of the Fortran D Programming System. In *Languages and Compilers for Parallel Computing, Fourth International Workshop*, Santa Clara, CA. Springer-Verlag.

Hirsch, C. 1990. *Numerical Computation of Internal and External Flows*, Volume 2, Wiley, New York.

Hoare, R.A. 1966. Problems of Heat Transfer in Lake Vanda, A Density Stratified Antarctic Lake, *Nature*, **210**, 787–789.

Holmes, P.J., Berkooz, G., and Lumley, J.L. 1996. *Turbulence, Coherent Structures, Dynamical Systems and Symmetry*, Cambridge University Press, New York.

Howard, J.H.G., Patankar, S.V., and Bordynuik, R.M. 1980. Flow Prediction in Rotating Ducts Using Coriolis-Modified Turbulence Models, *J. Fluids Eng.*, **102**, 456–461.

Howell, L.H. 1994. A Multilevel Adaptive Projection Method for Unsteady Incompressible Flow, from Xnetlib at *ftp.ornl.gov*.

Howes, F.A. and Whitaker, S. 1985. The Spatial Averaging Theorem Revisited, *Chem. Engng. Sci.*, **40**, 1387–1392.

Huang, P.G., Bradshaw, P., and Coakley, T.J. 1992. Assessment of Closure Coefficients for Compressible Flow Turbulence Models, *NASA TM-103882*.

Huang, P.G., Bradshaw, P., and Coakley, T.J. 1994. Turbulence Models for Compressible Boundary Layers, *AIAA J.*, **32**, 735–740.

Huang, S.-C. and Glicksman, M.E. 1981. Fundamentals of Dendritic Solidification I. Steady-State Tip Growth, *Acta Metall.*, **29**, 701–715, and II. Development of Side-Branch Structure, *Acta Metall.*, **29**, 717–734.

Huppert, H.E. 1990. The Fluid Mechanics of Solidification, *J. Fluid Mechanics*, **212**, 209–240.

Huppert, H.E. and Turner, J.S. 1980. Ice Blocks Melting Into a Salinity Gradient, *J. Fluid.*

Mech., **100**, 367–384.

Huppert, H.E. and Turner, J.S. 1981. Double-Diffusive Convection, *J. Fluid Mech.*, **106**, 299–329.

Hurle, D.T.J., Jakeman, E., and Wheeler, A.A. 1983. Hydrodynamics Stability of the Melt during Solidification of a Binary Alloy, *Phys. Fluids*, **26**, 624–626.

Hutchinson, B.R. and Raithby, G.D. 1986. A Multigrid Method Based on the Additive Correction Strategy, *Numer. Heat Transf.*, **9**, 511–537.

Hwang, K. 1993. *Advanced Computer Architecture: Parallelism, Scalability, and Programmability*. McGraw-Hill, New York.

Hwang, K. and Briggs, F.A. 1984. *Computer Architecture and Parallel Processing*, McGraw-Hill, New York, NY.

Hwang, T. and Parsons, I. 1994. Parallel Implementation and Performance of Multigrid Algorithms for Solving Eigenvalue Problems, *Comput. Struct.*, **50**(3), 325–336.

Issa, R.I. 1985. Solution of the Implicitly Discretised Fluid Flow Equations by Operator Splitting, *J. Comp. Phys.*, **61**, 40–65.

Jameson, A. 1993. Artificial Diffusion, Upwind Biasing, Limiters and their Effect on Accuracy and Multigrid Covergence in Transonic and Hypersonic Flows, *AIAA Paper No. 93-3359*.

Jespersen, D.C. and Levit, C. 1989. A Computational Fluid Dynamics Algorithm on a Massively Parallel Computer. *Inter. J. Supercomp. Appl.*, **3**(4), 9–27.

Johnson, R.A. and Belk, D.M. 1995. Multigrid Approach to Overset Grid Communication, *AIAA J.*, **33**, 2305–2308.

Johnston, J.P., Halleen, R.M., and Lezius, D.K. 1972. Effects of Spanwise Rotation on the Structure of Two-Dimensional Fully Developed Turbulent Channel Flow, *J. Fluid Mech.*, **56**, 533–557.

Jones, W.P. and Launder, B.E. 1972. The Prediction of Laminarization with a Two-Equation Model of Turbulence, *Int. J. Heat Mass Trans.*, **15**, 301–314.

Kallinderis, Y. 1992. Numerical Treatment of Grid Interfaces for Viscous Flows, *J. Comp. Phys.*, **98**, 129–144.

Kallinderis, Y., Khawaja, A., and McMorris, H. 1996. Hybrid Prismatic/Tetrahedral Grid Generation for Viscous Flow Around Complex Geometries, *AIAA J.*, **34**, 291–298.

Kamakura, K. and Ozoe, H. 1993. Experimental and Numerical Analyses of Double-Diffusive Natural Convection Heated and Cooled from Opposing Vertical Walls with an Initial Condition of a Vertically Linear Concentration Gradient, *Int. J. Heat Mass Transfer*, **36**(8), 2125–2134.

Kao, K.-H. and Liou, M.-S. 1995. Advances in Overset Grid Schemes: From Chimera to DRAGON Grids, *AIAA J.*, **33**, 1809–1815.

Kessler, D.A., Koplik, J., and Levine, H. 1988. Pattern Selection in Fingered Growth Phenomena, *Advances in Physics*, **37**, 255–339.

Kim, J. 1983. The Effect of Rotation on Turbulence Structure, *Proc. 4th Symp. on Turbulent Shear Flows, Karlsruhe*, 6.14–6.19.

Kim, J., Kline, S.J., and Johnston, J.P. 1978. Investigation of Separation and Reattachment of a Turbulent Shear Layer: Flow over a Backward-Facing Step, Report No. MD-37, Thermosciences Division, Dept. of Mechanical Engineering, Stanford University.

Kobayashi, R. 1993. Modeling and Numerical Simulations of Dendritic Crystal Growth, *Physica D*, **63**, 410–423.

Krishnamurty, V.S. 1996. *Effect of Compressibility on the Turbulence Structure and its Modeling*, Ph.D. Dissertation, University of Florida, Gainesville.

Krishnamurty, V.S. and Shyy, W. 1996. Study of K-ε Based Modeling of Compressible Turbulent Flows, to be published.

Kristoffersen, R. and Andersson, H.I. 1993. Direct Simulation of Low-Reynolds-Number Turbulent Flow in a Rotating Channel, *J. Fluid Mech.*, **256**, 163–197.

Kumar, V. and Singh, V. 1991. Scalability of Parallel Algorithms for the All-Pairs Shortest-Path Problem, *J. Parall. Distr. Comput.*, **13**, 124–138.

Kumar, V., Grama, A., Gupta, A., and Karypis, G. 1994. *Introduction to Parallel Computing, Design and Analysis of Algorithms*. Benjamin/Cummings, Redwood City, CA.

Kurz, W. and Fisher, D.J. 1989. *Fundamentals of Solidification*, Trans. Tech. SA, Switzerland.

Lai, Y.G., Jiang, Y., and Przekwas, A.J. 1993. An Implicit Multi-Domain Approach for the Solution of Navier-Stokes Equations in Body-Fitted-Coordinate Grids, *AIAA-93-0541*.

Landhal, M.T. and Mollo-Christensen, E. 1992. *Turbulence and Random Processes in Fluid Mechanics*, Cambridge University Press, Cambridge, U.K.

Langer, J.S. 1980. Instabilities and Pattern Formation in Crystal Growth, *Rev. Modern Phys.*, **52**, 1–28.

Launder, B.E. 1989. Second-Moment Closure and Its Use in Modeling Turbulent Industrial Flows, *Int. J. Numer. Methods Fluids*, **9**, 963–985.

Launder, B.E. and Sharma, B.I. 1974. Application of the Energy Dissipation Model of Turbulence to the Calculation of Flow near a Spinning Disc, *Lett. Heat Mass Transf.*, **1**, 1112–1128.

Launder, B.E. and Spalding, D.B. 1974. The Numerical Computation of Turbulent Flows, *Comp. Meth. Appl. Mech Eng.*, **3**, 269–289.

Launder, B.E., Reece, G., and Rodi, W. 1973. Progress in the Development of a Reynolds-Stress Turbulence Closure, *J. Fluid Mech.*, **68**, 537–566.

Launder, B.E., Priddin, C.H., and Sharma, B.I. 1977. The Calculation of Turbulent Boundary Layers on Spinning and Curved Surfaces, *J. Fluids Eng.*, **102**, 231–239.

Launder, B.E., Tselepidakis, D.P., and Younis, B.A. 1987. A Second-Moment Closure Study of Rotation Channel Flow, *J. Fluid Mech.*, **183**, 63–75.

Lee, J.W. and Hyun, J.M. 1991. Time-Dependent Double Diffusion in a Stably Stratified Fluid Under Lateral Heating, *Int. J. Heat Mass Transf.*, **34**, 2409–2421.

Lele, S.K. 1994. Compressibility Effects on Turbulence, *Ann. Rev. Fluid Mech.*, **26**, 211–254.

Lesieur, M. 1993. Advance and State of the Art on Large-Eddy Simulations, in *Refined Flow Modeling and Turbulence Measurements*, Proceeding of the 5th International Symposium, Paris, Sept. 7–10.

Laster, B.P. 1993. *The Art of Parallel Programming*. Prentice-Hall, Inc., Englewood Cliffs, NJ.

Levit, C. 1989. Grid communication on the Connection Machine: Analysis, Performance, and Improvements. In H.D. Simon, editor, *Scientific Applications of the Connection Machine*, 316–332, World Scientific, New York.

Levit, V.I., Bul, I.A., Hu, J., and Kaufman, M.J. 1996. High Tensile Elongation of β-NiAl Single Crystals at 293 K, submitted to *Scripta Metall. Mater.*

Lichtenberg, A.J. and Lieberman, M.A. 1982. *Regular and Stochastic Motion*, Springer, New York.

Lien, F.S. and Leschziner, M.A. 1991. Multigrid Convergence Acceleration for Complex Flow Including Turbulence. In W. Hackbusch and U. Trottenberg, editors, *Multigrid Methods III*, 277–288, Birkhauser, Boston.

Lien, F.S. and Leschziner, M.A. 1994. Assessment of Turbulence-Transport Models Including Non-Linear RNG Eddy-Viscosity Formulation and Second-Moment Closure for Flows over a Backward-Facing Step, *Comp. Fluids.*, **23**(8), 983–1004.

Linden, P.F. 1976. The Formation and Destruction of Fine-Structure by Double Diffusive Process, *Deep-Sea Res.*, **23**, 895–908.

Linden, J., Lonsdale, G., Steckel, B., and Stuben, K. 1990. Multigrid for the Steady-State Incompressible Navier-Stokes Equations: A Survey. In *International Conference for Numerical Methods in Fluids*, 57–68, Berlin, Springer-Verlag.

Linden, J., Lonsdale, G., Ritzdorf, H., and Schuller, A. 1993. Block-Structured Multigrid for the Navier-Stokes Equations: Experiences and Scalability Questions. In R.B. Pelz, A. Ecer, and J. Hauser, editors, *Parallel Computational Fluid Dynamics '92*, 267–278, Elsevier, Amsterdam, The Netherlands.

Linden, J., Lonsdale, G., Ritzdorf, H., and Schuller, A. 1994. Scalability Aspects of Parallel Multigrid, *Future Generation Computer Systems*, **10**(4), 103–122.

Liou, M.-S. and Steffen, C.J. 1993. A New Flux Splitting Scheme, *J. Comp. Phy.*, **107**, 23–29.

Liou, M.-S., van Leer, B., and Shuen, J.S. 1990. Splitting of Inviscid Fluxes for Real Gases, *J. Comp. Phys.*, **87**, 1–24.

Liu, J. and Shyy, W. 1996. Assessment of Grid Interface Treatments for Multi-Block Incompressible Viscous Flow Computation, *Computers Fluids*, **25**, 719–740.

Löhner, R. and Parikh, P. 1988. Generation of Three-Dimensional Unstructured Grids by the Advancing-Front Method, *Int. J. Numer. Methods Fluids*, **8**, 1135–1149.

Lonsdale, G. and Schuller, A. 1993. Multigrid Efficiency for Complex Flow Simulations on Distributed Memory Machines, *Parallel Computing*, **19**(1), 23–32.

Luchini, P. and Dalascio, A. 1994. Multigrid Pressure-Correction Techniques for the Computation of Quasi-Incompressible Internal Flows, *Inter. J. Numer. Meths. Fluids*, **18**(5), 489–507.

Lumley, J.L. 1978. Computational Modeling of Turbulent Flows. In C.S. Yih, editor, *Advances in Applied Mechanics* 18, 123, Academic Press, New York.

Lynch, D.R. 1982. Unified Approach to Simulation on Deforming Elements with Application to Phase Change Problem, *J. Comput. Phys.*, **47**, 387–441.

MacCormack, R. 1991. Algorithmic Trends in CFD in the 1990s for Aerospace Flow Field Calculations, in M.Y. Hussaini, A. Kumar, and M.D. Salas (eds), *Algorithmic Trends in Computational Fluid Dynamics*, 21–27, Springer-Verlag, Berlin.

Malvern, L.E. 1969. *Introduction to the Mechanics of a Continuous Medium*, Prentice-Hall, Englewood Cliffs, New Jersey.

McBryan, O.A., Frederickson, P.O., Linden, J., Schuller, A., Solchenbach, K., Stuben, K., Thole, C.A., and Trottenberg, U. 1991. Multigrid Methods on Parallel Computers – A Survey of Recent Developments, *Impact of Computing in Science and Engineering*, **3**, 1–75.

McComb, W.D. 1990. *The Physics of Fluid Turbulence*, Oxford University Press, Oxford, UK.

McCormick, S.F. 1989. *Multilevel Adaptive Methods for Partial Differential Equations*. Frontiers in Applied Mathematics, SIAM, Philadelphia.

McFadden, G.B. and Coriell, S.R. 1987. Non-Planar Interface Morphologies During Unidirectional Solidification, *J. Crystal Growth*, **84**, 371–388.

Meakin, R.L. 1994. On the Spatial and Temporal Accuracy of Overset Grid Methods for Moving Body Problems, *AIAA-94-1925*.

Meakin, R.L. and Street, R.L. 1988. Simulation of Environmental Flow Problems in Geometrically Complex Domain, Part 2: A Domain-Splitting Method, *Computer. Meths. Appl. Mech. Eng.*, **68**, 311–331.

Menter, F.R. 1994. Two-Equation Eddy-Viscosity Turbulence Models for Engineering Applications, *AIAA J.*, **32**, 1598–1605.

Meyer, R.E. and Parter, S.V. (eds.) 1980. *Singular Perturbations and Asymptotics*, Academic Press, New York.

Michielse, P. 1993. Parallel Multigrid Using PVM, *Supercomputer*, **10**(6), 10–23.

Michelson, J.A. 1991. Mesh-Adaptive Solution of the Navier-Stokes Equations. In W. Hackbusch and U. Trottenberg, editors, *Multigrid Methods III*, 301–312, Birkhauser, Boston.

Minion, M.L. 1994. Two Methods for the Study of Vortex Patch Evolution on Locally Refined Grids, Technical Report, Lawrence Berkeley Laboratory, LBL-35719, Berkeley, CA 94720, May.

Miles, J. 1984. Strange Attractors in Fluid Dynamics. In J.W. Hutchinson and T.Y. Wu, editors, *Advances in Applied Mechanics*, Academic Press, New York, **24**, 189–214.

Moin, P., Squires, K., Cabot, W., and Lee, S. 1991. A Dynamic Subgrid-Scale Model for Compressible Turbulence and Scalar Transport, *Phys. Fluids*, A3(11).

Mohammadi, B. and Pironneau, O. 1994. *Analysis of the K-Epsilon Turbulence Model*, John Wiley & Sons, New York.

Moon, F.C. 1987. *Chaotic Vibrations: An Introduction for Applied Scientists and Engineers*, Wiley, New York.

Moon, Y. and Liou, M.-S. 1989. Conservative Treatment of Boundary Interfaces for Overlaid Grids and Multi-Level Grid Adaptations, *AIAA-89-1980*.

Monin, A.S. and Yaglom, A.M. 1971 and 1975. *Statiscal Fluid Mechanics*, 2 volumes, The MIT Press, Cambridge, MA.

Morgan, K., Peraire, J., Perio, J., and Hassan, O. 1991. The Computation of Three-Dimensional Flows Using Unstructured Grids, *Comp. Methods Appl. Mech. Eng.*, **87**, 335–352.

Narusawa, U. and Suzukawa, Y. 1981. Experimental Study of Double-Diffusive Cellular Convection Due to a Uniform Lateral Heat Flux, *J. Fluid Mech.*, **113**, 387–405.

Ni, J. and Beckermann, C. 1991. A Volume-Averaged Two-Phase Model for Transport Phenomena During Solidification, *Metall. Trans. B*, **22B**, 349–361.

Nicol, D.M. and Saltz, J.H. 1990. An Analysis of Scattered Decomposition, *IEEE Trans. Computers*, **39**, 1337–1345.

Nicolaides, R.A. 1979. On Some Theoretical and Practical Aspects of Multigrid Methods, *Math. Comput.*, **33**(147), 933–952.

Nishimura, T., Imoto, T., and Miyashita, H. 1994. Occurrence and Development of Double-Diffusive Convection during Solidification of a Binary System, *Int. J. Heat Mass Transf.*, **37**(10), 1455–1464.

Noebe, R.D., Bowman, R.R., and Nathal, M.V. 1993. Physical and Mechanical Properties of the B2 Compound NiAl, *Int. Mat. Rev.*, **38**(4), 193–232.

Ochoa-Tapia, J.A. and Whitaker, S. 1995a. Momentum Transfer at the Boundary Between a Porous Medium and a Homogeneous Fluid – I. Theoretical Development, *Int. J. Heat Mass Transf.*, **38**(14), 2635–2646.

Ochoa-Tapia, J.A. and Whitaker, S. 1995b. Momentum Transfer at the Boundary between a Porous Medium and a Homogeneous Fluid – II. Theoretical Development, *Int. J. Heat Mass Transf.*, **38**(14), 2647–2655.

Oran, E.S., Boris, J.P., and Brown, E.F. 1990. Fluid-Dynamic Computations on a Connection Machine – Preliminary Timings and Complex Boundary Conditions, AIAA Paper 90-0335, 28th Aerospace Sciences Meeting and Exhibit, Reno, NV.

Orszag, S.A., Yakhot, V., Flannery, W.S., Boysan, F., Choudhury, D., Maruzewski, J., and Patel, B. 1993. Renormalization Group Modeling and Turbulence Simulations. In R.M.C. So, C.G. Speziale, and B.E. Launder, editors, *Near-Wall Turbulent Flows*, 1031–1047, Elsevier, Amsterdam, The Netherlands.

Osher, S. 1984. Riemann Solvers, the Entropy Condition and Difference Approximations, *SIAM J. Num. Anal.*, **21**, 217–235.

Ouyang, H. and Shyy, W. 1996. Multi-Zone Simulation of Bridgman Growth Process of β-NiAl Crystal, *Int. J. Heat Mass Transf.*, **39**, 2039–2051.

Overman, A. and Van Rosendale, J. 1993. Mapping Robust Parallel Multigrid Algorithms to Scalable Memory Architectures. In S. McCormick, editor, *Proceedings of the Third Copper Mountain Conference on Multigrid Methods*, Marcel Dekker, New York.

Part-Enander, E. and Sjogreen, B. 1994. Conservative and Non-Conservative Interpolation Between Overlapping Grids for Finite Volume Solutions of Hyperbolic Problems, *Comp. Fluids*, **23**, 551–574.

Patankar, S.V. 1980. *Numerical Heat Transfer and Fluid Flow*, Hemisphere, Washington, DC.

Patel, V.C., Rodi, W., and Scheuerer, G. 1985. Turbulence Models for Near-Wall and Low Reynolds Number Flows, *AIAA J.*, **23**, 1308–1319.

Perng, C.Y. and Street, R.L. 1991. A Coupled Multigrid-Domain-Splitting Technique for Simulating Incompressible Flow in Geometrically Complex Domains, *Int. J. Numer. Meths. Fluids*, **13**, 269–286.

Perry, R.H., Green D.W., and Maloney J.O. 1984. *Perry's Chemical Engineerings' Handbook (6th ed)*, McGraw-Hill, New York.

Poirier, D.R. and Nandapurkar, P. 1988. Enthalpies of a Binary Alloy during Solidification, *Metall. Trans A*, **19A**, 3057–3061.

Poirier, D.R., Nandapurkar, P., and Ganesan, S. 1991. The Energy and Solute Conservation Equations for Dendritic Solidification, *Metall. Trans B*, **22B**, 889–900.

Prakash, C. and Voller, V.R. 1989. On the Numerical Solution of Continuum Mixture Model Equations Describing Binary Solid-Liquid Phase Change, *Numer. Heat Transf. Part B*, **15**, 171–189.

Prescott, P.J. and Incropera, F.P. 1994. Convective Transport Phenomena and Macrosegregation of a Binary Metal Alloy: I – Numerical Predictions, *J. Heat Transf.*, **116**, 735–741.

Press, W.H., Teukolsky, S.A., Vetterling, W.T., and Flannery, B.P. 1992. *Numerical Recipes in Fortran, The Art of Scientific Computing*. Cambridge University Press, London, 2nd edition.

Rai, M.M. 1985. A Implicit, Conservative, Zonal-Boundary Scheme for Euler Equation Calculations, *AIAA-85-0488*.

Rai, M.M. 1986. A Conservative Treatment of Zonal Boundaries for Euler Equation Calculations, *J. Comp. Phys.*, **62**, 472–503.

Rappaz, M. 1989. Modeling of Microstructure Formation in Solidification Processes, *Int. Mater. Rev.*, **34**, 93–123.

Reynolds, O. 1895. On the Dynamical Theory of Incompressible Viscous Fluids and the Determination of the Criterion, *Phil. Trans. Royal Soc. A*, **186**, 123–164.

Rhie, C.M. 1989. A Pressure-Based Navier-Stokes Solver Using the Multigrid Method. *AIAA J.*, **27**, 1017–1018.

Ristorcelli, J.R. 1993. A Representation for the Turbulent Mass Flux Contribution to Reynolds Stress and Two-Equation Closures for Compressible Turbulence, *NASA CR-191569*.

Rodrigue, G. and Shah, S. 1989. Pseudo-Boundary Conditions to Accelerate Parallel Schwarz

Methods. In G. Carey, editor, *Parallel Supercomputing Methods, Algorithms and Applications*, Wiley, Chichester, England.

Roe, P.L. 1981. Approximate Riemann Solvers, Parameter Vectors and Difference Schemes, *J. Comp. Phy.*, **43**, 357–372.

Roe, P.L. 1991. Beyond the Riemann Problem, Part 1. In M.Y. Hussaini, A. Kumar, and M.D. Salas, editors, *Algorithmic Trends in Computational Fluid Dynamics*, 3410–3467, Springer-Verlag, Berlin.

Rogallo, R.S. and Moin, P. 1984. Numerical Simulation of Turbulent Flows, *Ann. Rev. Fluid Mech.*, **16**, 99–137.

Rubesin, M.W. 1990. Extra Compressibility Terms for Favre-Average Two-Equation Models for Inhomogeneous Turbulent Flows, *NASA CR-177556*.

SC-NET 1995. The Electronic Newsletter of the SIAM Supercomputing Activity Group, August.

Sani, R.L. and Gresho, P.M. 1994. Resume and Remarks on the Open Boundary Condition Mini Symposium, *Int. J. Numer. Meths. Fluids*, **18**, 983–1008.

Sarkar, S. 1992. The Pressure Dilatation Correlation in Compressible Flows, *Phys. Fluids A*, **4**, 2674–2682.

Sarkar, S., and Lakshmanan, B. 1991. Application of a Reynolds-Stress Turbulence Model to the Compressible Shear Layer, *AIAA J.*, **29**, 743–749.

Sarkar, S., Erlebacher, G., Hussaini, M.Y., and Kreiss, H.O. 1991. The Analysis and Modeling of Dilatational Terms in Compressible Turbulence, *J. Fluid Mech.*, **227**, 473–493.

Sathyamurthy, P. and Patankar, S.V. 1994. Block-Correction-Based Multigrid Method for Fluid Flow Problems, *Numer. Heat Transf., Part B*, **25**(4), 375–394.

Schladow, S.G., Thomas, E. and Koseff, J.R. 1992. The Dynamics of Intrusions Into a Thermohaline Stratification, *J. Fluid Mech.*, **236**, 127–165.

Schlichting, H. 1979. *Boundary-Layer Theory*, 7th edition, McGraw-Hill, New York.

Schneider, M.C. and Beckermann, C. 1995. Formation of Macrosegregation by Multicomponent Thermosolutal Convection during the Solidification of Steel, *Metall. Trans. A*, **26A**, 2373–2388.

Schreiber, R. 1990. An Assessment of the Connection Machine, Technical report, RIACS, NASA Ames Research Center, Mountain View, CA, April.

Schreiber, R. and Simon, H.D. 1992. Towards the Teraflops Capability for CFD. In H.D. Simon, editor, *Parallel Computational Fluid Dynamics: Implementations and Results*, chapter 16, 313–342, MIT Press, Cambridge, MA.

Schwabe, D. 1988. Surface-Tension Driven Fluid Flow in Crystal Growth Melts, *Crystals*, **11**, 75–112, Springer-Verlag, New York.

Sen, S. and Stefanescu, D.M. 1991. Melting and Casting Processes for High-Temperature Intermetallics, *JOM*, *May*, 30–32.

Shaw, G.J. and Sivaloganathan, S. 1988a. A Multigrid Method for Recirculating Flows, *Inter. J. Numer. Meths. Fluids*, **8**(4), 417–440.

Shaw, G.J. and Sivaloganathan, S. 1988b. On the Smoothing Properties of the SIMPLE Pressure-Correction Algorithm, *Inter. J. Numer. Meths. Fluids*, **8**(4), 441–461.

Sherman, B.S. and Imberger, J. 1991. Control of a Solar Pond, *Solar Energy*, **46**(2), 71–81.

Shih, T.-H. 1996. Developments in Computational Modeling of Turbulent Flows, *NASA Contract Report 198458 (ICOMP-96-04)*, NASA Lewis Research Center.

Shih, T., Liou, W.W., Shabbir, A., Yang, Z., and Zhu, J. 1995. A New K-ε Eddy Viscosity Model for High Reynolds Number Turbulent Flows, *Comp. Fluids.*, **24**(3), 227–238.

Shih, T.I.-P., Rimlinger, M.J., and Chyu, W.J. 1993. Three-Dimensional

Shock-Wave/Boundary-Layer Interactions with Bleed, *AIAA J.*, **31**, 1819–1826.

Shimomura, Y. 1993. Turbulence Modeling Suggested by System Rotation. In R.M.C. So, C.G. Speziale, and B.E. Launder, editors, *Near-Wall Turbulence*, 115–123, Elsevier, Amsterdam, The Netherlands.

Shyy, W. 1987. Effects of Open Boundary Condition on Incompressible Navier-Stokes Flow Computation: Numerical Experiments, *Numer. Heat Transf.*, **12**, 157–178.

Shyy, W. 1989. A Unified Pressure Correction Algorithm for Computing Complex Fluid Flows, *Recent Advances in Computational Fluid Mechanics*, C.C. Chao, S.A. Orzag, and W. Shyy (eds.), Lecture Notes in Engineering, Springer-Verlag, New York, **43**, 135–147.

Shyy, W. 1991a. Structure of an Adaptive Grid Computational Method from the Viewpoint of Dynamic Chaos, *Appl. Numer. Math.*, **7**, 263–285.

Shyy, W. 1991b. Structure of an Adaptive Grid Computational Method from the Viewpoint of Dynamic Chaos, Part II: Grid Addition and Probability Distribution, *Appl. Numer. Math.*, **7**, 523–545.

Shyy, W. 1994. *Computational Modeling for Fluid Flow and Interfacial Transport*, Elsevier, Amsterdam, The Netherlands.

Shyy, W. and Sun, C.S. 1993. Development of a Pressure-Correction/Staggered-Grid Based Multigrid Solver for Incompressible Recirculating Flows, *Computers Fluids*, **22**(1), 51–76.

Shyy, W. and Thakur, S.S. 1994a. A Controlled Variation Scheme in a Sequential Solver for Recirculating Flows. Part I. Theory and Formulation, *Numer. Heat Transf., Part B*, **25**(3), 245–272.

Shyy, W. and Thakur, S.S. 1994b. A Controlled Variation Scheme in a Sequential Solver for Recirculating Flows. Part II. Applications, *Numer. Heat Transf., Part B*, **25**(3), 273–286.

Shyy, W. and Vu, T.C. 1991. On the Adoption of Velocity Variable and Grid System for Fluid Flow Computation in Curvilinear Coordinates, *J. Comp. Phys.*, **92**, 82–105.

Shyy, W. and Vu, T.C. 1993. Modeling and Computation of Flow in a Passage with 360 Degree Turning and Multiple Airfoils, *J. Fluids Eng.*, **115**, 103–108.

Shyy, W., Tong, S.S., and Correa, S.M. 1985. Numerical Recirculating Flow Calculation Using a Body-Fitted Coordinate System., *Numer. Heat Transf.*, **8**, 99–113.

Shyy, W., Thakur, S., and Wright, J. 1992a. Second-Order Upwind and Central-Difference Schemes for Recirculating Flow Computations, *AIAA J.*, **30**, 923–931.

Shyy, W., Chen, M.-H., and Sun, C.-S. 1992b. A Pressure-Based FMG/FAS Algorithm for Flow at All Speeds, *AIAA Paper 92-0548*, 30th Aerospace Sciences Meeting and Exhibit, Reno, NV, also *AIAA J.*, **30**, 2660–2669.

Shyy, W., Pang, Y., Hunter, G.B., Wei, D.Y., and Chen, M.-H. 1992c. Modeling Turbulent Transport and Solidification during Continuous Ingot Casting, *Int. J. Heat Mass Transf.*, **35**, 1229–1245.

Shyy, W., Thakur, S.S., and Udaykumar, H.S. 1993. A High Accuracy Sequential Solver for Simulation and Active Control of a Longitudinal Combustion Instability, *Computing Systems in Eng.*, **4**(1), 27–41.

Shyy, W., Liu, J., and Wright, J. 1994. Pressure-Based Viscous Flow Computation Using Multiblock Overlapped Curvilinear Grid, *Numer. Heat Transf.*, **25**, 39–59.

Shyy, W., Udaykumar, H.S., Rao, M.M., and Smith, R.W. 1996a. *Computational Fluid Dynamics with Moving Boundaries*, Taylor and Francis, Washington, DC.

Shyy, W., Liu, J. and Ouyang, H. 1996b. Multi-Resolution Computations for Fluid Flow and Heat/Mass Transfer. *Advances in Numerical Heat Transfer*, W.J. Minkowycz and E.M. Sparrow (editors), **1**, 201–239, Taylor and Francis, Washington, DC.

Simon, H.D. 1991. Partitioning of Unstructured Problems for Parallel Processing, *Computing Systems in Engineering*, **2**(2), 135–148.

Simon, H.D., Van Dalsem, W.R., and Dagum, L. 1992. Parallel CFD: Current status and future requirements. In H.D. Simon, editor, *Parallel Computational Fluid Dynamics: Implementations and Results*, chapter 1. MIT Press, Cambridge, MA.

Slattery, J.C. 1967. Flow of Viscoelastic Fluids Through Porous Media, *A.I.Ch.E.*, **13**, 1066–1071.

Smith, R.A. and Weiser, A. 1992. Semicoarsening multigrid on a hypercube, *SIAM J. Sci. Stat. Comput.*, **13**(6), 1314–1329.

Sockol, P.M. 1993. Multigrid solution of the Navier-Stokes equations on highly stretched grids with defect correction. In S. McCormick, editor, *Proceedings of the Third Copper Mountain Conference on Multigrid Methods*. Marcel Dekker, New York.

Speich, G.R. and Fisher, R.M. 1966. Recrystallization of a Rapidly Heated 31/4% Silicon Steel, Recrystallization. In *Grain Growth and Textures*, ASM, Metal Park OH, 563–598.

Speziale, C.G. 1991. Analytical Methods for the Development of Reynolds Stress Closures in Turbulence, *Ann. Rev. Fluid Mech.*, **23**, 107–157.

Speziale, C.G. 1995. A Review of Reynolds Stress Models for Turbulent Shear Flows, *NASA CR-195054*.

Speziale, C.G. and Thangam, S. 1992. Analysis of a RNG Based Turbulence Model for Separated Flows, *Int. J. Engng. Sci.*, **30**, 1379–1388.

Speziale, C.G., Abid, R., and Anderson, E.C., 1992. Critical Evaluation of Two-Equation Models for Near-Wall Turbulence, *AIAA J.*, **30**, 324–331.

Steger, J.L. 1991. Thoughts on the Chimera Method of Simulation of Three-Dimensional Viscous Flow, in Proceedings, *Computational Fluid Dynamics Symposium on Aeropropulsion*, Cleveland, Ohio, *NASA CP-3078*, 1–10.

Steger, J.L. and Warming, R.F. 1981. Flux Vector Splitting of the Inviscid Gas-Dynamic Equations with Applications to Finite Difference Methods, *J. Comp. Phys.*, **40**, 263–293.

Stone, H.S. 1993. *High-Performance Computer Architecture: Third Edition*, Addison-Wesley, Reading, MA.

Strikwerda, J.C. and Scarbnick, C.D. 1993. A Domain Decomposition Method for Incompressible Viscous Flow, *SIAM J. Sci. Comput.*, **14**(1), 49–67.

Swartztrauber, P.N. 1987. Multiprocessor FFTs, *Parallel Computing*, **5**, 197–210.

Szekely, J. 1979. *Fluid Flow Phenomena in Metals Processing*, Academic Press, New York.

Takasugi, T., Watanabe, S., and Hanada, S. 1992. The Temperature and Orientation Dependence of Tensile Deformation and Fracture in NiAl Single Crystals, *Mat. Sci. Eng.*, **149**, 183–193.

Tanny, J. and Tsinober, A.B. 1988. The Dynamics and Structure of Double-Diffusive Layers in Sidewall-Heating Experiments, *J. Fluid Mech.*, **196**, 135–156.

Tatsumi, S., Martinelli, L., and Jameson, A. 1995. Flux-Limited Schemes for the Compressible Navier-Stokes Equations, *AIAA J.*, **33**(2), 252–261.

Tennekes, H. and Lumley, J.L. 1972. *A First Course in Turbulence*, The MIT Press, Cambridge, MA.

Thakur, S.S. and Shyy, W. 1992. Unsteady, One-Dimensional Gas Dynamics Computations Using a TVD Type Sequential solver, *28th AIAA/SAE/ASME/ASEE Joint Propulsion Conference*, Paper No. 92-3640.

Thakur, S.S. and Shyy, W. 1993. Development of High Accuracy Convection Schemes for Sequential Solvers, *Numer. Heat Transf., Part B*, **23**, 175–199.

Thakur, S., Wright, J., Shyy, W., Liu, J., Ouyang, H., and Vu, T. 1996. Development of

Pressure-Based Composite Multigrid Method for Complex Fluid Flows, *Prog. Aero. Sci.*, **32**, 313–375.

Thakur, S.S., Wright, J., and Shyy, W. 1996. A Pressure-Based Composite Multigrid Method with Conservative Interface Treatment, AIAA Paper No. 96-0298, *34th Aerospace Sciences Meeting*, Reno, Nevada.

Thinking Machines Corporation, Cambridge, MA. *CM-5 Technical Summary*, November 1992.

Thompson, J.M.T. and Stewart, H.B. 1986. *Nonlinear Dynamics and Chaos*, Wiley, New York.

Thompson, M.C. and Ferziger, J.H. 1989. An Adaptive Multigrid Technique for the Incompressible Navier-Stokes Equations, *J. Comp. Phys.*, **82**, 94–121.

Thorpe, S.A., Hutt, P.K., and Soulsby, R. 1969. The Effect of Horizontal Gradients on Thermohaline Convection, *J. Fluid Mech.*, **38**, 375–400.

Tiller, W.A. 1991. *The Science of Crystallization: Microscopic Interfacial Phenomena*, Cambridge University Press, Cambridge, U.K.

Tritton, D.J. 1992. Stabilization and Destabilization of Turbulent Shear Flow in a Rotating Fluid, *J. Fluid Mech.*, **241**, 503–523.

Tu, J.Y. and Fuch, L. 1992. Overlapping Grids and Multigrid Methods for Three-Dimensional Unsteady Flow Calculations in IC Engines, *Int. J. Numer. Meths. Fluids*, 15, 693–714.

Tucker, P.K. and Shyy, W. 1993. A Numerical Analysis of Supersonic Flow over an Axisymmetric Afterbody, *AIAA Paper* No. 93-2347.

Tuminaro, R.S. and Womble, D.E. 1993. Analysis of the Multigrid FMV Cycle of Large-Scale Parallel Machines. *SIAM J. Sci. Comput.*, **14**(5), 1159–1173.

Turner, J.S. 1974. Double-Diffusive Phenomena, *Ann. Rev. Fluid Mech.*, **6**, 37–57.

Turner, J.S. 1979. *Buoyancy Effects in Fluids*, Cambridge University Press, London.

Turner, J.S. 1985. Multicomponent Convection, *Ann. Rev. Fluid Mech.*, **17**, 11–44.

Udaykumar, U.S. and Shyy, S. 1995. Development of a Grid-Supported Marker Particle Scheme for Interface Tracking, *Numer Heat Transf.*, Part B, **27**, 127–153.

Udaykumar, H.S., Shyy, W., and Rao, M.M. 1996. *ELAFINT*: A Mixed Eulerian-Lagrangian Method for Fluid Flow with Complex and Moving Boundaries, *Int. J. Numer. Meths. Fluids*, **22**, 691–712.

Van Dyke, M. 1975. *Perturbation Methods in Fluid Mechanics*, annotated edition, Parabolic Press, Stanford, CA.

Van Leer, B. 1982. Flux Vector Splitting for Euler Equations, In *Proceedings of the 8th International Conference on Numerical Methods in Fluid Dynamics*, Springer-Verlag, New York.

Vanka, S.P. 1986. Block-Implicit Multigrid Solution of Navier-Stokes Equations in Primitive Variables, *J. Comput. Phys.*, **65**, 138–158.

Vedula, K., Pathare, V., Aslandis, I., and Titran, R. 1985. Alloys Based on NiAl for High Temperature Applications, *Mat. Res. Soc. Symp. Proc.*, **39**, 411–421.

Venkatakrishnan, V. 1996. Perspective on Unstructured Grid Flow Solvers, *AIAA J.*, **34**, 533–547.

Viegas, J.R., Rubesin, M.W., and Horstman, C.C. 1985. On the Use of Wall Functions as Boundary Conditions for Two-Dimensional Separated Compressible Flows, *AIAA Paper* 85-0180.

Viskanta, R. 1990. Mathematical Modeling of Transport Processes During Solidification of Binary Systems, *JSME Int. J.*, **33**(3), 409–423.

Voller, V. 1985. Implicit Finite-Difference Solutions of the Enthalpy Formulation of Stefan

Problems, *IMA J.*, **5**, 201–214.

Voller, V.R. and Peng, S. 1994. An Enthalpy Formulation Based on an Arbitrarily Deforming Mesh for Solution of the Stefan Problem, *Comput. Mech.*, **14**, 492–502.

Voller, V.R. and Prakash, C. 1987. A Fixed Grid Numerical Modeling Methodology for Convection–Diffusion Mushy Region Phase-Change Problems, *Int. J. Heat. Mass. Transf.*, **30**(8), 1709–1719.

Voller, V.R., Brent, A.D., and Prakash, C. 1989. The Modeling of Heat, Mass and Solute Transport in Solidification Systems, *Int. J. Heat. Mass. Transf*, **32**(9), 1719–1731.

Vu, T.C. and Shyy, W. 1988. Navier-Stokes Computation of Radial Inflow Turbine Distributor, *J. Fluids Eng.*, **110**, 29–32.

Vu, T.C. and Shyy, W. 1990. Navier-Stokes Flow Analysis for Hydraulic Turbine Draft Tubes, *J. Fluids Eng.*, **112**, 199–204.

Wada, Y. and Liou, M.-S. 1994. A Flux Splitting Scheme with High-Resolution and Robustness for Discontinuities, *AIAA Paper No. 94-0083*.

Walston, W.S. and Darolia, R. 1995. Effect of Alloying on Physical Properties of NiAl, *MRS Symp. Proc.*, **28**, 237–242.

Wang, J.C. and Widhopf, G.F. 1987. A High-Resolution TVD Finite Volume Scheme for the Euler Equations in Conservation Form, *J. Comp. Phys.*, **84**, 145–173.

Wang, Z.J. and Yang, H.Q. 1994. A Unified Conservative Zonal Interface Treatment for Arbitrarily Patched and Overlapped Grids, *AIAA-94-0320*.

Wang, Z.J., Buning, P., and Benek, J. 1995. Critical Evaluation of Conservative and Non-Conservative Interface Treatment for Chimera Grids, *AIAA-95-0077*.

Warfield, M.J. and Lakshminarayana, B. 1987. Computation of Rotating Turbulent Flow with an Algebraic Reynolds Stress Model, *AIAA J.*, **25**, 957–964.

Wesseling, P. 1987. Linear Multigrid Methods. In S.F. McCormick, editor, *Multigrid Methods*, chapter 2. SIAM, Philadelphia.

Wesseling, P. 1991. A Survey of Fourier Smoothing Analysis Results. In W. Hackbusch and U. Trottenberg, editors, *Multgrid Methods III*, 105–127. Birkhauser, Boston.

Wey, T.C. 1994. Development of a Mesh Interface for Overlapped Structured Grids, *AIAA Paper No. 94-1924*.

Wheeler, A.A., Murray, B.T., and Schaefer, R.J. 1993. Computation of Dendrites using a Phase Field Model, *Physica D*, **66**, 243–262.

Whitaker, S. 1967. Diffusion and Dispersion in Porous Media, *A.I.Ch.E.*, **13**, 420–427.

Wilcox, D.C. 1993. *Turbulence Modeling for CFD*, DCW Industries, Inc., La Canada, CA.

Wirtz, R.A. and Liu, L.H. 1975. Numerical Experiments on the Onset of Layered Convection in a Narrow Slot Containing a Stably Stratified Fluid, *Int. J. Heat Mass Transf.*, **18**, 1299–1305.

Womble, D.E. and Young, B.C. 1990. Multigrid on Massively-Parallel Computers. In *Proceedings of the Fifth Distributed Memory Computing Conference*, 559–563, Charleston, SC, IEEE Computer Society Press.

Wright, J.A. and Shyy, W. 1993. A Pressure-Based Composite Grid Method for the Navier-Stokes Equations, *J. Comp. Phys.*, **107**(2), 225–238.

Wright, J.A. and Shyy, W. 1996. Numerical Simulation of Unsteady Convective Intrusion in a Thermohaline Stratification, *Int. J. Heat Mass Transf.*, **39**, 1183–1201.

Wright, N. and Gaskell, P. 1995. An Efficient Multigrid Approach to Solving Highly Recirculating Flows, *Computers Fluids*, **24**(1), 63–79.

Yakhot, V. and Orszag, S.A. 1986. Renormalization Group Analysis of Turbulence, I: Basic Theory, *J. Sci. Comput.*, **1**, 3–51.

Yakhot, V. and Smith, L.M. 1992. The Renormalization Group, the ε-Expansion and Derivation of Turbulence Models, *J. Sci. Comput.*, **3**, 35.

Yakhot, V., Orszag, S.A., Thangam, S., Gatski, T.B., and Speziale, C.G. 1992. Development of Turbulence Models for Shear Flows by a Double Expansion Technique, *Phys. Fluids.*, **4A**(7), 1510–1520.

Yavneh, I. 1993. A Method for Devising Efficient Multigrid Smoother for Complicated PDE Systems. *SIAM J. Scient. Comp.*, **14**, 1437–1463.

Yee, H.C. 1986. Linearized Form of Implicit TVD Schemes for the Multidimensional Euler and Navier-Stokes Equations, *Comp. Math. Appls.*, **12A**, 413–432.

Yee, H.C. 1987. Construction of Explicit and Implicit Symmetric TVD Schemes and Their Applications, *J. Comp. Phys.*, **68**, 151–179.

Yee, H.C., Warming, R.F., and Harten, A. 1985. Implicit Total Variation Diminishing (TVD) Schemes for Steady-State Calculations, *J. Comp. Phys.*, **57**, 327–360.

Yung, C., Keith, Jr. T.G., and De Witt, K.J. 1989. Numerical Simulation of Axisymmetric Turbulent Flow in Combustors and Diffusers, *Int. J. Numer. Meths. Fluids*, **9**, 167–183.

Younnis, B.A. 1993. Prediction of Turbulent Flows in Rotating Rectangular Ducts, *J. Fluids Eng.*, **115**, 646–652.

Zeman, O. 1990. Dilatation Dissipation: The Concept and Application in Modeling Compressible Mixing Layers, *Phys. Fluids A*, **2**, 178–188.

Zeman, O. 1993. New Model for Super/Hypersonic Turbulent Boundary Layers, *AIAA Paper No. 93-0897*.

Zeng, S. and Wesseling P. 1994. Multigrid Solution of the Incompressible Navier-Stokes Equations in General Coordinates, *SIAM J. Numer. Anal.*, **31**, 1764–1784.

Index

314